Direct Digital Control Systems

C[3] INDUSTRIAL CONTROL, COMPUTERS AND COMMUNICATIONS SERIES

Series Editor: **Professor Derek R. Wilson**
University of Westminster, England

3. Dataflow Architecture for Machine Control
   Bogdan Lent

4. RISC Systems
   Daniel Tabak

6. Linear Control Systems
   VOLUME 1 – ANALYSIS OF MULTIVARIABLE SYSTEMS
   T. Kaczorek

7. Linear Control Systems
   VOLUME 2 – SYNTHESIS OF MULTIVARIABLE SYSTEMS AND MULTIDIMENSIONAL SYSTEMS
   T. Kaczorek

8. Transputers in Real-Time Control
   Edited by **G. W. Irwin** *and* **P. J. Fleming**

9. Parallel Processing in Cellular Arrays
   Y. I. Fet *(Forthcoming)*

10. Direct Digital Control Systems
    J. B. Knowles

Direct Digital Control Systems

J. B. Knowles

RESEARCH STUDIES PRESS LTD.
Taunton, Somerset, England

JOHN WILEY & SONS INC.
New York · Chichester · Toronto · Brisbane · Singapore

RESEARCH STUDIES PRESS LTD.
24 Belvedere Road, Taunton, Somerset, England TA1 1HD

Marketing and Distribution:

Australia and New Zealand:
Jacaranda Wiley Ltd.
GPO Box 859, Brisbane, Queensland 4001, Australia

Canada:
JOHN WILEY & SONS CANADA LIMITED
22 Worcester Road, Rexdale, Ontario, Canada

Europe, Africa, Middle East and Japan:
JOHN WILEY & SONS LIMITED
Baffins Lane, Chichester, West Sussex, England

North and South America:
JOHN WILEY & SONS INC.
605 Third Avenue, New York, NY 10158, USA

South East Asia:
JOHN WILEY & SONS (SEA) PTE LTD.
37 Jalan Pemimpin 05-04
Block B Union Industrial Building, Singapore 2057

Library of Congress Cataloging-in-Publication Data

Available

British Library Cataloguing in Publication Data

A catalogue record for this book
is available from the British Library.

ISBN 0 86380 167 6 (Research Studies Press Ltd.)
ISBN 0 471 95147 1 (John Wiley & Sons Inc.)

Typeset by Jennifer Sparham, Portland, Dorset, England
Printed in Great Britain by SRP Ltd., Exeter

To

Jim Nightingale

sportsman, creative genius, and friend

Editorial Preface

Following the massive expansion of the British University system in 1993, the basic professorial question has become - what do you profess? Brian Knowles' unambiguous answer is - the design of Direct Digital Control Systems. In today's jargon, he is undisputedly a world class expert. I can unhesitatingly recommend all my Professorial colleagues in Europe and the USA to invite him to lecture in their departments. Students will certainly enjoy his infectious enthusiasm as well as gain considerable insight into the subject.

The book contains a rigorous exposition of direct digital control (DDC) system design techniques. In addition, lucid detailed analyses of system dynamics with random or deterministic inputs are included, together with new material on the statistical characteristics of quantisation errors. Despite the mathematical maturity, the overall style of presentation makes it an easy book to digest. It will certainly become a standard reference work for engineers, scientists and students alike.

The key element of the text, the successful engineering design of DDC systems, is set within the framework of a modern design environment such as a PC with a commercial software design package. Here, the machine algorithms for computing system responses to frequency or time domain stimuli circumvent the otherwise tedious manual mechanics of the design process, so that a designer is able to concentrate on the intellectual principles clearly described herein.

The author has unique credentials for writing on this subject. His interest was first awakened at Manchester in the late 1950's by an exceptional symbiotic combination of academic and industrial excellence. Manchester University was home to the Royal Society's Digital Computer Laboratory, which lays some claim to the development of the first digital computer with a stored programme. In addition, the city hosted the Ferranti Company who with considerable foresight incorporated digital technology into advanced industrial and missile control systems, as well as in mainframe machines. It was in this remarkably stimulating environment that Brian initially became involved with DDC systems design. This book is written by a truly master craftsman, who has not only contributed to the intellectual development of the subject, but who has also applied it to the solution of real industrial problems. It is therefore with great pleasure that I welcome his contribution to the series.

D.R. Wilson
London

January 1994

Preface

Digital systems can be engineered with constructional and operational advantages over their counterparts operating with continuous data provided that the effects of data sampling and amplitude quantisation are understood. This book examines the fundamental aspects of single-input DDC systems in terms of rigorously formulated mathematical relationships, whose physical interpretation and practical significance are also carefully presented. Deterministic and stochastic input data are considered with an objective of demonstrating the close mathematical similarities between results derived by the z- and Laplace Transformations. Apart form the pedagogical benefits of such an approach (actually exploited in an early UMIST control engineering course), the accuracy of a DDC system must usually be assessed as a function of continuous time, and the two transformation techniques are required for this purpose. The losses of control accuracy due to the finite sampling frequency and wordlength of a practical system are quantified by methods developed during my career at UMIST. A comprehensive, yet computationally undemanding, design procedure is described that embodies the preceding theory. Additional topics considered include: antialiasing filters, notch networks, and the use of coarse-fine control modes to combat the destabilising effect of overflow. Because digital controllers are devised in practice on the basis of linearised plant models, it cannot be overemphasised that a design should always be vindicated by an adequate non-linear simulation.

Innovation is the hallmark of good engineering design, and I believe it stems more from incisive physical insight than mathematical 'flute music'. Accordingly, it is recommended that the book is read first without too much regard for the mathematical rigour. Qualitative explanations are provided wherever possible for this purpose, and much of the mathematical detail is relegated to appendices. A busy practising engineer might even start by reading Chapter 7 just 'to get his design off the ground' and to establish priorities for subsequent more detailed attention. The required design calculations can generally be implemented with presently available personal computer software. References to formulae derived in earlier Sections are included there with these ideas in mind. However, frequently during my own career the resolution of a problem has required a thorough understanding of both the underlying mathematics and physics. If needed, the reader should find here the mathematical rigour sufficient for most such awkward occasions.

Lecturing always provides me with considerable personal enjoyment and stimulus because students are largely free of various preconceptions. As well as interrogating the logic and lucidity of my arguments, bright students have on occasion discovered that "the emperor wasn't wearing any clothes". The expositions herein have therefore benefited from the remarks of many of my past control engineering students, and I look forward to continuing these discussions with future generations.

I am also particularly appreciative of the efforts of my former research colleagues at UMIST; especially Drs Rod. Edwards and Ender Olcayto. Shortly after publication of our finite wordlength control error analysis in the Proceedings of the Institute of Electrical Engineers I received written congratulations from Dr Jim Kaiser of Bell System Laboratories, Murray Hill NJ. Since then he has kept me in touch with the evolution of digital filter technology in the USA, and a close friendship has developed. His unselfish efforts in encouraging others over the years merit recognition. I have pleasure in recording my appreciation as well of Miss Jennifer Sparham who, apart from her splendid typing, remained so cheerful and helpful in spite of our many set-backs.

My first meeting with (now Professor) Jim Nightingale was in 1954 as a first year undergraduate at UMIST. As my later research supervisor and personal friend, he taught me a great deal. For this, and the kindness of himself and Joy over the years, I am deeply indebted. This book endeavours to approach the intellectual penetration and practical utility that he demanded of himself and of others. With these thoughts in mind, it is respectfully dedicated.

Finally to all of you, mentioned or not mentioned above, thank you for your friendship and intellect which have so enriched my career.

J.B. Knowles
'Riverhouse'
Caters Place, Dorchester

November 1993

Contents

Nomenclature xv

1 SAMPLING AND BINARY CODES 1

1.1 Introduction 1

1.2 Some Mathematical Preliminaries and Observations 4

1.3 The Periodic Sampling Process 18

1.4 Some Aspects of Binary Arithmetic 26

1.5 Data Converters 34

2 DETERMINISTIC INPUT SIGNAL ANALYSIS 43

2.1 The Single-Sided z-Transformation 43

2.2 The Pulse Transfer Function for a Discrete Data System 49

2.3 Programming Structures for Pulse Transfer Functions 55

2.4 The Pulse Transfer Function for a Continuous Data Component 61

2.5 Intersample Behaviour and the Modified z-Transformation 72

3 STABILITY, ROOT LOCI AND NYQUIST DIAGRAMS 77

3.1 Stability 77

3.2 Root Loci 84

3.3 Real Frequency Response 93

3.4 Time and Frequency Domain Behaviour 97

3.5 Nyquist Diagrams 105

3.6 Inverse Nyquist Diagrams 115

3.7 The Bilinear Transformation 117

4 STOCHASTIC INPUT SIGNAL ANALYSIS 121

4.1 Relevant Aspects of Mathematical Probability 121

Table of Contents

4.2 A Geometrical Insight into Correlation Analysis 127
4.3 Power Spectra 130
4.4 Pulse Power Spectra 134
4.5 The Pulse Spectrum of a Sampled Continuous Data Process 140
4.6 Input-Output Relationships 149
4.7 Multiple Uncorrelated Inputs 159
4.8 Shaping Filters and Mean Square Value Calculations 166

5 THE CHOICE OF SAMPLING FREQUENCY 173
5.1 A Perspective 173
5.2 Sampling Continuous Stochastic Data 179
5.3 Output Ripple Power Analysis 187
5.4 An Initial Choice of Sampling Frequency and Practical Considerations 193
5.5 Experimental Validation and Pulsed RC Networks 199

6 FINITE WORDLENGTH EFFECTS 206
6.1 Introducing the Statistical Approach 206
6.2 The Idealised Statistics of Amplitude Quantisation Errors 210
6.3 Actual Statistics of Amplitude Quantisation Errors 216
6.4 An Upper-Bound on the Control Error due to a Finite Wordlength 231
6.5 A Statistical Analysis of Finite Wordlength Control Errors 240
6.6 Coefficient Quantisation and Real Frequency Response 252

7 NYQUIST BASED DESIGN PRINCIPLES 261
7.1 Preliminaries 261
7.2 Prefiltering and the Choice of Sampling Frequency 264
7.3 Gain-Phase Curves for Frequency Domain Compensation 269
7.4 Structural Resonances 276
7.5 Compensation of Class-0 or Regulator Systems 283
7.6 Time Domain Synthesis of DDC Systems 286
7.7 Overflow 292

8 MULTIRATE AND SUBRATE SYSTEMS 297
8.1 Stability and Compensation of Multirate DDC Systems 297
8.2 Ripple in Multirate Systems 307
8.3 Rounding Errors in Multirate Systems 318
8.4 Aspects of Subrate DDC Systems 326

9 PRINCIPAL CONCLUSIONS AND FINAL REMARKS 341

APPENDICES 349

A2.1 The Limit in Equation (2.1.12) 349
A2.4 The Normal Summability of $\left\{H(s)(z^{-1}e^{sT})^n\right\}$ over [c - j∞, c + j∞] 351
B2.4 Convergence around Infinite Semicircles 353
A3.3 The Spectrum of a Sequence 357
A4.6 Poles, Analytic Regions and Power Series 359
A4.8 The Sum of the Squares of a Weighting Sequence 362
A6.2 The First Order Characteristic Function for Quantisation Errors 364
A6.3 The Bandwidth of a Second Order Characteristic Function 369
A7.2 The Non-Filterable Input or Output Situation 371
A7.3 Gain-Phase Curves for Digital Compensator Design 374
A8.1 Extracting a Deterministically Embedded z-Transformation 382
A8.3 Derivation of a Multiplication Pattern 385

REFERENCES 389

INDEX 401

Nomenclature

Perhaps St Cyril (9th Cent.) was right after all? At least there are insufficient Roman letters to provide a unique symbolism for each entity in this book. However, there should be no real ambiguity because symbols are defined quite locally in most cases and because the limited duplication arises in widely disparate contexts. Throughout the text an upper case letter generally denotes the z- or Laplace etc transformation of the function represented by the corresponding lower case letter. Apart from this convention, the following list constitutes the principal nomenclature.

$\triangleq$	-	equality by definition, so not deducible
$\Vert\ \Vert$	-	norm of a vector or function
t	-	time
$\frac{dG}{dt}$, $\dot{G}$	-	total temporal derivative of an arbitrary function G(t)
e, exp	-	exponential function
$\vert\ \vert$	-	modulus of a complex number
$\Re$, $\Im$	-	real, imaginary, part of a complex number
$\bar{z}$	-	conjugate of a complex number z
f	-	real frequency (Hz)
ω	-	angular frequency (rad/s)

T	-	sampling period or occasionally the period of a periodic function
ω_0	-	angular sampling frequency ($2\pi/T$)
i,k,m,n,p	-	integer variables; likewise upper case letters
s	-	complex variable of the Laplace Transformation
z	-	complex variable of the z- Transformation
j	-	$\sqrt{-1}$
$\delta(t)$	-	delta function
δ_{ko}	-	single pulse sequence
$U(t)$	-	unit step function
$x(t), y(t)$	-	continuous input, output, data of a dynamical system
x_k , y_k	-	periodically sampled input, output, of a dynamical system
e, f	-	error, plant forcing, function of a feedback control system
$p(t)$	-	a periodic train of sampling pulses
τ	-	time constant, or temporal displacement variable, of a correlation function
$a,b,c,$ α,β,γ	-	constants with locally defined values
$[a,b]$	-	closed line segment defined by the arbitrary numbers a and b
q	-	width of quantisation
M	-	wordlength of a digital device
$\oplus$	-	two's complement binary addition
$\mathcal{S}$	-	a defined region of a complex plane
$\mathcal{C}$	-	a defined contour of integration
Γ	-	unit circle contour in the z- plane
λ	-	scalar gain of a digital controller
$D(z)$	-	pulse transfer function of a digital controller
$G(z)$	-	pulse transfer function of a linearised plant model
$H(z)$	-	pulse transfer function of data hold (extrapolator) and plant

K(z) - overall pulse transfer function of a closed loop system

G(s), H(s), K(s) - as above but Laplace transfer functions

B(s) - Laplace transfer function of a feedback network

θ - angular displacement or argument of a complex number

ψ - phase displacement of a sinusoid

R - rank of a pulse or Laplace transfer function; usually of a plant model

M_T - maximum overshoot of a system's step response

$\tilde{t}$, $\tilde{n}T$ - time required to achieve M_T

ω_B - bandwidth of a system; if appropriate

M_p - maximum (resonant) amplitude of a real frequency response function

ω_p - resonant frequency corresponding to M_p

P_X - probability density function of an arbitrary, possibly vector, ensemble X

F_X - characteristic function of an arbitrary, possibly vector, ensemble X

ϕ_X - autocorrelation function or sequence of an arbitrary stochastic ensemble X

$\langle x \rangle$ - doubly infinite temporal average of an arbitrary realisation x(t) or x_k

P_{XY} - joint probability density function of arbitrary ensembles X and Y

ϕ_{XY} - cross correlation function or sequence of arbitrary stochastic ensembles X and Y

Φ - (pulse) power spectrum corresponding to a correlation function (sequence) ϕ

χ - shaping filter corresponding to a spectrum Φ derived from an autocorrelation function (sequence)

ω_s - bandwidth of a (pulse) power spectrum

ϕ_s - control error component due to the baseband of a sampled input

ϕ_r - control error component due to the sidebands of a sampled input (the ripple component)

ϕ_{fw} - control error component due to finite wordlength effects

ρ_X - correlation coefficient of an arbitrary stochastic ensemble X

σ_X - standard deviation of an arbitrary ensemble X

$\hat{\epsilon}$ - upper-bound of a finite wordlength control error

E(k) - computational error of a finite wordlength controller at kT

$\mathcal{E}(z)$ - z-transformation of the sequence {E(k)}

μ - number of 'significant' multiplications in realising a pulse transfer function

A(z),B(z) - numerator, denominator, of a pulse transfer function

$D_I(z)$ - ideal pulse transfer function granted an infinite wordlength

σ^2_ω - mean square deviation in a real frequency response due to coefficient quantisation

z_N, $r_N z_N$ - zero, pole, of a specialised digital notch filter

$\overline{B}$, ω_N - ideal bandwidth, centre frequency, of a notch filter

ΔB, $\Delta\omega_N$ - deviation in bandwidth, centre frequency, of a notch filter due to coefficient quantisation

$\underline{x}_k$ - fast rate sequence with the slow rate sequence x_p embedded

z_n - complex variable of the z_n-Transformation

$Z\left[C(z_n)\right]$ - z-transformation of the slow rate sequence embedded in an arbitrary fast rate sequence represented by $C(z_n)$

CHAPTER 1

Sampling and Binary Codes

A wise man, which built his house upon a rock. - St Matthew

1.1 INTRODUCTION

As a result of a sizeable threshold between the logic 0 and 1 states, external interference and ageing induce less performance degradation on digital circuits than their analogue counterparts. In addition, the on/off nature of the active circuit elements enables a more efficient utilisation of power, and thereby a greater device-density in very large scale integrated electronic circuits. Moreover, because such digital circuits are constructed from relatively few different types of 'building blocks' which do not generally require on-the-chip trimming, their relative fabrication-costs are even further reduced. Special digital controllers for missile guidance were considered viable as early as 1957. From this platform the Ferranti Company developed their successful range of Argus process control computers, which were first used on a chemical plant in 1962. However, even in the late 1970's, enclaves of innate conservatism advocated analogue controllers often on the basis of initial costs without a fair consideration of the following advantages of the digital alternative:

i) transmitted data have greater noise immunity, which can be further enhanced by coding techniques (1)

ii) interconnection of distributed controllers to achieve greater performance, flexibility and reliability (2,3)

iii) auxiliary off-line data processing to achieve improvements in overall plant-performance and accountability

iv) tamper resistance (4)

Direct digital control systems now find widespread applications due to the production of inexpensive micro-controllers and data converters, as well as the wider engineering appreciation of digital techniques. The presently insignificant cost of such micro-processors is illustrated by their increasingly frequent application in domestic equipment during the past decade. Indeed, the development of validated software can now be more expensive than the corresponding hardware.

The above advantages of direct digital control systems and signal processing systems are gained only at a price. The eminent system analyst Hendrick Bode (5) clearly identifies, if somewhat overstates, the general situation "It is as though we were Nature's tenants on a share basis. We must surrender half the crop for the privilege of farming the land". Digital systems must inherently operate on samples taken from continuous data at generally regular intervals, in order to allow time for data conversions (A-D and D-A) and arithmetic operations. In addition, physical or economic constraints dictate that these processes are implemented within a restricted accuracy that is synonymous with wordlength or bits. The analyses in Chapters 5 and 6 enable these degenerate effects to be engineered small enough for the advantages of digital control and signal processing systems to be realised at a very 'fair rent'.

Although some control systems respond to a single input variable, many industrial processes involve several independent input variables. This book largely examines the more easily analysed single-input systems, so that its existence requires perhaps some justification. Firstly, many single-input system design concepts re-appear with similar roles in multi-input analyses embodied in a matrix format. Single-input theory forms therefore an ideal educational stepping-stone into the mathematically more-demanding multi-input scene. Secondly, the effects of sampling-rate and wordlength are considered sufficiently complicated in themselves to merit an introduction with the simpler single-input analysis. Finally, single-input theory can still be applied successfully to multi-input control problems when experience suggests that the significant independent plant-variables have widely disparate response times, and when performance requirements are somewhat lax; that is to say, under conditions when the open loop transfer function matrix would be already strongly diagonally dominant (6, 7).

System dynamics that are governed by ordinary linear differential equations with constant coefficients form a general type that admits an analytical solution. More significantly for the control engineer, the inputs and outputs of these linear systems are related in the Laplace Transformation domain by functions of a complex variable. The mathematical theory of such functions enables the stability of linear feedback control systems to be established, and then improved as necessary by additional components or feedback loops. However, non-linear effects eventually become significant in all systems, and the corresponding differential equations are generally amenable to only ad hoc numerical solution. Consequently, there are no exact and generally applicable design techniques for non-linear feedback control systems. Nevertheless, experience shows that many successful control schemes for non-linear processes can be designed using linear theory, provided that the important non-linearities exhibit continuous input-output relationships. Under these conditions, sets of linearised differential equations or real frequency responses can be derived that form reasonable descriptions of plant dynamics for small enough perturbations about an intuitively adequate number of plant-states to characterise the parameter variations. A single set of controller parameters (eg gains and time constants) can then often be designed to achieve a satisfactory, albeit not optimum, control at all the selected states. Personal experience is that the normally generous stability margins are sufficient to coerce adequately damped responses as the plant variables traverse quite gross non-linear trajectories across the plant-states. Admittedly, there are contrived non-linear systems which are stable for small enough perturbations about any operating point, but which are globally unstable*. Although these peculiar systems are not representative of practical problems, the dynamics of a non-linear plant with linearly-designed linear controllers should always be validated by a proper non-linear simulation. It is concluded that control system design on the basis of linear dynamics is generally the only viable practical route. The exclusive use of linear plant models in this book is therefore considered justified, but the reader must always bear in mind the real gap to be bridged before trial on an actual plant.

The diffusion of heat down a temperature gradient is one example of a so-called thermodynamically irreversible process (9). As demonstrated by a refrigerator, this process is actually physically reversible by the expenditure of mechanical work. However, the sampling of data in continuous time is totally irreversible, in that once effected the initial continuous data can never generally be perfectly reconstructed again - at least by linear filtering. The control engineer must therefore understand the irrevocable nature of the sampling process as a first step towards designing a digital control system with data reconstruction that is adequate to meet the required accuracy-specification. For this purpose, the concepts of Fourier Series, Fourier Integrals and Transfer Functions are briefly reviewed in the next section.

1.2 SOME MATHEMATICAL PRELIMINARIES AND OBSERVATIONS

A ***norm*** of a vector or function x is written as $\|x\|$, and it has the properties (10):

$$\|x\| \text{ is a real number,} \quad \|x\| \geq 0 \quad \text{for every } x \text{ in the set}$$

---

*These are discussed in the literature under the heading of the Aizerman Conjecture (8)

$\|x\| = 0$ is equivalent to $x = 0$

$\|\lambda x\| = |\lambda| \ \|x\|$ for any scalar x

$\|x_1 + x_2\| \leq \|x_1\| + \|x_2\|$ for any two elements x_1 and x_2

If a function g of a real variable is bounded over the interval [a,b], then one possible definition of its norm is:

$$\|g\| = \mathrm{Sup}\left\{ |g(t)| \,\middle|\, a \leq t \leq b \right\}$$

where Sup denotes the least upper-bound. It is frequently possible to define several different norms as illustrated for real two dimensional vectors {(x,y)} by:

$$\|(x,y)\| = \sqrt{x_2 + y_2} \quad ; \quad \|(x,y)\| = \mathrm{Max}\left\{ |x| \ ; \ |y| \right\}$$

When norms are *equivalent* in the sense that real numbers a < b exist for which:

$a \|x\|_1 \leq \|x\|_2 \leq b \|x\|_1$ for all x in the set

then much of the arbitrariness disappears because continuity, convergence etc with respect to one norm implies the same with respect to the other (10). An infinite series of functions $\{g_n\}$ is *normally summable* if and only if

$$\lim_{N \to o} \left[\sum_{n=o}^{N} \|g_n\| \right] \tag{1.2.1}$$

exists. With functions of a real or complex variable, normal summability implies the *uniform convergence* of the series. That is, there exists a limit function G such that:

$$\text{Sup}\left\{ \left| G(t) - \sum_{n=o}^{N} g_n(t) \right| \right\}$$

can be made vanishingly small by the choice of any N that is large enough. Various tests, such as Abels' lemma (10, 12), are available to establish the normal summability of an inverse power series $\{a_n z^{-n}\}$ where $\{a_n\}$ is an infinite sequence of constants and z is a complex variable. These tests generally prove that a particular series is normally summable provided that

$$|z| > R$$

Within such a region of convergence, the integral (differential) of a function:

$$G(z) = \sum_{n=o}^{\infty} a_n z^{-n} \tag{1.2.2}$$

can be derived from a term by term integration (differentiation) of the series (10, 12):

$$\left.\begin{aligned} \int G(z)dz &= \sum_{n=o}^{\infty} a_n \int z^{-n} dz \\ \frac{dG}{dz} &= \sum_{n=o}^{\infty} a_n \frac{d}{dz}(z^{-n}) \end{aligned}\right\} \tag{1.2.3}$$

The above properties of normally summable series are exploited later in the z- Transformation analysis of digital systems. As an illustration, consider the inverse power series $\{e^{sn} z^{-n}\}$. Setting

$$b = \exp(-s)$$

then every term of the series $\{|e^{sn} b^n|\}$ is bounded by the one value of 1*. The condition of Abel's lemma is thefore satisfied, so that the series $\{e^{sn} z^{-n}\}$ is normally summable for all $|z| > |b|^{-1}$ and from first principles it is easily verified that:

$$\sum_{n=o}^{\infty} e^{sn} z^{-n} = 1 / (1 - e^{S} z^{-1}) \tag{1.2.4}$$

An infinite set of integrable functions over a real interval [a,b] is *orthogonal* if and only if (11):

$$\left.\begin{array}{ll} \int_a^b \phi_p \, \bar{\phi}_q \, dx = 0 & \text{for integers } p \neq q \\ \text{and} & \\ \int_a^b |\phi_p|^2 \, dx = \|\phi_p\| & \text{for every integer } p \end{array}\right\} \tag{1.2.5}$$

where $\bar{\phi}_p$ denotes the complex conjugate of ϕ_p. A set of orthogonal functions is *not complete* if and only if there exists a non-zero function that is orthogonal to every function in the set. Given a function g that is integrable and absolutely square integrable over the real interval [a,b] together with a corresponding complete set of orthogonal functions, the quest is for a set of coefficients $\{\alpha_n\}$ in:

$$g_N = \sum_{n=-N}^{N} \alpha_n \, \phi_n \tag{1.2.6}$$

which would minimise the mean square error quantity:

---

* any constant value is sufficient

$$\epsilon_N = \int_a^b |g - g_N|^2 dt$$

Substituting equation (1.2.6) into the above yields:

$$\epsilon_N = \int_a^b |g|^2 dt + \sum_{n=-N}^{N} \left\{ |\alpha_n|^2 - \alpha_n \bar{\beta}_n - \bar{\alpha}_n \beta_n \right\} \|\phi_n\|^2 \qquad (1.2.7)$$

where by definition:

$$\beta_n = \frac{1}{\|\phi_n\|^2} \int_a^b g \, \bar{\phi}_n \, dt \qquad (1.2.8)$$

By simple manipulation of equation (1.2.7), it follows that

$$\epsilon_N = \int_a^b |g|^2 dt + \sum_{n=-N}^{N} |\alpha_n - \beta_n|^2 \|\phi_n\|^2 - \sum_{n=-N}^{N} |\beta_n|^2 \|\phi_n\|^2 \qquad (1.2.9)$$

so that choosing:

$\alpha_n = \beta_n$ for all integer n

effects the required minimisation of the mean square error, and $\{\beta_n \phi_n\}$ is the Fourier Series expansion of the function g over the interval [a,b]. It follows from equation (1.2.9) that each coefficient (α_n) can be adjusted to its optimum (β_n) independently of all the others, and this property finds application in curve-fitting and plant-identification. As long as the series of orthogonal functions is complete and a function is integrable and absolutely square integral, then it may be shown that ***Parceval's theorem:***

$$\int_a^b |g|^2 dt = \sum_{-\infty}^{\infty} |\beta_n|^2 \|\phi_n\|^2 \qquad (1.2.10)$$

is generally true (11).

The uniform convergence (or otherwise) of a Fourier Series can be established by a variety of tests (10, 12). To investigate the limit value of a uniformly convergent series define:

$$w = \sum_{-\infty}^{\infty} \beta_n \phi_n - g$$

Multiplying both sides of the above equation by the conjugate of any element ϕ_k in the complete set $\{\phi_n\}$, and then integrating over the interval [a, b] yields:

$$\int_a^b w \, \bar{\phi}_k \, dt = \int_a^b \left\{ \sum_{-\infty}^{\infty} \beta_n \phi_n \bar{\phi}_k \right\} dx - \beta_k \left\| \phi_k \right\|^2$$

Because the infinite series is assumed uniformly convergent, the order of integration and summation in the above can be interchanged to give

$$\int_a^b w \, \bar{\phi}_k \, dt = 0$$

As the set of orthogonal functions is taken to be complete, then

$$w = 0$$

and so

$$g = \sum_{-\infty}^{\infty} \beta_n \phi_n$$

A uniformly convergent Fourier Series therefore converges at every point to its 'parent' function. In practice, Fourier Series often lose their uniform convergence as a result of analytical idealisations of a function. For example, the exponentially rising and falling edges of electronically generated pulses are frequently replaced by abrupt transitions to produce a so-called square-wave. When such a square-wave is expanded in terms of the complete orthogonal set $\{e^{j2\pi nt/T}\}$ over its period T, Fig 1.2.1 illustrates that the finite partial sums of its Fourier Series exhibit oscillatory overshoots at each transition, and these maintain a constant amplitude whilst decaying in the limit to zero width (13). This behaviour of non-uniformly convergent Fourier Series exemplifies the so-called *Gibbs Phenomenon.* However, it is of no relevance in the analysis of digital systems because the zero-width Gibbs-spikes of an infinite sum generate zero power over the period, and because they are absent for practical signals anyway. A more convincing argument is perhaps that with an increasingly large number of terms, the transitional oscillations become progressively faster and therefore more effectively smoothed out by the stray capacitances, inertias etc of practical systems.

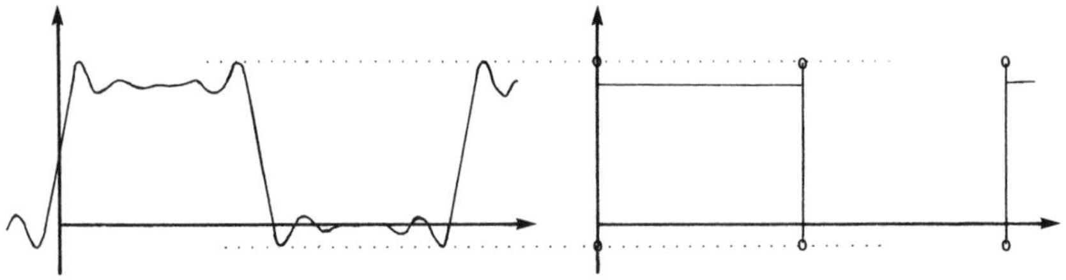

PARTIALLY SUMMED SERIES FULLY SUMMED SERIES

Fig 1.2.1 Illustrating the Gibbs Phenomenon

A system is defined as *linear* if and only if for every input x_1 and x_2 with corresponding outputs y_1 and y_2, it obtains that:

$$\left.\begin{array}{l}\text{Input } (x_1 + x_2) \text{ gives the Output } (y_1 + y_2) \\ \text{Input } (\kappa x_1) \text{ gives the Output } (\kappa y_1) \\ \qquad\qquad \text{for all values of the scalar } \kappa\end{array}\right\} \quad (1.2.11)$$

One important example of a linear system is one whose time dependent real input and output functions are related by an ordinary linear differential equation with constant coefficients:

$$\sum_{m=o}^{M} b_m \frac{d^m y}{dt^m} = \sum_{m=o}^{M} a_m \frac{d^m x}{dt^m} \tag{1.2.12}$$

Direct substitution confirms that the input and output functions:

$$x(t) = e^{j\omega t} \quad ; \quad y(t) = \left\{ \sum_{m=o}^{M} a_m (j\omega)^m \Big/ \sum_{m=o}^{M} b_m (j\omega)^m \right\} e^{j\omega t} \tag{1.2.13}$$

satisfy equation (1.2.12) for all values of ω. An eigen-function z and corresponding eigen-value λ of a linear transformation A have the property (14):

$$Az = \lambda z$$

Physically, therefore, a linear transformation alters its eigen-functions in amplitude and phase, but not in 'shape'. From equation (1.2.13), $\exp(j\omega t)$ for any value of ω is an eigen-function of any ordinary linear differential equation with constant coefficients. Because this type of equation occurs frequently in the analysis of electrical and control engineering systems, the particular Fourier Series expansion:

$$\left.\begin{aligned} g &= \sum_{-\infty}^{\infty} \beta_n \exp(j2\pi nt/T) \\ \beta_n &= \frac{1}{T} \int_{-T/2}^{T/2} g(t) \exp(-j2\pi nt/T)\,dt \end{aligned}\right\} \tag{1.2.14}$$

is especially important for present purposes. Noting in this case that Parceval's theorem becomes:

$$\frac{1}{T}\int_{-T/2}^{T/2} |g|^2 dt = \sum_{-\infty}^{\infty} |\beta_n|^2 \qquad (1.2.15)$$

and that by De Moivre's theorem:

$$2\cos \omega_0 t = \exp(j\omega_0 t) + \exp(-j\omega_0 t) \qquad (1.2.16)$$

the description of $\{|\beta_n|^2\}$ as the *line power spectrum* of a periodic function g becomes transparently justified.

If the following limits exist for a function g of a real variable:

$$\lim_{T\to\infty}\int_{-T/2}^{T/2} g\,dt \quad ; \quad \lim_{T\to\infty}\int_{-T/2}^{T/2} |g|^2 dt \qquad (1.2.17)$$

then from equation (1.2.14) there exists the Fourier Series:

$$g(t) = \sum_{-\infty}^{\infty} (1/T)T\beta_n \exp(j2\pi nt/T) \qquad (1.2.18)$$

with

$$T\beta_n = \int_{-T/2}^{T/2} g(t) \exp(-j2\pi nt/T)dt \qquad (1.2.19)$$

over any arbitrary interval [-T/2 ; T/2]. Defining in terms of equation (1.2.19) the function:

$$G(n/T) = T\beta_n$$

and the reciprocal of time as frequency in the manner:

$$\delta f = 1/T \quad \text{and} \quad f = n/T$$

then in the limit as the interval [-T/2 ; T/2] becomes arbitrarily large, the right-hand-side of equation (1.2.19) identifies with the Riemann integral (10):

$$g(t) = \int_{-\infty}^{\infty} G(f)\ \exp(j2\pi ft)df \tag{1.2.20}$$

where the existence of

$$G(f) = \int_{-\infty}^{\infty} g(t)\ \exp(-j2\pi ft)dt \tag{1.2.21}$$

is assured from statement (1.2.17). Equations (1.2.20) and (1.2.21) define the Fourier Transformation, but a rigorous development would require Lebesgue integrals to deduce ***Plancherel's theorem (11).*** This affirms the existence of G(f), whose integral transformation according to equation (1.2.20) produces a function that in general equates with g everywhere except for a set of zero measure (15). In fact, when g is continuous, there is a complete point-by-point correspondence. However, when g is discontinuous, 'Gibbs-type' spikes can appear at the abrupt transitions, but for the reasons presented earlier these are largely irrelevant in engineering analysis. If g_1 and g_2 have Fourier Transformations G_1 and G_2, then Parceval's theorem for Fourier Integrals reads as (11):

$$\int_{-\infty}^{\infty} G_1(f)\overline{G_2(f)}df = \int_{-\infty}^{\infty} g_1(t)g_2(t)dt \tag{1.2.22}$$

from which:

$$\int_{-\infty}^{\infty} |g(t)|^2 dt = \int_{-\infty}^{\infty} |G(f)|^2 df \tag{1.2.23}$$

so that $|G(f)|^2$ is called the *energy spectrum* of the signal g(t).

Moreover, if:

$$H(f) = G_1(f)G_2(f) \qquad \text{with} \qquad h(t) = \int_{-\infty}^{\infty} H(f)e^{j2\pi ft}df$$

than the *convolution integral* for real valued time domain functions is derived directly from equation (1.2.22) as:

$$h(t) = \int_{-\infty}^{\infty} g_1(\varsigma)g_2(t-\varsigma)d\varsigma = \int_{-\infty}^{\infty} g_1(t-\varsigma)g_2(\varsigma)d\varsigma \tag{1.2.24}$$

Similarly if:

$$h(t) = g_1(t)g_2(t) \qquad \text{with} \qquad H(f) = \int_{-\infty}^{\infty} h(t)e^{-j2\pi ft}dt$$

then

$$H(f) = \int_{-\infty}^{\infty} G_1(\varsigma)G_2(f-\varsigma)d\varsigma = \int_{-\infty}^{\infty} G_1(f-\varsigma)G_2(\varsigma)d\varsigma \tag{1.2.25}$$

Control system studies involve singularity functions like the Heaviside unit step function:

$$\left.\begin{aligned} U(t) &= 1 \qquad \text{for } t > 0 \\ &= 0 \qquad \text{otherwise} \end{aligned}\right\} \tag{1.2.26}$$

but such functions, and clearly many others, do not possess a Fourier Transformation because the conditions in equation (1.2.16) are not satisfied. However, the product of such functions and exp(-ct).U(t) can generally be transformed provided the real parameter c is sufficiently positive. In this way set:

$$h(t) = \exp(-ct)g(t).U(t)$$

then:

$$H(f,c) = \int_{-\infty}^{\infty} g(t)\ U(t)\ \exp(-c-j2\pi ft)dt$$

$$h(t) = \int_{-\infty}^{\infty} H(f,c)\ \exp(j2\pi ft)df$$

Defining the complex variable:

$$s = c + j2\pi f$$

then straightforward manipulation of the above equations gives the well-known ***Laplace Transformation*** integrals:

$$G(s) = \int_{0+}^{\infty} g(t)\ \exp(-st)dt \qquad (1.2.27)$$

$$g(t).U(t) = \frac{1}{2\pi j} \int_{c-j\infty}^{c+j\infty} G(s)\ \exp(st)ds \qquad (1.2.28)$$

In equation (1.2.27) the strictly positive interval of integration which is implied by the nomenclature o+, is far from academic*. It is absolutely necessary so that equation (1.2.28) can be evaluated by the ***Method of Residues*** (10, 12). Provided the linear system in equation (1.2.12) is initially at rest with:

$$\left.\frac{d^m y}{dt^m}\right|_{t=0} = 0 \qquad \text{for} \qquad m < M$$

then its Laplace Transformation yields:

---

*o+ denotes an arbitrary small positive real number. See Appendix B2.4 for its deeper analytical significance

$$Y(s) = G(s)\ X(s) \tag{1.2.29}$$

where:

$$G(s) = \sum_{m=o}^{M} a_m s^m \Big/ \sum_{m=o}^{M} b_m s^m \tag{1.2.30}$$

is called the *transfer function* of the system. The output y(t) can be derived numerically from the convolution integral (1.2.24) as:

$$y(t) = \int_{o+}^{\infty} g(\varsigma)x(t - \varsigma)U(t - \varsigma)d\varsigma = \int_{o+}^{t} g(\varsigma)x(t - \varsigma)d\varsigma \tag{1.2.31}$$

but the analytical route by the inversion integral (1.2.28) and the method of residues (10, 12) gives:

$$y(t) = \sum \text{Residues of } Y(s)e^{st} \text{ at its poles within the Bromwich Contour}$$

For a qth order pole at s = p:

Residue of Y(s)exp(st) at p

$$= \frac{1}{(q-1)!} \lim_{s \to p} \left[\frac{d^{q-1}}{ds^{q-1}} (s - p)^q Y(s)\exp(st) \right]$$

Because y(t) is a real valued function for <u>all</u> time, the exponential term must be real or have a complex conjugate partner*. That is, the poles of Y(s) must be real or complex conjugate pairs. In many cases of practical interest, a Laplace transformation can be written as:

---

*so together their sum involves a cosine or sine function

$$Y(s) = A(s) / B(s)$$

where A(s) and B(s) are rational or irrational functions; the latter arising from partial differential equations. If Y(s) has a real pole at p, then A(p) is necessarily real and therefore:

$$A(\bar{p}) = \overline{A(p)} \qquad (1.2.32)$$

Similarly, if Y(s) possesses a complex pole pair p and $\bar{p}$, then a real contribution to the time domain function is made only if equation (1.2.32) is satisfied. For the special case when A(s) is a finite polynomial, equation (1.2.32) implies that its coefficients are real, so that its roots (zeros of Y(s)) exist (10, 12) and they are real or complex conjugate pairs (29). Theorems concerning the z-Transformation for the analysis of digital data systems are shown in Chapter 2 to parallel those derived here for the Laplace Transformation.

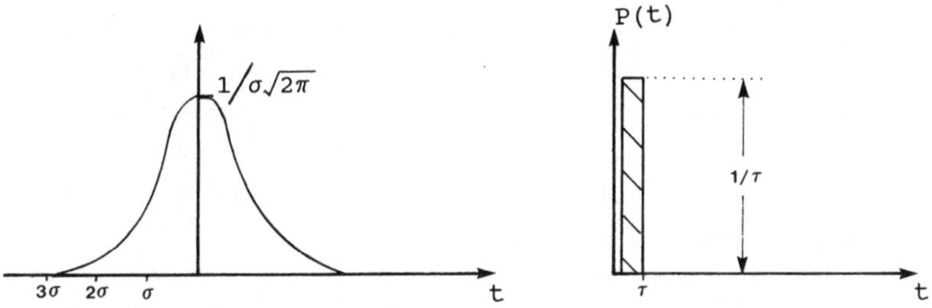

$$\delta(t) = \lim_{\sigma \to 0} \exp\left[-t^2/2\sigma^2\right] / \sigma\sqrt{2\pi}$$

Dirac's Delta Function

$$\delta(t) = \lim_{\tau \to 0} P(t)$$

Engineers' Delta Function

Fig 1.2.2 Delta Functions

The δ- *function* or impulse function was devised by the celebrated atomic physicist Dirac, whose early career was actually in electrical engineering. Fig 1.2.2 illustrates the δ-function used by electrical or control engineers in the context of the Laplace Transformation, and that proposed by Dirac in the context of

Fourier Integrals. Although these two functions differ in form, they share the essential elements:

$$\left.\begin{aligned} &\int_{-\infty}^{\infty} \delta(t)dt = 1 \\ &\int_{-\infty}^{\infty} g(t)\delta(t - a)dt = g(a) \end{aligned}\right\} \qquad (1.2.33)$$

Delta-functions facilitate the manipulation of Fourier and Laplace Transformations, but until the rigorous development by Lighthill (16) they were generally considered by mathematicians as a 'poor man's choice'.

It may well be felt that the frequency domain analysis of continuous data systems has been over-emphasised for a book on DDC systems. However, experience has evoked many examples of unsatisfactory designs that have evolved as a result of an inadequate appreciation of the data sampling process. In the time-domain, "We see through the glass darkly", but the frequency domain analysis in the next section reveals the process of periodically sampling continuous data in total clarity.

1.3 THE PERIODIC SAMPLING PROCESS

A real function of continuous time is referred to in this text as *continuous data*. The relationship between the Fourier transformations of the input and output functions for the linear continuous data system in equation (1.2.12) is derived similarly to equation (1.2.29) as:

$$Y(f) = G(f)X(f) \qquad (1.3.1)$$

where

$$G(f) = G(s)\Big|_{s=j2\pi f} \tag{1.3.2}$$

defines the *real frequency response function.* Equation (1.3.1) shows for example that an input spectral component (power or energy) would be intensified in the output if the real frequency response function has a relatively large magnitude in that frequency range. Because of the 1-1 correspondence between a time domain function and its Fourier transformation in practical situations, linear systems can sometimes offer therefore a conceptually simple method of promoting the significance of a particular additive component in continuous data. As illustrated in Fig 1.3.1, a 'wanted' component which occupies a different spectral frequency band to other 'unwanted' additive components can be relatively enhanced by a suitably engineered real frequency response function. Systems which deliberately promote some spectral components of data relative to others are generically termed *filters.*

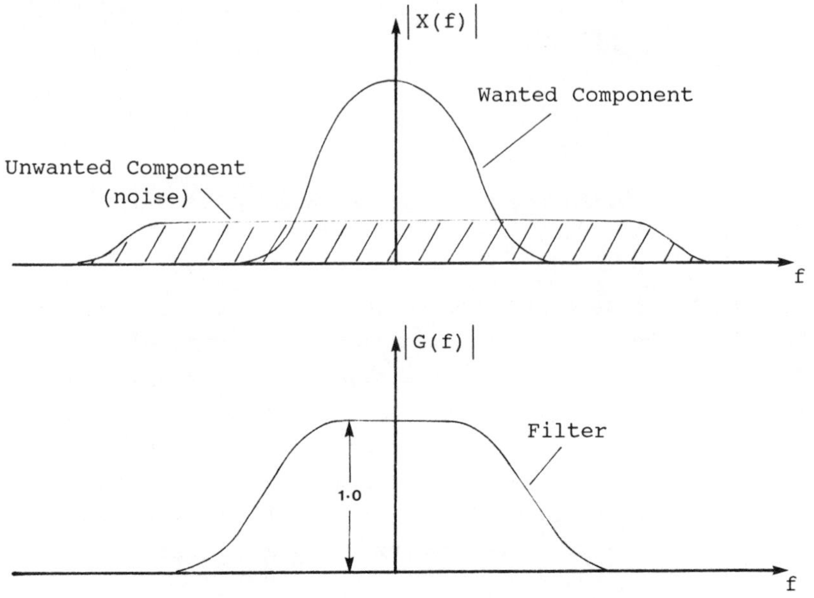

Fig 1.3.1 Attenuation of an Additive Unwanted Signal-Component by Linear Filtering

By virtue of equation (1.3.1), the time domain output (y) of a linear continuous data system is related to its input (x) by the convolution integral (1.2.24) according to:

$$y(t) = \int_{-\infty}^{\infty} g(t - \varsigma)x(\varsigma)d\varsigma \qquad (1.3.3)$$

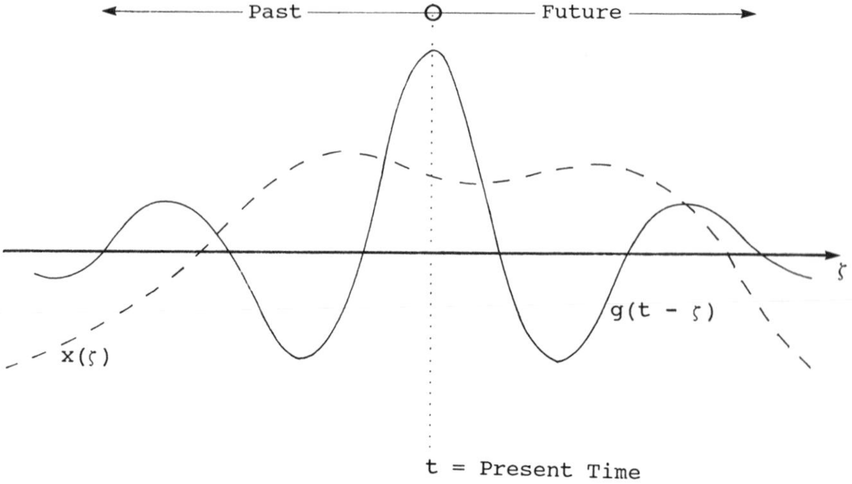

Fig 1.3.2 Illustrating the Convolution Integral (1.3.3)

Noting that $g(t - \varsigma)$ is $g(\varsigma)$ reversed in time (ς) and shifted to the right by t, then terms in the integrand of equation (1.3.3) are as illustrated in Fig 1.3.2. The significance of past and future values of an input in determining the present output is seen to be governed by g; which is therefore described as the *weighting function* of the linear continuous data system. When the timescale corresponds to that of the real world, so-called *real time*, the inability to anticipate the future demands that:

$$g(t) = 0 \qquad \text{for} \qquad t \leq 0 \qquad (1.3.4)$$

Indeed, this view is re-enforced by equations (1.2.33) and 1.3.3) which imply that a weighting function is also the response of the linear system to an impulse (δ-) function applied at time zero*. Practical control systems must evidently be real time systems. The idealised low pass filter:

$$\left. \begin{aligned} G(f) &= 1 \quad \text{for } |f| \leq F/2 \\ &= 0 \quad \text{otherwise} \end{aligned} \right\} \tag{1.3.5}$$

with weighting function:

$$g(t) = F \sin(\pi Ft)/(\pi Ft) \tag{1.3.6}$$

evidently exemplifies a non-real time filter for continuous data. Although this particular filter is of little more than academic interest, non-real time filters based on digital techniques find many applications when the associated time delays are inconsequential (17, 18, 19). While the basic analytical techniques for non-real and real time digital filters apply also to digital controllers, their respective synthesis procedures and application-problems are radically different. Consequently, from Chapter 7 onwards this book becomes specifically orientated towards digital controllers.

Continuous data are transformed into the operational domain of digital computers by means of analogue to digital (A-D) converters, which are usually of electronic (20, 21 22), or occasionally of electro-mechanical (23), construction. Fundamentally, these devices effect the two distinct operations of:

i) successively sampling the continuous data at discrete instants of time

---

*For this reason, a weighting function is sometimes referred to as the Impulse Response. It is also described as the Green's Function or Kernel. Strictly, zero should be replaced by o+ - see later

ii) transforming the amplitudes of these samples into a suitably coded format (eg binary)

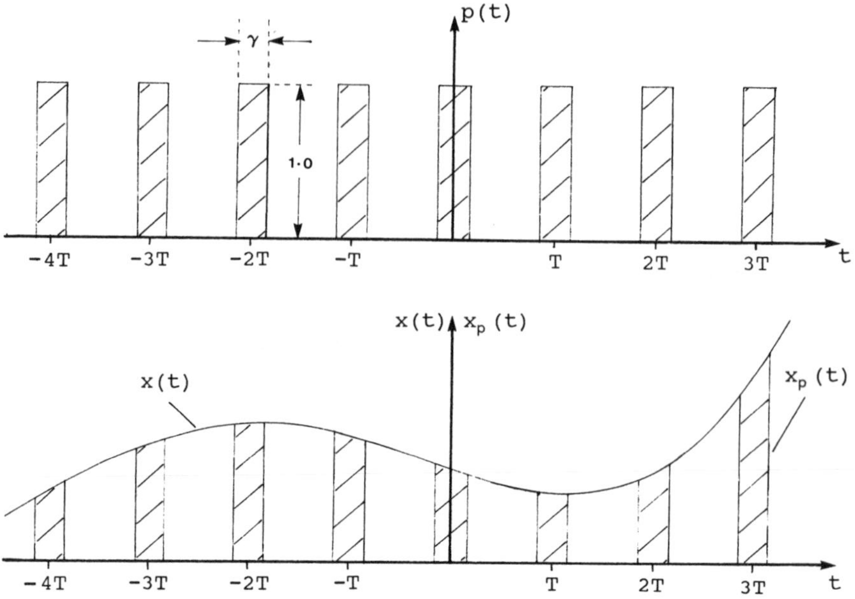

Fig 1.3.3 Periodic Sampling as Idealised Pulse Amplitude Modulation

Some essential properties of various coding schemes are described later in Sections 1.4 and 1.5, but the 1-1 correspondence between code elements and the real numbers enables the following separate analysis of the sampling operation. In many cases it is expedient to sample the continuous data periodically, and under these circumstances the process is closely approximated by the amplitude modulation process in Fig 1.3.3. Samples (x_p) of the continuous data (x) are thus specified by:

$$x_p(t) = p(t).x(t) \tag{1.3.7}$$

where elements of the pulse-train p(t) each have unit height, a narrow width γ, and a mutual separation of T. Expressing the idealised pulse-train as the exponential Fourier series:

$$p(t) = \sum_{-\infty}^{\infty} P_n \exp(j2\pi nt/T)$$

with

$$P_n = \frac{1}{T} \int_{-\gamma/2}^{\gamma/2} \exp(-j2\pi nt/T)dt = (\gamma/T)\sin(\pi n\gamma/T)\Big/(\pi n\gamma/T) \qquad (1.3.8)$$

then provided that the continuous data is Fourier transformable as X(f), the sampled-data has the Fourier transform:

$$X_p(f) = \int_{-\infty}^{\infty} \left\{ \sum_{-\infty}^{\infty} P_n \exp(j2\pi nt/T) \right\} x(t) \exp(-j2\pi ft)dt \qquad (1.3.9)$$

Terms in the above series of functions $\left\{P_n \exp\left[-j2\pi(f-n/T)t\right].x(t)\right\}$ are evidently bounded by $\left\{\left|x(t)\right|/\pi n\right\}$. Consequently, the series is not normally summable with respect to time (t) even with a bounded signal (x), because the series {1/n} is divergent (38, 46). Although an interchange of the order of integration and summation in the above equation is presently desirable, it cannot therefore be justified. In practice, perfectly rectangular pulses cannot be generated due to intrinsic stray capacitance, whose effect can be represented by processing the idealised pulses with the simple lag circuit:

$$G(s) = 1/(1 + s\tau_R)$$

Because electronic circuits are engineered with:

$$\tau_R \ll \gamma$$

actual pulses are a good approximation to the rectangular idealisation, but, mathematically, it is necessary now to allow for their finite rates of change. By equation (1.2.29) the actual Fourier Series coefficients are:

$$P'_n = P_n(1 + j2\pi n\tau_R/T)^{-1} \qquad (1.3.10)$$

No matter how small the realised time constant τ_R, the sequence $\{|P'_n|\}$ eventually converges to zero as fast as $1/n^2$. It follows that this modification renders the series of functions in equation (1.3.9) normally summable, and therefore in practice:

$$X_p(f) = \sum_{-\infty}^{\infty} P'_n X(f - n/T) \qquad (1.3.11)$$

Equation (1.3.11) establishes that the spectrum of the sampled-data (X_p) is in fact a series of spectra. As depicted on the next page in Fig 1.3.4a, each term of this series is the spectrum of the continuous data (X) shifted by an integer multiple of $\pm$ 1/T and weighted by the corresponding Fourier Series coefficient (P'_n).

The spectral component of X_p that is centred about the origin is termed the *base-band*, and the others are referred to as *sidebands*. By virtue of the unique relationship:

$$x(t) = \int_{-\infty}^{\infty} X(f) \exp(j2\pi ft)df \qquad (1.3.12)$$

perfect reconstruction of the continuous data (x) from its samples (x^*) can be achieved only by the perfect separation of the base-band from the sidebands. As illustrated on the next page by Fig 1.3.4b, some sideband energy is inevitably recovered to distort the filtered (reconstructed) data, because real time continuous data systems cannot posses the real frequency response function that is exactly unity over the base-band and zero

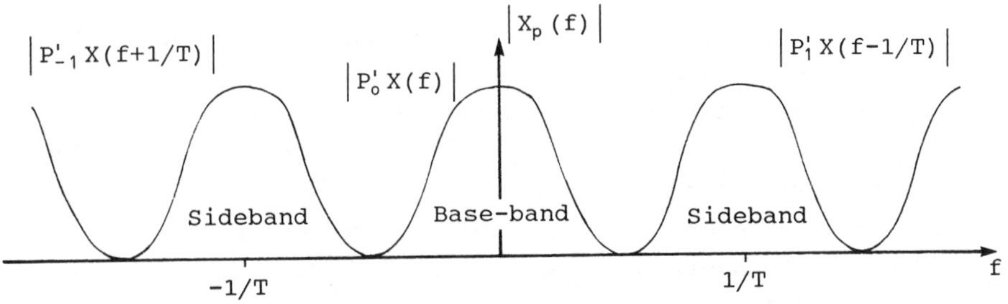

a) Idealised Spectrum of Sampled-Data (Band-limited)

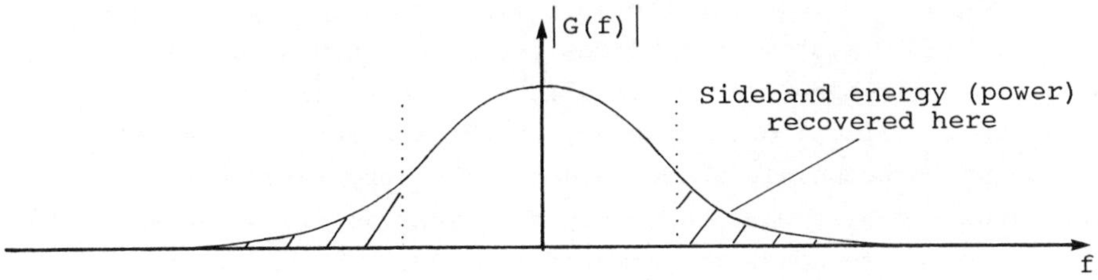

b) Real time Filter for Data Reconstruction

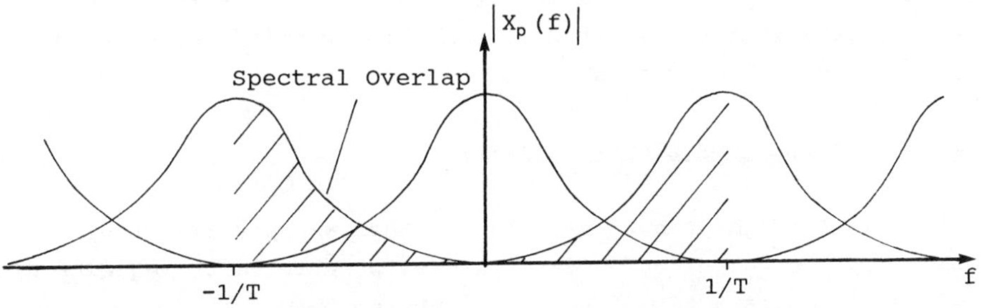

c) Broadening of Pre-sampled Spectrum and Spectral Overlap

Fig 1.3.4 Sources of Distortion in the Reconstruction of Sampled-Data

elsewhere (5)*. In addition, some overlap of the base-band and sidebands occurs in practice due to an extension of pre-sampled spectral bandwidth by:

---

*See also equations 1.3.5 and 1.3.6

i) unwanted components (noise) from a variety of sources
ii) the abrupt start, and possibly termination, of the continuous data
iii) intermodulation by non-linearities and interactions*

Consequently, continuous data is usually pre-filtered to reduce unwanted components prior to sampling, in order to reduce distortion in its subsequent reconstruction. Such filters, like that in Fig 1.3.1, are termed *anti-aliasing* though clearly their operation can never be perfect. With the inevitability of imperfect data reconstruction in real time, the sampling rate (1/T) of a DDC system must be designed high enough to reject sufficient sideband-energy (power) as is necessary to meet its accuracy specification. However, unlike the design of a filter for signal processing, whose structure is very largely arbitrary, a control engineer must take very much what is given in terms of the low-pass frequency response of a plant. The analysis in Chapter 5 places a lower-bound on the sampling rate in this context, but as described in Chapter 7 wordlength considerations can somewhat surprisingly decide its upper-bound. Next in this chapter, attention is directed at some binary coding schemes that are employed in DDC systems.

1.4 SOME ASPECTS OF BINARY ARITHMETIC

Given any integer r not less than 2, then any positive real number (b) can be uniquely identified with an infinite sequence of positive integers β_0 β_1 β_2 β_3 ...... where:

β_0 - positive integer ; $0 \le \beta_m \le r - 1$ for $m \ge 1$

$$\text{and} \quad b = \beta_0 + \sum_{m=1}^{\infty} \beta_m 2^{-m} \tag{1.4.1}$$

---

*Expand a non-linear gain characteristic as a Maclaurin Series, and suppose its input is $\sin\omega t$, and then $\sin\omega_1 t + \sin\omega_2 t$

Although everyday life is concerned with decimal numbers ($r = 10$), digital computers perform arithmetic operations with binary numbers ($r = 2$), in which the on-off states of electronic devices represent the 0 and 1 *bits* of the sequence $\{\beta_m\}$. With early mainframe computers like the Ferranti Pegasus, mathematical equations required pre-scaling so that computed variables had magnitudes less than unity* in order to achieve compatibility with their *fixed-point arithmetic units.* Since then, developments in digital techniques and solid state electronics have enabled the realisation of *floating-point arithmetic units* for numbers with magnitudes larger than unity. These circuits have greatly eased the programming of complex scientific problems, and software equivalents have enabled the now widespread public use of personal computers. On the other hand, digital controllers and filters involve relatively simple equations whose variables are easily pre-scaled by their professional designers, so that the less expensive fixed-point units currently offer the most cost-effective realisations. Accordingly, this text concerns itself exclusively with fixed-point arithmetic, in which positive real numbers less than unity are approximated by finite binary sequences of the form $0.\ \beta_1\ \beta_2\ \beta_3\ \ldots\ldots\ \beta_{M-1}$ where:

. – *binary point* ; M – *wordlength* ; β_{M-1} – *least significant bit*

and

$$2^{-M+1} = q \tag{1.4.2}$$

is the *width of quantisation* . Two established conventions extend the above representation to positive and negative real numbers in the range [-1, 1]. Adopting the *Sign-Magnitude convention,* the binary sequence $\beta_0\ \beta_1\ \beta_2\ \ldots\ldots\ \beta_{M-1}$ uniquely equates with the real number:

---

*This was once described as the 'Scientific Programming Convention'. Because financial institutions accounted to the nearest £1, the ascending power series was termed the 'Commercial Convention'

$$b = (1 - 2\beta_0)\left\{\sum_{m=1}^{M-1} \beta_m 2^{-m}\right\} \qquad (1.4.3)$$

so that for example:

$$0.11 = 3/4 \quad \text{and} \quad 1.11 = -3/4$$

In the *Two's-Complement convention*, the binary sequence $\beta_0 \; \beta_1 \; \beta_2 \; \cdots\cdots \; \beta_{M-1}$ equates uniquely with the real number:

$$b = -\beta_0 + \sum_{m=1}^{M-1} \beta_m 2^{-m} \triangleq V(\beta) \qquad (1.4.4)$$

so that for example:

$$0.11 = 3/4 \quad \text{and} \quad 1.11 = -1/4$$

Because

$$\sum_{m=1}^{M-1} 2^{-m} = 1 - 2^{-M+1} = 1 - q \qquad (1.4.5)$$

the ranges of real numbers accommodated by the fixed-point sign-magnitude and two's-complement conventions are [-1+q, 1-q] and [-1, 1-q] respectively. *Overflow* occurs when an arithmetic operation produces a corresponding real number outside the machine-range (eg 1/2 + 1/2). Unlike fixed-point sign-magnitude arithmetic, the two's complement addition of three or more numbers is shown later to be immune to *intermediate overflows* provided their total lies in machine range. Another advantage of two's complement arithmetic is the simpler electronic design of addition and subtraction circuits (24, 25). Consequently, digital controllers and filters are usually implemented with fixed-point two's-complement circuits (30), and subsequent analysis relates to this particular representation.

As with decimal numbers, the fixed-point two's-complement addition of two binary sequences proceeds as a sequential bit-by-bit with carry process starting at the least-significant bit. However, due to the finite structure of the representation, a carry beyond the ***most-significant bit*** **(β_0)** is deliberately discarded. For example,

$$3/4 + (-1/4) = 1/2$$

discarded

$$011 \oplus 111 \quad = 1\,010 \;\; (= 1/2)$$

where the $\oplus$ is used to emphasise this distinction between the addition process for conventional decimals and that for finite two's-complement binary sequences. Subtraction of two real numbers a and b with corresponding binary sequences α and β is implemented as the addition process:

$$a - b = a + (-b)$$

where in fixed-point two's-complement representation, it is readily verified that:

$$-b = -\bar{\beta}_0 + \sum_{m=o}^{M-1} \bar{\beta}_m 2^{-m} + 2^{-M+1}$$

with

$$\beta_m + \bar{\beta}_m = 1 \qquad \text{for all m} \qquad (1.4.6)$$

Because any bit (β_m) is the output from one of two coupled semiconductor devices forming a bistable pair (flip-flop), its complement ($\bar{\beta}_m$) is available as the output of the other device.

It is now shown that the fixed-point two's-complement addition of two machine-numbers $V(\alpha)$ and $V(\beta)$ provides the correct total $V(\alpha) + V(\beta)$; provided this also lies in the machine-range $[-1, 1-q]$. In symbolic terms, the assertion is that if:

$$\left.\begin{array}{l} -1 \leq V(\alpha) + V(\beta) \leq 1 - q \\ \text{then} \\ V(\alpha \oplus \beta) = V(\alpha) + V(\beta) \end{array}\right\} \quad (1.4.7)$$

Towards this end, define the fractional part of any machine-sequence γ as:

$$\gamma_f = 0\ \gamma_1\ \gamma_2\ \gamma_3\ \ldots\ldots\ \gamma_M \quad (1.4.8)$$

so that:

$$0 \leq V(\gamma_f) \leq 1 - q \quad (1.4.9)$$

Two alternative situations can occur in the addition process:

$$V(\alpha_f) + V(\beta_f) \leq 1 - q \qquad \text{no carry}$$

or

$$V(\alpha_f) + V(\beta_f) > 1 - q \qquad \text{a carry}$$

In the former:

$$V((\alpha \oplus \beta)_f) = V(\alpha_f) + V(\beta_f) \quad (1.4.10)$$

while in the latter:

$$V((\alpha \oplus \beta)_f) = -1 + V(\alpha_f) + V(\beta_f) \quad (1.4.11)$$

If there is no carry then -

i) $\alpha_0 = \beta_0 = 0$ implies $V(\alpha) = V(\alpha_f)$, $V(\beta) = V(\beta_f)$ and $V(\alpha \oplus \beta) = V((\alpha \oplus \beta)_f)$.
Hence from equation (1.4.10), $V(\alpha \oplus \beta) = V(\alpha) + V(\beta)$

ii) $\alpha_0 = 1$ and $\beta_0 = 0$ implies $V(\alpha) = -1 + V(\alpha_f)$, $V(\beta) = V(\beta_f)$ and $V(\alpha \oplus \beta) = -1 + V((\alpha \oplus \beta)_f)$.
Hence from equation (1.4.10), $V(\alpha \oplus \beta) = V(\alpha) + V(\beta)$

iii) $\alpha_0 = \beta_0 = 1$ implies as above that $V(\alpha) + V(\beta) = -2 + V(\alpha_f) + V(\beta_f) = -2 + V((\alpha \oplus \beta)_f)$.
Because there is no carry, then $V((\alpha \oplus \beta)_f) \leq 1 - q$ so that $V(\alpha) + V(\beta) < -1$. This inequality contradicts the hypothesis that the sum $V(\alpha) + V(\beta)$ lies in machine-range, so that $\alpha_0 = \beta_0 = 1$ is untenable and the case can be discounted (29).

If there is a carry so that equation (1.4.11) obtains, then exactly similar logic demonstrates the validity of equation (1.4.7) for this situation as well. Moreover, it is readily extended to an arbitrary finite number of machine-numbers by mathematical induction.

Although the above proof may appear somewhat academic, even possibly obvious, the same technique can be used to establish the important immunity of two's-complement addition to intermediate overflow. Symbolically, the assertion is that if:

$$\left.\begin{array}{l} -1 \leq V(\alpha) + V(\beta) + V(\gamma) \leq 1 - q \\ \text{then} \\ V((\alpha \oplus \beta) \oplus \gamma) = V(\alpha) + V(\beta) + V(\gamma) \end{array}\right\} \qquad (1.4.12)$$

If intermediate overflow does not occur during the addition of three machine-numbers $V(\alpha)$, $V(\beta)$ and $V(\gamma)$, then it follows directly from equation (1.4.7) that:

$$(V(\alpha) + V(\beta)) + V(\gamma) = V(\alpha \oplus \beta) + V(\gamma) = V((\alpha \oplus \beta) \oplus \gamma))$$

An intermediate overflow can occur only if the numbers $V(\alpha)$ and $V(\beta)$ have the same sign. If these numbers are both positive, then $V(\gamma)$ must be negative to ensure a total within machine-range. Thus:

$$V(\alpha) + V(\beta) + V(\gamma) = V(\alpha_f) + V(\beta_f) - 1 + V(\gamma_f)$$

and as

$$V(\alpha_f) + V(\beta_f) = 1 + V((\alpha \oplus \beta)_f)$$

then

$$V(\alpha) + V(\beta) + V(\gamma) = V((\alpha \oplus \beta)_f) + V(\gamma_f)$$

By hypothesis the left-hand-side of the above equation is in the machine-range, and likewise for each of the fractional parts on the right-hand-side, so that by equation (1.4.7):

$$V(\alpha) + V(\beta) + V(\gamma) = V((\alpha \oplus \beta)_f \oplus \gamma_f)$$

Under the assumed conditions:

$$\alpha + \beta = 1.(\alpha \oplus \beta)_f \qquad \text{and} \qquad \gamma = 1.\gamma_f$$

then as required:

$$V(\alpha) + V(\beta) + V(\gamma) = V((\alpha \oplus \beta) \oplus \gamma)$$

Exactly similar logic proves the result for when $V(\alpha)$ and $V(\beta)$ are both negative and $V(\gamma)$ necessarily positive. Unlike the situation without overflow, equation (1.4.12) is not so readily extended by mathematical induction. Nevertheless, it can be established for an arbitrary finite number of terms by a more recondite analysis based on algebraic congruences and finite fields (29, 33). Linear time invariant digital controllers and filters are characterised by finite difference equations of the form:

$$\sum_k b_k y_{n-k} = \sum_k a_k x_{n-k}$$

where

$\{x_k\}$ - input sequence ; $\{y_k\}$ - output sequence ;

$\{a_k\}$ and $\{b_k\}$ - finite sets of constants

In this context, it should be observed that multiplication of machine-numbers cannot cause overflow, and that division is not normally involved in the realisation of such systems because the recurrence relation can always be scaled to make:

$$b_0 = 1$$

Overflow, which can occur in implementing the summations, destroys linearity, and the performance of a system would be degraded or even destabilised by the induced distortion. The simplification afforded by equation (1.4.12) in *pre-scaling* the input data to prevent such overflows is therefore of considerable practical importance (35, 36).

Multiplication of two fixed-point two's-complement binary sequences is usually effected in digital controllers and filters by Booth's algorithm (32, 34). It suffices to note for present purposes that the double-length product is produced in two coupled registers, which contain its most- and least-significant halves. However, the least-significant half must be eventually discarded in order to achieve compatibility with subsequent machine-operations. As a result, a *multiplicative rounding error* is incurred and in this respect multiplication differs from the addition operation which is performed perfectly without error. If the least-significant half is totally rejected as for example:

$$001\ 110 \longrightarrow 001$$

the product is said to be *unrounded*. Whereas, if half of one least-significant bit is added before its rejection as for example:

$$\left.\begin{array}{r} 001\ 110 \\ \\ +\ 100 \end{array}\right\} \longrightarrow 010$$

the product is said to be *rounded*. The operation of multiplying two machine-numbers a and b is conveniently denoted by:

$$ab' = ab + \epsilon \tag{1.4.13}$$

where

ab'- actual single-length value retained by the machine

ab - true double-length value

ϵ - multiplicative rounding error

It follows from the above discussion that multiplicative rounding errors are bounded by:

$$\left.\begin{array}{ll} -q < \epsilon \leq 0 & \text{for unrounded multiplication} \\ -q/2 < \epsilon \leq q/2 & \text{for rounded multiplication} \end{array}\right\} \tag{1.4.14}$$

Multiplication of input or output samples by the generally constant parameters of a digital controller or filter can be viewed therefore as an unavoidable introduction of noise sources into these systems. These noise sources evidently degenerate the accuracy of control or the signal to noise ratio of a filter to some extent, and their mean square value is shown in Chapter 6 to be a key parameter. Consequently, rounded multiplication is usually implemented in practice because its symmetric error distribution has intrinsically the smaller mean square value. In that the rounding error noise in digital systems is caused by amplitude quantisation into an integral numbers of least-significant bits, it is analogous to noise in continuous data electronic systems that originates in part from the discreteness of electrical charge.

1.5 DATA CONVERTERS

Data is transformed into the operational domain of a digital computer by means of an analogue-digital (A-D) converter, and a digital-analogue (D-A) converter effects the inverse operation. Comprehensive technical information on these devices, rather than just sales rhetoric, is available in literature (20, 21, 26, 27) published by several United States companies which include: Digital Equipment, Analogic, Burr-Brown and Hewlett-Packard. The significant educational value of these and other similar reports merits commendation. Certain types of A-D and D-A converters are particularly cost-effective for digital control systems and these are now described quite briefly in order to provide a background for later theoretical developments.

D-A converters are described first of all because they form part of a frequently deployed type of A-D converter. Fig 1.5.1 on the next page illustrates the so-called parallel-bit voltage-steering D-A converter (22), in which current-sources proportional to the weight of each bit are constructed from a reference voltage-source and a binary-scaled resistor network. According to the value of a bit (0 or 1), its particular current-source is *steered* into the input of the operational amplifier, which always remains at essentially zero potential because an enormous gain($\simeq$100 dB) is available over the range of signal frequencies ($\lesssim$ 100 Hz). Consequently, the output of the amplifier is given by:

$$V_0 = \left[-\beta_0 + \sum_{m=1}^{M} \beta_m 2^{-m} \right] I_0 R_0 \tag{1.5.1}$$

where

$$I_0 = E_{ref}/R \tag{1.5.2}$$

Because currents drawn from the reference source are steered rather than switched on and off, the input voltage to the binary-scaled resistor network remains very largely constant apart from small transient *glitches* (22), and small more easily compensated thermal effects. Another somewhat more accurate design of D-A converter steers very high impedance weighted current-sources into the input of an operational amplifier or ground (22). Realisations of these designs as integrated circuits achieve the ideal

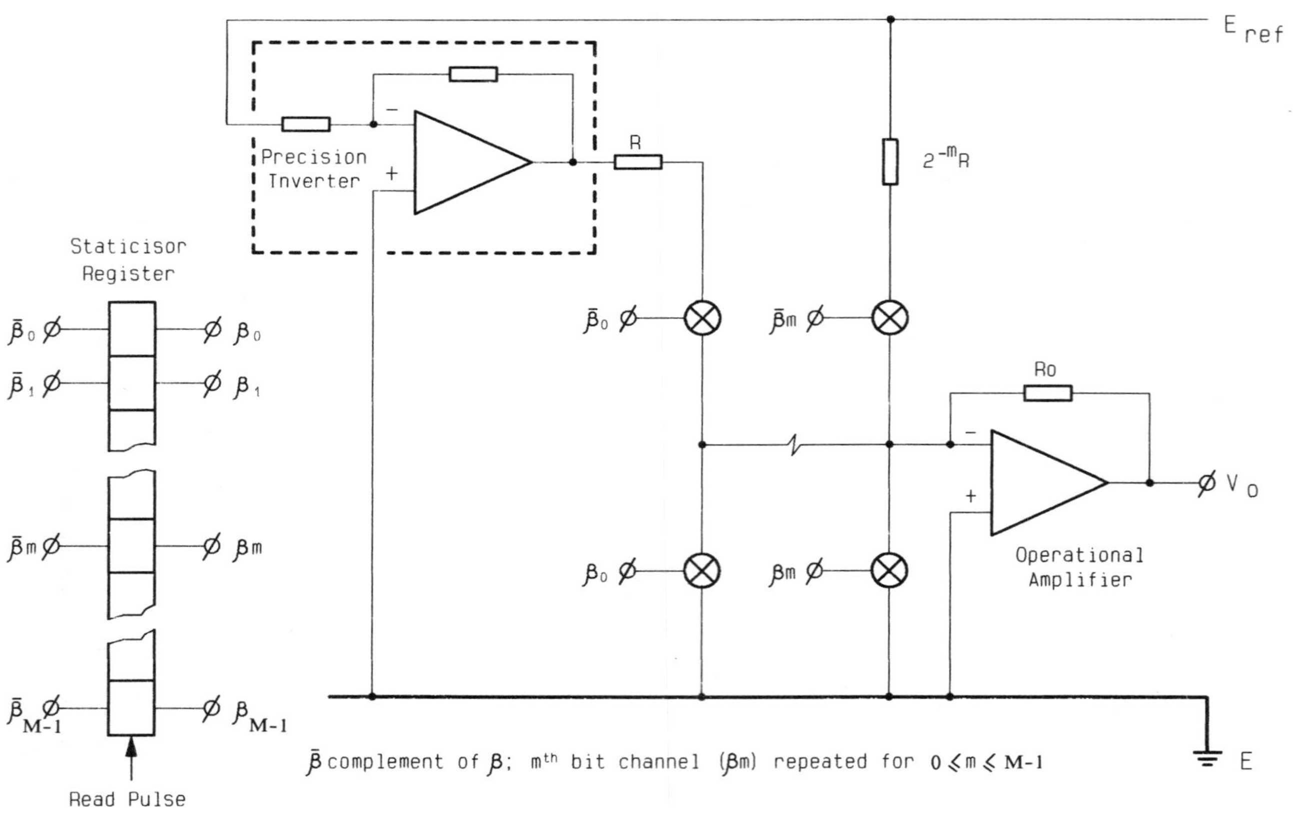

$\bar{\beta}$ complement of β; m[th] bit channel (βm) repeated for $0 \leqslant m \leqslant M-1$

Fig 1.5.1 Parallel-bit Voltage-steering D-A Converter

linearity of equation (1.5.1) within 12 to 18 bit accuracies and 0.05 to 10 μs settling times.

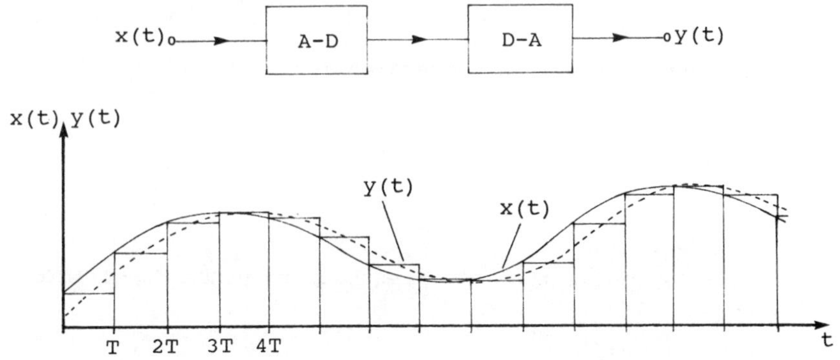

Fig 1.5.2 Illustrating the Phase-Lag associated with D-A Conversion

For present purposes, the electronic circuit design of D-A converters is secondary to their dynamic behaviour as control system elements. In this context, a binary number from a digital controller regularly replaces that already held in the staticisor register of its D-A converter every output period. These sample values are converted into a regular staircase function which represents a partial reconstruction of the sampled-data. As can be visualised from the cascaded A-D then D-A conversion processes in Fig 1.5.2, the D-A conversion introduces a phase-lag which must be quantified for the design of closed loop control systems. Derivation of the transfer function for a D-A converter is conceptually complicated by the mixed arrangement of digital and analogue elements. However, an easily analysed model that reproduces the same response as an actual device is shown on the next page in Fig 1.5.3. Here the imaginary **δ-** *unit* delivers a δ-function of current or voltage etc with an area that is proportional to the binary number entered into the output staticisor (register) of a digital controller. By regarding the δ-unit and staticisor as integral parts of a D-A converter, its Laplace transfer function is readily derived from Fig 1.5.3 as:

$$H(s) = K_{DA}\left[1 - \exp(-sT)\right]/s \qquad (1.5.3)$$

where

K_{DA} - gain constant of D-A conversion (eg mA/l.s. bit)

Setting

$$s = j2\pi f$$

in equation (1.5.3) yields the real frequency response function of the converter as:

$$H(f) = K_{DA}T\ \left|\sin(\pi fT)/\pi fT\right|\ \exp(-j\pi fT) \qquad (1.5.4)$$

from which its intrinsic phase-lag is self-evident. Although implementations of linear, quadratic or higher order interpolations of the output samples are all quite practicable, these create proportionately greater phase-lag (28). Because phase-lag aggravates the stabilisation of a feedback control system, it is usual practice to deploy the above interpolation provided by a D-A converter, and this simplest scheme is variously referred to as: a *zero-order hold* , a *clamp* or a *box-car circuit.*

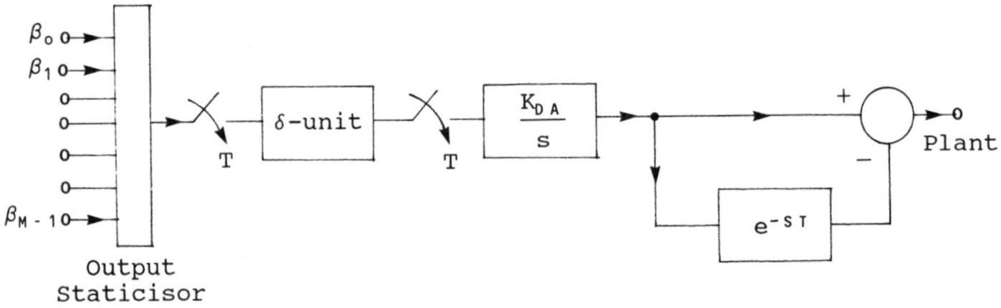

Fig 1.5.3 A Mathematical Model of a D-A Converter

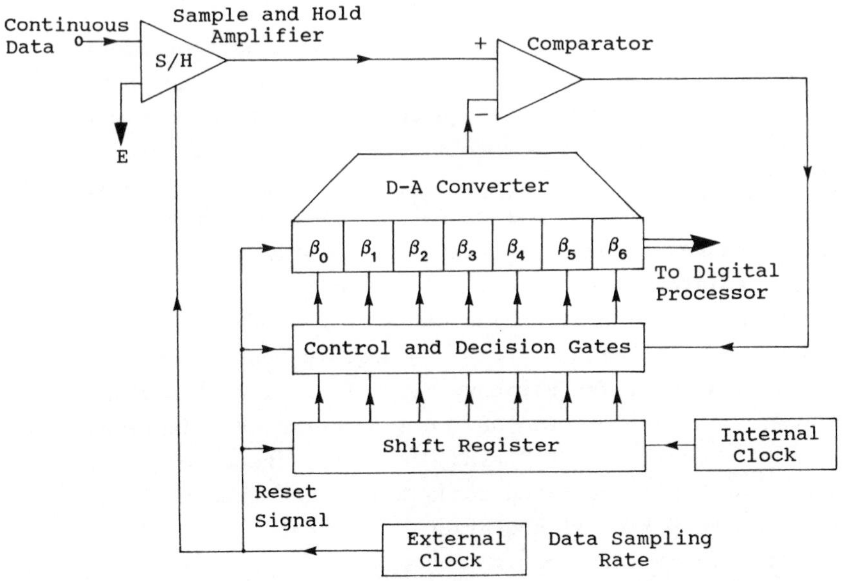

Fig 1.5.4 Illustrating a Successive Approximation A-D Converter

Many transducers convert a physical quantity such as pressure, temperature or mass flow etc into a closely proportional electrical voltage or current. For subsequent digital processing, this continuous data must be sampled and transformed into a sequence of binary numbers. The shift-programmed successive approximation architecture is perhaps the best known and most commonly employed A-D converter for this purpose (22). Although flash converters (20, 22) offer the higher sampling rates necessary for video signals, and although dual-slope and voltage to frequency converters are less expensive, the successive approximation device in Fig 1.5.4 is often the most cost-effective for digital control applications. In operation, the *sample and hold amplifier* provides virtually instantaneous sampling* and then storage during the subsequent conversion process. Though not

*Typical acquisition times are currently 0.02 to 10 μs

an essential component of an A-D converter, a sample and hold amplifier can increase its bandwidth by as much as four orders of magnitude (26). After initiating the acquisition of an input sample, the shift register (25) and output register are reset to binary 100...0 and 000...0 respectively. If the comparator output is high, the most significant bit (β_0) remains at 0 ; otherwise it is reset to 1. Following this decision, a high repetition rate internal clock moves the shift register contents to 010...0 and the output register to $\beta_0$10...0. If the output of the comparator is again high, the bit β_1 remains set at 1 but otherwise it is reset to 0. This trial and error sequence is repeated with successive bits until the ultimate resolution of better than q/2 is achieved. In essence, the device implements the classical mathematical technique of Dedekind Cuts (29); except of course that a real number is resolved to just a finite number of bits after the finite number of operations. Typical contemporary devices offer 12 to 16 bits with conversion times of 1 to 25 μs. Such delays are generally negligible from the viewpoint of designing a digital control system, and an A-D converter is characterised by a simple gain constant (K_{AD}) whose dimensions are l.s. bits per mA or mV. However, as described in Chapter 6, the limited resolution of amplitude can be regarded as equivalent to an additive noise process, which under certain circumstances can initiate a significant degeneration in the signal to noise ratio of a digital filter or the accuracy of a digital control system.

The *Cyclic Permuting* (CP) or *Gray code* forms the basis of an early electro-mechanical type of A-D converter for directly encoding the angular position of a rotating shaft (23). Positive binary numbers are translated into CP code by the algorithm:

$$\left.\begin{array}{lll} \beta_m \rightarrow \beta_m & \text{if} & \beta_{m+1} = 0 \\ \beta_m \rightarrow \bar{\beta}_m & \text{if} & \beta_{m+1} = 1 \end{array}\right\} \qquad (1.5.5)$$

and vice-versa according to:

$$\left.\begin{array}{llll} \beta_o \rightarrow \beta_o \quad ; \quad \beta_m \rightarrow 1 & \text{if} & \sum_{k=o}^{m} \beta_k & \text{odd} \\ \qquad\qquad\quad \beta_m \rightarrow 0 & \text{if} & \sum_{k=o}^{m} \beta_k & \text{even} \end{array}\right\} \qquad (1.5.6)$$

Fraction	Binary	CP
0	000	000
1/8	001	001
1/4	010	011
3/8	011	010
1/2	100	110
5/8	101	111
3/4	110	101
7/8	111	100

Table 1.5.1 Three bit Binary and Cyclic Permuting Codes

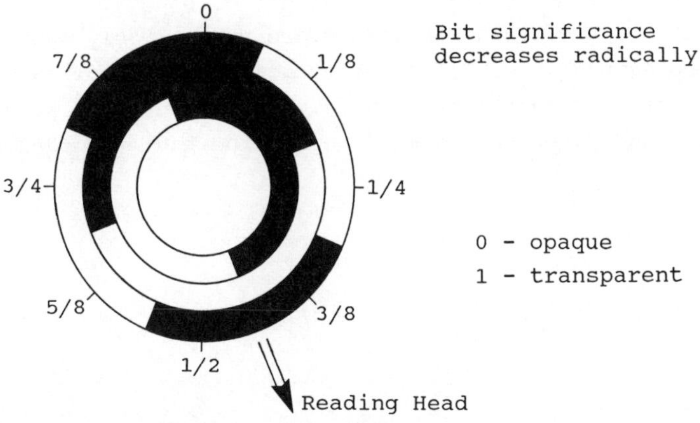

Fig 1.5.5 A Three bit Cyclic Permuting or Gray code Shaft Position Encoder

Table 1.5.1 lists the three bit binary and CP coded numbers that correspond to the real fractions 0, 1/8, 1/4, ... 7/8. For illustrative purposes an optical three bit CP coded disc, which defines just eight radial positions, is shown on the previous page in Fig 1.5.5. Due to inherently imperfect rejection of random load disturbances like wind gusts, sea waves etc, the output shaft of a servo-system 'dithers' slightly. Also small manufacturing tolerances exist on the locations of the transparent and opaque segments on a disc. In this practical environment, the relative advantages of CP over raw binary coding can be appreciated by considering the read-out head to be positioned just inside the 3/8 segment and closely adjacent to the 1/2 as in Fig 1.5.5. Variations in the output of the CP disc due to dither can create outputs of 011 or 010, which represent an uncertainty of about half the width of quantisation (q/2). On the other hand, Table 1.5.1 reveals that all the possible output states {000, 001, 010, 011, 100, 101, 111} could occur with binary coding under the same circumstances. Barker (23) provides a thorough analysis of static and dynamic reading errors in a CP encoder, and he also suggests a novel read-out scheme (V-scan) for further reducing the errors. After reading the encoded shaft position, it can be translated into two's complement format for easier arithmetic processing by means of a specific integrated circuit (31). Optical shaft encoders are still widely deployed, but the more extensive use of stepping motor drives now favours the lower cost incremental type (27), which measures relative displacement and direction of motion from the last recorded position.

CHAPTER 2

Deterministic Input Signal Analysis

It is not certain that everything is uncertain. - B Pascal

2.1 THE SINGLE-SIDED z- TRANSFORMATION

Chapter 1 describes briefly various electronic and electro-mechanical interfaces that enable the translation of continuous data into the coded sampled-data format of a digital computer; and vice versa. Also the sampling rate (1/T) is identified as a particularly crucial design parameter, but a design procedure for its prudent selection is deferred until Chapter 5 to allow development of the requisite analytical techniques. Towards this end, attention now focuses on the ***single-sided z- Transformation*** as an important design tool for linear sampled-data systems with deterministic inputs. A ***deterministic sequence*** takes values that are not subject to the uncertainties of probability. Simple examples are:

$$x_k = 1 \quad \text{for} \quad k \geq 0$$
$$= 0 \quad \text{otherwise} \qquad \text{unit step sequence} \qquad (2.1.1)$$

and

$$x_k = \sin\omega kT \quad \text{for} \quad k \geq 0$$
$$= 0 \quad \text{otherwise} \qquad \text{sinusoidal sequence} \qquad (2.1.2)$$

The z- Transformation X(z) of an arbitrary single-sided sequence $\{x_k\}$ is defined by:

$$X(z) = \sum_{k=0}^{\infty} x_k \ z^{-k} \qquad (2.1.3)$$

where

z- complex variable

As described in Section 1.2, Abel's lemma and other tests (10, 12, 38) are available to establish the normal summability of such series; and therefore the existence of a particular z-Transformation. However, it is often unnecessary to apply such tests in the engineering design of digital controllers or filters. The relevant sampled-data sequences generally originate from continuous data with well-known Laplace transformations, that possess just finite order isolated singularities (poles) in any finite region of the complex s- plane*. The convergence and existence of their corresponding z- Transformations is examined in detail later in Section 2.4. For the moment, it suffices to assume that the series in equation (2.1.3) is normally summable for:

$$|z| > R > 0 \qquad (2.1.4)$$

and that this corresponds to a region (S) of the z- plane in which X(z) has no poles. Because a normally summable series is always point-wise (or simply) convergent (10), the existence of X(z) is then assured with:

$$x_0 = X(z)\Big|_{z^{-1} = 0} \qquad (2.1.5)$$

which is sometimes written as (28, 37):

$$x_0 = \lim_{z \to \infty} X(z) \qquad (2.1.6)$$

and termed the *Initial Value Theorem.* The initial response (x_0) of a practical digital controller or filter is generally zero due

---

* so-called 'meromorphic functions'

to computational delays, so that equation (2.1.5) or (2.1.6) can be applied as a check on the algebraic manipulation of z-transformations.

By applying the principle of mathematical induction to finite sub-series, two normally convergent inverse power series can be shown to define the same function if and only if their respective coefficients are identical (38). A sequence and its z-transformation are therefore uniquely related. The following analysis provides an integration technique for recovering a sequence $\{x_m\}$ from its z- transformation X(z), and thereby offers as well an alternative proof of this 1-1 correspondence. Multiplying a summable series $\{x_k z^{-k}\}$ by z^{n-1}, where n is a constant integer, can be effected term by term, and its region of normal summability (S) is unaltered (10, 12, 38). Furthermore, integration of a function defined by a normally summable series can also be implemented as a term by term operation (10, 12), so that for <u>any</u> closed contour (C) within S:

$$\int_C X(z)\, z^{n-1} dz = \sum_{k=o}^{\infty} x_k \int_C z^{n-1-k} dz \tag{2.1.7}$$

By virtue of equation (2.1.4) the contour C encloses the origin, so that (10, 12):

$$\left.\begin{aligned} \int_C z^{n-1-k} dx &= 2\pi j \quad \text{for} \quad k = n \\ &= 0 \text{ otherwise} \end{aligned}\right\} \tag{2.1.8}$$

Thus equation (2.1.7) reduces to:

$$x_n = \frac{1}{2\pi j} \int_C X(z) z^{n-1} dz \tag{2.1.9}$$

which constitutes the *Inversion Integral* of the z-Transformation. Cauchy's theory of residues (10, 12) enables the above integral to be evaluated as:

$$x_n = \sum_m \text{Residue of } \left[X(z)z^{n-1}\right] \text{ at } p_m \qquad (2.1.10)$$

where

$\{p_m\}$ - the set of poles for $X(z)z^{n-1}$

and if p_m is a qth order pole then:

Residue of $\left[X(z)z^{n-1}\right]$ at p_m

$$= \frac{1}{(q-1)!} \lim_{z \to p_m} \left\{ \frac{d^{q-1}}{dz^{q-1}} \left[(z - p_m)^q \, X(z)z^{n-1} \right] \right\} \qquad (2.1.11)$$

Because $\{x_n\}$ is a real valued sequence, it follows from equation (2.1.11) by the same argument as in Section 1.2 that the poles of $X(z)$ are real or complex conjugate pairs. Also for the special case when its numerator is a finite polynomial, the zeros of $X(z)$ exist and they too are real or complex conjugate pairs.

To demonstrate the application of these results consider the sequence $\{z^{-k} \exp(sk)\}$, which is shown in Section 1.2 to be normally summable for:

$$|z| > |e^s|$$

with:

$$X(z)z^{n-1} = z^{n-1}/(1 - e^s z^{-1}) = z^n/(z - e^s)$$

This particular function clearly possesses just one first-order pole at

$$z = e^s$$

which can obviously be located within an admissible contour of integration (C). Its corresponding residue is derived from equation (2.1.11) as:

$$\text{Residue of } \left[\frac{z^n}{z - e^s}\right] \text{ at } e^s = \lim_{z \to e^s} \left\{ (z - e^s) \frac{z^n}{z - e^s} \right\} = e^{sn}$$

so that according to equation (2.1.10):

$$x_n = e^{sn}$$

which matches the terms of the original sequence.

Although the above example demonstrates the recovery of a sequence from its z- Transformation by the Inversion Integral, the numerical procedure described in the following Section 2.2 is generally far more effective. Indeed, the principal application of the Inversion Integral lies in the formulation of analytical relationships, such as that now derived between the behaviour of a sequence $\{x_k\}$ and the pole locations $\{p_m\}$ of its z-Transformation. Apart from possibly the case of n = 0, the poles of $X(z)\ z^{n-1}$ are those of X(z), and equation (2.1.10) shows that each of these poles makes its own particular contribution to the value of the terms in the corresponding sequence $\{x_k\}$. Furthermore, equation (2.1.11) and the classical formula for differentiating a product of functions implies that the growth or decay of the contribution from a qth order pole is governed by factors of the form $k^q p_m^k$.

It is shown in Appendix A2.1 that:

$$\lim_{k \to \infty} k^q p_m^k = 0 \quad \text{for} \quad |p_m| < 1 \qquad (2.1.12)$$

Hence, if all the poles $\{p_m\}$ of X(z) have magnitudes strictly less than unity, equation (2.1.10) and (2.1.12) establish that the sequence $\{x_k\}$ converges to zero. Symbolically:

$$\left\{ |p_m| < 1 \text{ for all } m \right\} \text{ implies } \left\{ \lim_{k \to \infty} x_k = 0 \right\} \qquad (2.1.13)$$

On the other hand, if there exists even one pole with a magnitude not strictly less than unity, its contribution and therefore the sequence cannot converge to zero. Symbolically:

$$\left\{ |p_m| \nless 1 \text{ for some } m \right\} \text{ implies } \left\{ \lim_{k \to \infty} x_k \neq 0 \right\} \qquad (2.1.14)$$

Statements (2.1.12) and (2.1.14) together assert that an arbitrary sequence $\{x_k\}$ converges to zero if and only if the poles of its z-transformation lie strictly within the unit circle $|z| = 1$.

In the special circumstances when a z- Transformation has all its poles within the unit circle, except for a first order pole at $1\angle 0$, the above discussion shows that the corresponding sequence converges to the contribution provided by just this particular pole. Accordingly, it follows from equation (2.1.10) and (2.1.11) that

$$\lim_{k \to \infty} x_k = \lim_{z \to 1} \left\{ (z - 1)X(z)z^{k-1} \right\}$$

which clearly reduces to:

$$\lim_{k \to \infty} x_k = \lim_{z \to 1} \left\{ (z-1)X(z) \right\} \qquad (2.1.15)$$

Equation (2.1.15) constitutes the ***Final Value Theorem*** of the z-Transformation, which is useful in design as a means of confirming the correct algebraic manipulation of digital control system equations by an evaluation of their d.c. gain or stiffness.

These initial results reveal that the z- Transformation and Laplace Transformation are mathematically very similar. Subsequent theoretical developments re-enforce this viewpoint of the z- Transformation which is shown to enable:

- a structured method for manipulating the dynamical equations of series, parallel and feedback configurations of sampled-data systems

- insight into system dynamics without recourse to a detailed inversion of a z- Transformation into the sequence (time) domain

- a systematic improvement of digital control system stability margins or performance by the inclusion of additional compensating elements

- realisation of special frequency response characteristics for signal processing applications (17, 18, 19).

Accordingly, the next section formulates the concept of the pulse transfer function for a digital controller or filter.

2.2 THE PULSE TRANSFER FUNCTION FOR A DISCRETE DATA SYSTEM

As described in Section 1.1, the design of control systems and filters is generally implemented in terms of linear fixed parameter equations. Accordingly, the output number sequence $\{y_n\}$ of a digital controller or filter is related to its input number sequence $\{x_n\}$ by:

$$y_n = \sum_{k=1}^{K} a_k x_{n-k} - \sum_{k=1}^{K} b_k y_{n-k} \qquad (2.2.1)$$

where:

$\{a_k\}$, $\{b_k\}$ - families of constant real coefficients

K - *order* of the digital controller or filter

Because a digital controller and some filters must necessarily operate in real time, an output value (y_n) at any instant is a combination of past inputs and outputs only. On the other hand, the processing of pre-recorded data for example can be carried out in non-real time, for which equation (2.2.1) becomes:

$$y_n = \sum_{-K}^{K} a_k x_{n-k} - \sum_{k=1}^{K} b_k y_{n-k} \qquad (2.2.2)$$

A difference equation, like (2.2.1) or (2.2.2), is termed *recursive* if at least one of the coefficients $\{b_k\}$ is a non-zero. If all the coefficients $\{b_k\}$ are zero, the difference equation is termed *non-recursive.* The same nomenclature is used to describe the digital controllers or filters themselves. Due to the feedback of multiplicative rounding errors, the output sequence of a recursive digital filter is susceptible to non-linear limit cycle oscillations. The oscillations obviously cannot occur with non-recursive filters, whose 'quietness' is a factor that decides their deployment in digital telephone systems (39). Furthermore, non-real time non-recursive digital filters can be designed with zero phase (time) shift, which considerably assists the interpretation of filtered experimental data (40).

Equation (2.2.1) is regarded for immediate purposes as a special case of equation (2.2.2) with:

$$a_k = 0 \quad \text{for} \quad k < 1$$

Defining the z- transformations of the input and output sequences respectively as:

$$X(z) = \sum_{k=o}^{\infty} x_k z^{-k} \qquad \text{and} \qquad Y(z) = \sum_{k=o}^{\infty} y_k z^{-k} \tag{2.2.3}$$

then from equation (2.2.2):

$$\lim_{N \to \infty} \sum_{N=o}^{N} \left\{ \sum_{k=o}^{K} b_k y_{n-k} \right\} z^{-n} = \lim_{N \to \infty} \sum_{n=o}^{N} \left\{ \sum_{-K}^{K} a_k x_{n-k} \right\} z^{-n} \tag{2.2.4}$$

where always

$$b_o = 1$$

and due to computational delays, it is usual in real time systems that the coefficient a_o is zero. Because multiplication is distributive, and addition is both associative and commutative for finite arrangements of complex numbers (29), equation (2.2.4) can be re-ordered as:

$$\lim_{N\to\infty} \sum_{k=o}^{K} b_k z^{-k} \left[\sum_{n=k}^{N} y_{n-k} z^{-n+k} \right]$$

$$= \lim_{N\to\infty} \sum_{-K}^{K} a_k z^{-k} \left[\sum_{n=k}^{N} x_{n-k} z^{-n+k} \right] \qquad (2.2.5)$$

If each sequence in a finite family of infinite sequences has a limit, the limit of the sum of sequences is the sum of their individual limits (10, 46), so that equation (2.2.5) reduces to:

$$\sum_{k=o}^{K} b_k z^{-k} Y(z) = \sum_{-K}^{K} a_k z^{-k} X(z)$$

Hence the z- transformations of the input and output sequences of a linear discrete data system are related by:

$$Y(z) = D(z)X(z) \qquad (2.2.6)$$

where

$$D(z) \triangleq \left[\sum_{-K}^{K} a_k z^{-k} \right] \Big/ \left[\sum_{k=1}^{K} b_k z^{-k} \right] \quad \text{with} \quad b_o = 1 \qquad (2.2.7)$$

is termed the *pulse transfer function*. These equations are exactly analogous to the Laplace Transformation relationships for continuous data systems in equations (1.2.29) and (1.2.30).

Noting that the z- transformation of the single pulse sequence:

$$\left.\begin{aligned} \delta_{ko} &= 1 \quad \text{for} \quad k = 0 \\ &= 0 \quad \text{otherwise} \end{aligned}\right\} \qquad (2.2.8)$$

is

$$\Delta(z) = 1 \tag{2.2.9}$$

then it follows from equation (2.2.6) that rational z-transformations of arbitrary complexity can be inverted numerically by solving the difference equation (2.2.2) with the coefficients $\{a_k\}$ and $\{b_k\}$ corresponding to those in equation (2.2.7). For instance, the now familiar example of the z-transformation:

$$D(z) = 1/(1 - e^s z^{-1})$$

can be inverted as the step by step solution of:

$$d_n = \delta_{no} + e^s d_{n-1}$$

Explicitly:

$$d_o = 1 + 0 = 1$$

$$d_1 = 0 + (e^s)1 = e^s$$

$$d_2 = 0 + (e^s)e^s = e^{2s}$$

etc

More complicated difference equations, that are derived from more complicated transforms, are easily solved by a simply constructed generalised computer programme.

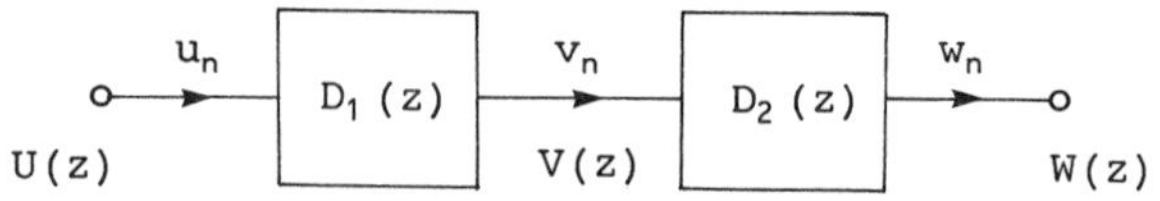

Fig 2.2.1 Pulse Transfer Functions in Cascade

Fig 2.2.1 depicts a cascaded arrangement of two discrete data systems with pulse transfer functions $D_1(z)$ and $D_2(z)$. If the z-transformation of the input sequence exists as $U(z)$, then the output sequence $\{w_n\}$ of the combination is derived directly from equation (2.2.6) as:

$$W(z) = D_2(z)\left[D_1(z)U(z)\right] = D_2(z)D_1(z)U(z) \qquad (2.2.10)$$

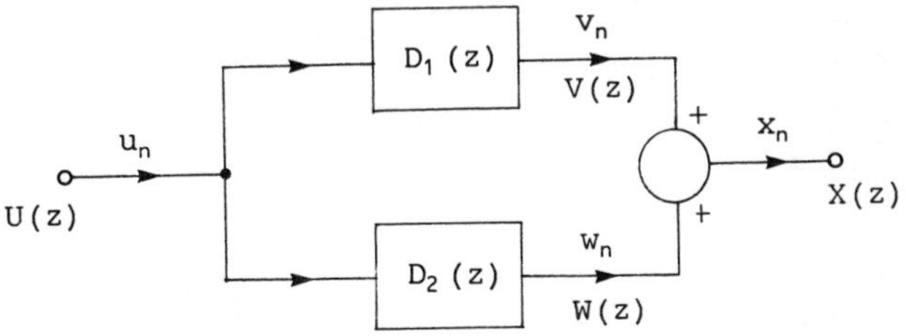

Fig 2.2.2 Pulse Transfer Functions in Parallel

The combined output sequence $\{x_n\}$ of the parallel arrangement of discrete data systems in Fig 2.2.2 is:

$$x_n = v_n + w_n$$

where $\{v_n\}$ and $\{w_n\}$ are their separate responses to a common input sequence with z- transformation $U(z)$. By definition:

$$X(z) = \lim_{N\to\infty} \sum_{n=o}^{N} x_n z^{-n} = \lim_{N\to\infty} \sum_{n=o}^{N} (v_n z^{-n} + w_n z^{-n})$$

Because the separate limits $V(z)$ and $W(z)$ exist from equation (2.2.6) as:

$$V(z) = D_1(z)U(z) \quad \text{and} \quad W(z) = D_2(z)U(z)$$

the previous equation simplifies to the sum of the separate limits:

$$X(z) = V(z) + W(z)$$

or

$$X(z) = \left[D_1(z) + D_2(z)\right] U(z) \tag{2.2.11}$$

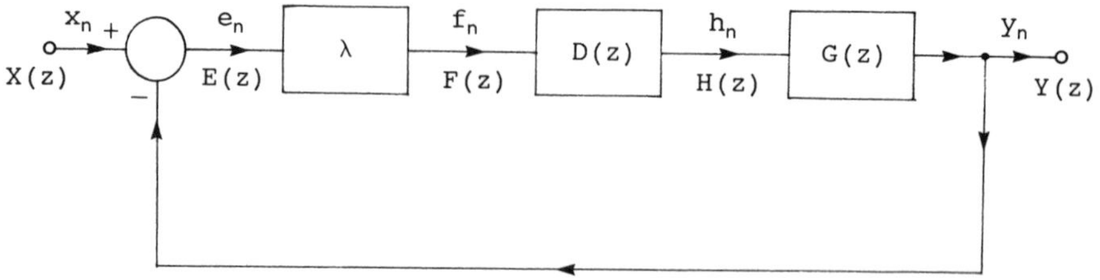

Fig 2.2.3 A Unity Feedback System with a Series Compensator (Controller)

Similarly, if the z- transformation of an input sequence to the feedback system in Fig 2.2.3 exists as X(z), then equation (2.2.10) gives:

$$Y(z) = \left[\lambda D(z)G(z)\right] E(z)$$

However, because:

$$E(z) = X(z) - Y(z)$$

it follows that:

$$E(z) = \left[\frac{1}{1 + \lambda D(z)G(z)}\right] X(z) \tag{2.2.12}$$

and

$$Y(z) = \left[\frac{\lambda D(z)G(z)}{1 + \lambda D(z)G(z)} \right] X(z) \qquad (2.2.13)$$

The functions:

$$1 + \lambda D(z)G(z) \quad \text{and} \quad \frac{\lambda D(z)G(z)}{1 + \lambda D(z)G(z)}$$

which figure prominently in subsequent analysis, are termed respectively the *loop-* and *overall pulse transfer functions* of the feedback system. Equations (2.2.10) to (2.2.13) again form exact parallels with formulae for the analysis of continuous data systems by the Laplace Transformation, and likewise they are readily extended to an arbitrary finite number of components. Jury (37) tabulates the overall pulse transfer functions for a variety of topologically different pulse transfer function arrangements, and these could well be used as worked examples to gain familiarity with the above results.

2.3 PROGRAMMING STRUCTURES FOR PULSE TRANSFER FUNCTIONS

Design procedures for digital controllers or filters create pulse transfer functions for realisation by a digital computer. With the so-called *direct programming technique*, a pulse transfer function is first expressed in inverse powers of z as in equation (2.2.7). The coefficients of the corresponding linear difference equation are then derived from a comparison with equation (2.2.1) or (2.2.2). To illustrate the procedure, consider the simple digital controller:

$$D(z) = 0.5(z - 0.9)/z(z - 0.4)$$

which is re-written as:

$$D(z) = (0.5\, z^{-1} - 0.45z^{-2})/(1 - 0.4\, z^{-1})$$

and its realisation is effected as:

$$y_n = 0.5x_{n-1} - 0.45x_{n-2} + 0.4y_{n-1}$$

where $\{x_n\}$ and $\{y_n\}$ denote the input and output sequences respectively. In this example, coefficients in the difference equation have been contrived to have magnitudes less than unity, so that they can be directly represented in a fixed point two's complement machine. Generally, however, some coefficients of a pulse transfer function have magnitudes greater than unity, so that various combinations of shift-left operations and repeated addition are necessary for their representation in machine format. For example:

$$10x_n = 2^4 \left\{ \frac{5}{8} x_n \right\}$$ with four left-shifts

or

$$10x_n = 2^3 \left\{ \frac{5}{8} x_n + \frac{5}{8} x_n \right\}$$ one repeated addition of $\frac{5}{8} x_n$ with three left-shifts

etc.

Digital controllers frequently have low order pulse transfer functions with often no more than a quadratic numerator and denominator, and consequently direct programming is the only option. On the other hand, higher order pulse transfer functions are frequently deployed in signal processing. These can be realised by a variety of programming techniques (28, 37), each of which affords specific advantages in terms of: multiplicative rounding error noise (41, 42), coefficient sensitivity (43, 44), or dynamic range without overflow (35, 36).

For realisation by the *parallel programming technique*, a pulse transfer function is first written as:

$$D(z) = \left[\lambda + \left\{ \sum_{k=o}^{K-1} a_k z^k \Big/ \sum_{k=o}^{K} b_k z^k \right\} \right] z^{-p} \qquad (2.3.1)$$

where z^{-p} represents the particular computational delay of p sampling periods. Because its numerator polynomial is at least one degree lower than its denominator polynomial, the rational function can then be expanded into partial fractions. When for example there are just simple poles*, this expansion takes the form:

$$D(z) = \left[\lambda + \sum_{i=1}^{I} \frac{\alpha_i}{z + \beta_i} + \sum_{q=1}^{Q} \frac{A_{1q} z + A_{2q}}{z_2 + B_{1q} z + B_{2q}} \right] z^{-p} \qquad (2.3.2)$$

Real Poles   Complex Pole Pairs

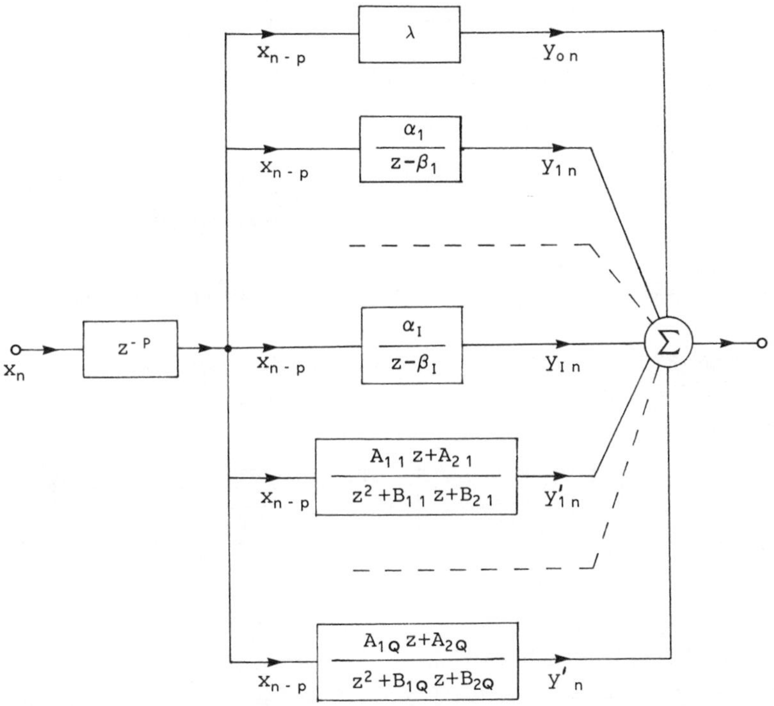

Fig 2.3.1 Illustrating the Parallel Programming Technique

---

*discounting of course the pth order pole of z^{-p} at the origin of the z- plane

Denoting the input sequence to the filter D(z) by $\{x_n\}$, then each first- and second-order pulse transfer function in equation (2.3.2) is directly programmed as:

$$\left.\begin{aligned} y_{on} &= \lambda\, x_{n-p} \\ y_{in} &= \alpha_i x_{n-1-p} - \beta_i y_{i\,n-1} \qquad \text{for } 1 \le i \le I \\ y'_{qn} &= A_{1q} x_{n-1-p} + A_{2q} x_{n-2-p} - B_{1q} y'_{q\,n-1} - B_{2q} y'_{q\,n-2} \\ &\qquad\qquad \text{for } 1 \le q \le Q \end{aligned}\right\} \quad (2.3.3)$$

By adding together all the individual output sequences $\{y_{on}\}$, $\{y_{in}\}$ and $\{y_{qn}\}$ in the manner of Fig 2.3.1, the addition rule for transfer functions in parallel implies that the procedure realises the pulse transfer function D(z).

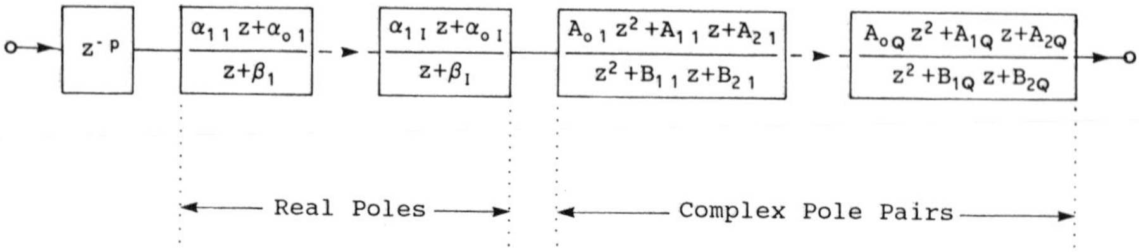

Fig 2.3.2 Illustrating the Cascade Programming Technique

Cascade programming is another realisation technique for which a pulse Transfer function is first factored as:

$$D(z) = z^{-p} \underbrace{\prod_{i=1}^{I} \left(\frac{\alpha_{1i} z + \alpha_{oi}}{z + \beta_i} \right)}_{\text{Real Poles of } D(z)} \cdot \underbrace{\prod_{q=1}^{Q} \left(\frac{A_{oq} z^2 + A_{1q} z + A_{2q}}{z^2 + B_{1q} z + B_{2q}} \right)}_{\text{Complex Pole Pairs of } D(z)} \qquad (2.3.4)$$

where z^{-p} again represents the particular computational delay of p sampling periods. Each of the above first- and second-order pulse transfer functions is directly programmed as for parallel programming, but the output sequence from one component now forms the input sequence to the next as illustrated by Fig 2.3.2. It follows from the multiplication rule for pulse transfer functions in cascade that the procedure realises the pulse transfer function D(z).

Consider an output sequence $\{y_n\}$ that is generated from an input sequence $\{x_n\}$ by the following set of linear difference equations with constant coefficients $\{a_k\}$ and $\{b_k\}$:

$$\left.\begin{aligned} v_{1\,n+1} &= x_n - \sum_{k=1}^{K} b_k v_{k\,n} \\ v_{k\,n+1} &= v_{k-1\,n} \qquad \text{for} \qquad 2 \le k \le K \\ y_n &= a_o x_n + \sum_{k=1}^{K} a_k v_{k\,n} - a_o \sum_{k=1}^{K} b_k v_{k\,n} \end{aligned}\right\} \qquad (2.3.5)$$

These equations are usually derived from the theory of linear multi-input multi-output systems, and the vector of sequences $\{(v_{1\,n}\ v_{2\,n}\ \cdots\ v_{k\,n})\}$ is the so-called *state vector*. However, for present purposes, it suffices to prove that they characterise the pulse transfer function relationship in equations (2.2.6) and (2.2.7). In this context, the z- transformations of equations (2.3.5) are evidently:

$$zV_1(z) = X(z) - \sum_{k=1}^{K} b_k V_k(z) \qquad (2.3.6)$$

$$V_k(z) = z^{-k+1} V_1(z) \qquad (2.3.7)$$

$$Y(z) + a_o X(z) + \sum_{k=1}^{K} a_k V_k(z) - a_o \sum_{k=1}^{K} b_k V_k(z) \qquad (2.3.8)$$

By substituting equation (2.3.7) into equations (2.3.6) and (2.3.8), the elimination of $V_1(z)$ yields

$$Y(z) = \left[a_0 + \left(\sum_{k=1}^{K} a_k z^{-k} \right) \Big/ \left(\sum_{k=0}^{K} b_k z^{-k} \right) \right.$$

$$\left. - a_0 \left(\sum_{k=1}^{K} a_k z^{-k} \right) \Big/ \left(\sum_{k=0}^{K} b_k z^{-k} \right) \right] X(z) \qquad (2.3.9)$$

in which by definition:

$$b_0 = 1$$

Hence, in terms of equations (2.2.6) and (2.2.7), the above expression simplifies as required to:

$$Y(z) = D(z)X(z)$$

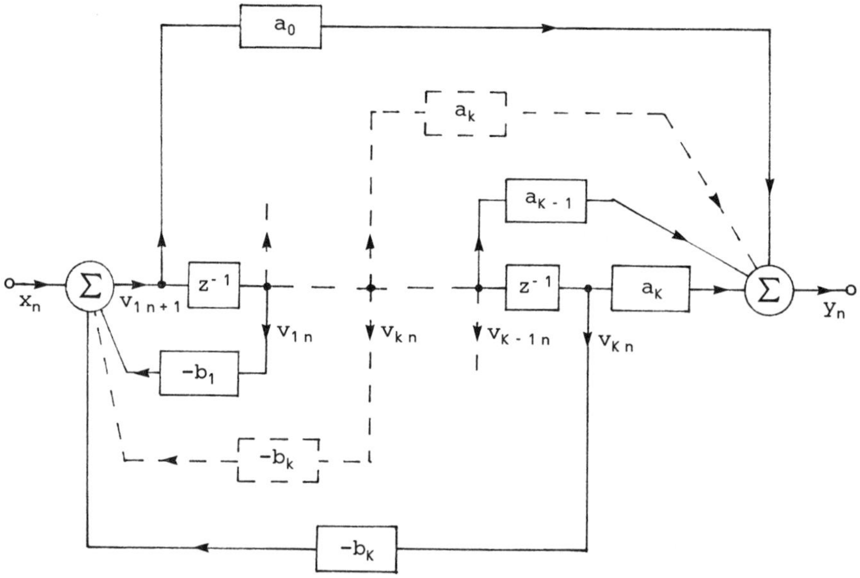

Fig 2.3.3 Illustrating the Canonical Programming Technique

If a computational delay of p sampling periods is incurred, the corresponding factor z^{-p} is first factored out of D(z) to define the coefficients $\{a_k\}$ and $\{b_k\}$, then as before x_{n-p} replaces x_n in equation (2.3.5). The realisation of a pulse transfer function by equation (2.3.5) is described as ***canonical programming***, and Fig 2.3.3 illustrates the technique applied to the real time version of equation (2.2.7). A personal interest in canonical programming arose with regard to a fixed point realisation of the 22nd order real time bandstop filter of Golden and Kaiser (45). Although the output noise of the canonically programmed filter was very much less than for other realisations, the ratio:

$$\frac{\text{rms rounding error noise}}{\text{maximum input amplitude without actual overflow}}$$

was over twelve decades larger (35). That is to say, the severely restricted range of the canonical realisation rendered it impractical. Subsequent experience and other research (36) confirms that parallel programming generally provides the most satisfactory realisations of high order recursive filters, but the amplitude of limit cycle oscillations arising from the feedback of rounding error noise depends on the circuit topology of their biquadratic 'building blocks' (47, 48, 49).

To conclude, Fig 2.3.4 on the next page depicts the realisation of the digital controller:

$$D(z) = (z - 0.9)(z - 0.95) \Big/ z(z - 0.7)(z - 0.8) \qquad (2.3.10)$$

by the various programming techniques described above. It is well-worth a reader's effort to repeat the same for other pulse transfer functions of his or her choice; proficiency is awarded for completing the worked example in reference 43.

2.4 <u>THE PULSE TRANSFER FUNCTION FOR A CONTINUOUS DATA COMPONENT</u>

Despite the analytical complications posed by the mixed arrangement of digital and analogue components in a D-A converter,

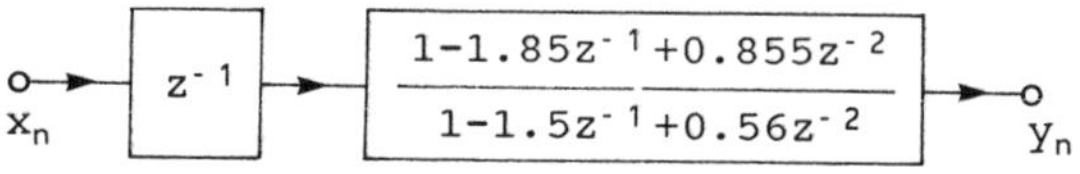

Direct Programming

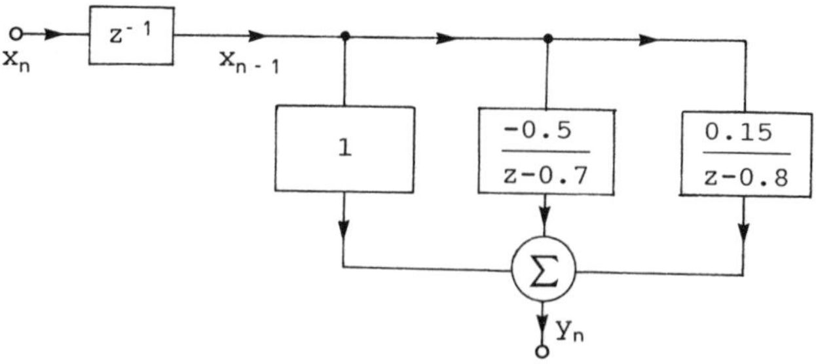

Parallel Programming

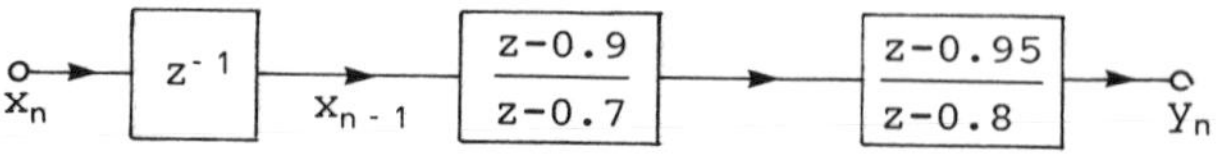

Cascade Programming

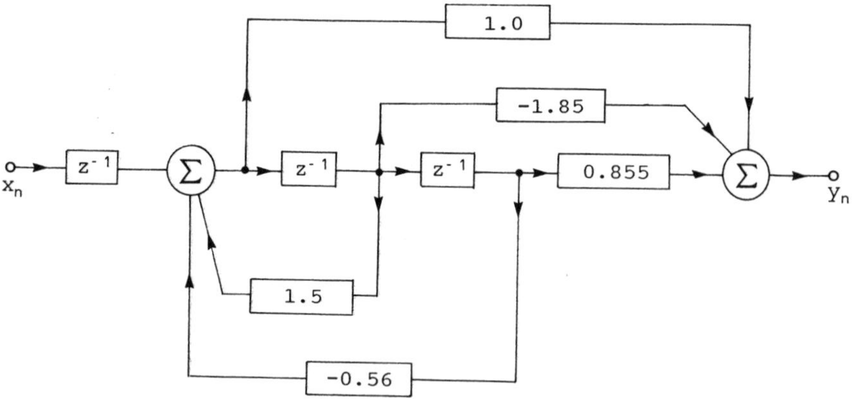

Canonical Programming

Fig 2.3.4 Some Realisations of the Pulse Transfer Function in Equation (2.3.10)

its Laplace transfer function is derived in Section 1.5 in terms of a conceptual model that reproduces the same output response. As illustrated by Fig 1.5.3, this model involves an imaginary δ-unit which delivers a δ-function of current or voltage etc whose area is proportional to the binary number that is entered into the output register (staticisor) of a digital controller. The design of DCC systems generally centres around the response of a linearised plant model that is driven by a D-A converter and digital controller in the manner of Fig 2.4.1. It is now shown that the dynamics of such hybrid systems can be characterised by the z- Transformation, provided that attention is artificially restricted to the sampling instants set by the digital controller. That is to say, the response of the plant is actually preserved, but it is *fictitiously sampled* to enable a systematic analysis of the overall system.

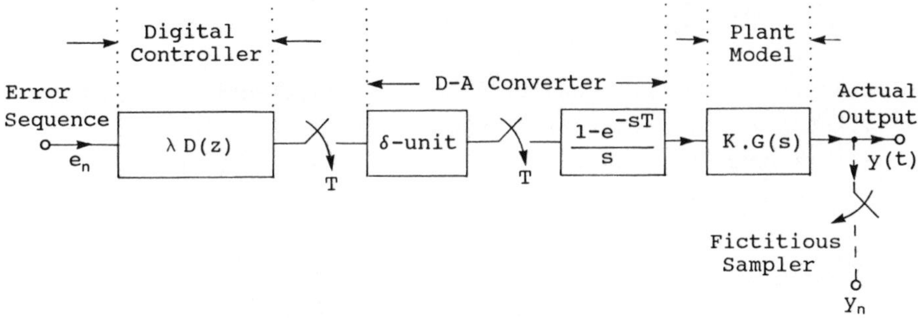

Fig 2.4.1 The z- Transformation Analysis of a DDC System by means of a Fictitious Output Sampler

The Laplace transfer function of the D-A converter and plant model in Fig 2.4.1 is denoted by:

$$H(s) = K \left[\frac{1 - e^{-sT}}{s} \right] G(s) \qquad (2.4.1)$$

which corresponds to a weighting function or impulse response of h(t). After an arbitrary time t from the first output (f_0) of the digital controller, the response of the plant is evidently $f_0 h(t)$.

Because the modelled D-A converter and plant are linear, then superposition applies and the response to all outputs from the digital controller at time t is therefore:

$$y(t) = f_0 h(t) + f_1 h(t - T) + \dots + f_k h(t - kT)$$

where k is the largest integer for which $kT \le t$. Hence the output sequence $\{y_n\}$ is specified by:

$$y_n = \sum_{k=0}^{n} h_k f_{n-k} \qquad (2.4.2)$$

so that by definition:

$$Y(z) = \lim_{n \to \infty} \left\{ \sum_{k=0}^{n} h_k f_{n-k} z^{-n} \right\} \qquad (2.4.3)$$

To establish sufficient conditions for the existence of the above limit, suppose that the series $\{h_n z^{-n}\}$ and $\{f_n z^{-n}\}$ are both normally summable over a region S of the z- plane, so that there are real numbers β_1 and β_2 such that:

$$\sum_{0}^{\infty} \left\| h_n z^{-n} \right\|_S = \beta_1 \quad \text{and} \quad \sum_{0}^{\infty} \left\| f_n z^{-n} \right\|_S = \beta_2$$

Any finite subseries of $\left\{ \left\| h_k f_{n-k} z^{-n} \right\|_S \right\}$, as it appears in equation (2.4.3), is therefore bounded by $\beta_1 \beta_2$. Consequently, the series is normally summable, and its limit Y(z) exists in S. Moreover, the series can be summed in any order (10, 38), and in particular by summing first over:

$$q = n-k$$

equation (2.4.3) becomes:

$$Y(z) = \sum_{k=0}^{\infty} h_k z^{-k} \left[\sum_{q=0}^{\infty} f_q z^{-q} \right]$$

which simplifies to:

$$Y(z) = H(z)F(z) \tag{2.4.4}$$

To design a DDC system by the z- Transformation, a closed algebraic form is now seen to be required for:

$$H(z) = \sum_{0}^{\infty} h_n z^{-n} \tag{2.4.5}$$

where

$$h_n = \frac{1}{2\pi j} \int_{c-j\infty}^{c+j\infty} H(s)e^{snT} ds \tag{2.4.6}$$

As depicted in Fig 2.4.2, the line of integration from c-j∞ to c+j∞ lies strictly to the right of all the poles of H(s). The summation of series forms a classic application of complex variable theory, and in the present case, equation (2.4.6) is substituted into (2.4.5) to yield:

$$H(z) = \lim_{N \to \infty} \sum_{0}^{N} \left\{ \frac{1}{2\pi j} \int_{c-j\infty}^{c+j\infty} H(s)e^{snT} \right\} z^{-n}$$

A sufficient condition for the existence of H(z), and for interchanging the order of integration and summation, is that the series of functions is normally summable over the line c-j∞ to c+j∞(10). Provided that:

$$\left| z \right| > \exp\left[\mathbf{R}\, sT \right] = e^{cT} \tag{2.4.7}$$

the above series is shown in Appendix A2.4 to be normally summable so that:

$$H(z) = \frac{1}{2\pi j} \int_{c-j\infty}^{c+j\infty} H(s) \left\{ \sum_{0}^{\infty} \left[e^{sT} z^{-1} \right]^n \right\} ds \qquad (2.4.8)$$

The poles of the infinite series term in the above integrand evidently satisfy:

$$e^{sT} = z$$

Denoting these s- plane locations by:

$$s = \sigma + j\omega \qquad (2.4.9)$$

and writing:

$$z = re^{j\Theta}$$

gives:

$$\sigma = \frac{1}{T} \log |z| \quad \text{and} \quad \omega = (\Theta \pm 2n\pi)/T \quad \text{for integer } n \qquad (2.4.10)$$

It follows from equations (2.4.7) and (2.4.10) that:

$$\sigma > c$$

Consequently, a countably infinite number of poles are associated with the infinite series term in equation (2.4.8), and these lie strictly to the right of the line [c−j∞, c+j∞] as shown in Fig 2.4.2 on the next page.

Evaluation of the integral in equation (2.4.8) is accomplished by Cauchy's Theory of Residues (10, 12) after closing the line of

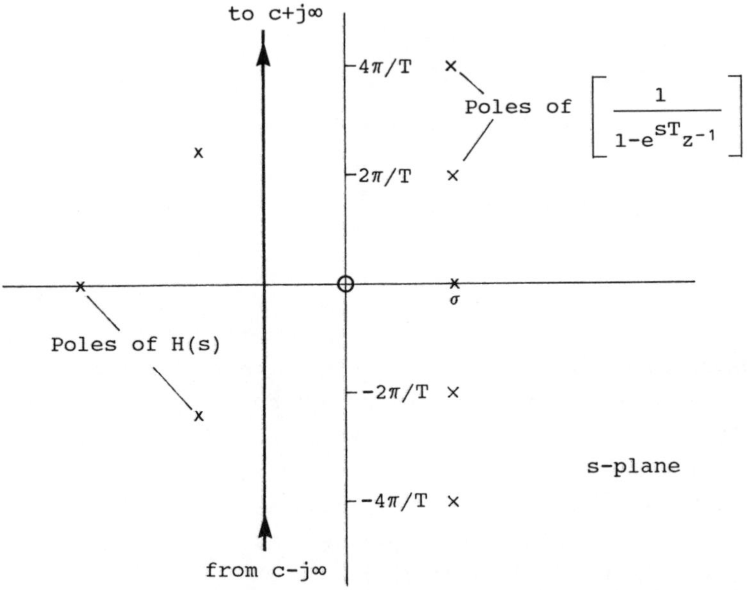

Fig 2.4.2 Pole Locations for the Integrand of Equation 2.4.8

integration* with an infinite semicircle in the left- or right-half s- plane. Stray capacitance, inertia etc ensure that the Laplace transfer functions of continuous data systems always behave as:

$$\lim_{s\to\infty} \left| sH(s) \right| = \alpha \qquad \text{a real constant} \tag{2.4.11}$$

which according to Appendix B2.4 represents a sufficient condition for the integral around either infinite semicircle to be zero.

---

*The dynamics represented by some partial differential equations have irrational Laplace transfer functions which involve for example $\sqrt{s}$ terms. In such cases, this contour must be contorted so as to lie on one sheet of the Riemann Surface associated with the necessary Branch Cut(s)

Thus equation (2.4.8) reduces to:

$$H(z) = \sum \text{Residues of} \left\{ \frac{H(s)}{1 - e^{sT}z^{-1}} \right\} \text{ at poles of } H(s) \tag{2.4.12}$$

for closure by a left-semicircle, and to:

$$H(z) = -\sum \text{Residues of} \left\{ \frac{H(s)}{1 - e^{sT}z^{-1}} \right\} \text{ at poles of } \frac{1}{1 - e^{sT}z^{-1}} \tag{2.4.13}$$

for closure by a right-semicircle. Equation (2.4.12) directly reveals that the z-transformation of a product $H_1(s)H_2(s)$ is not generally equal to $H_1(z)H_2(z)$. It also forms the principal technique for deriving z- transformations from Laplace transformations. For example, the familiar sequence $\exp(\beta nT)$ is the periodically sampled continuous data function $\exp(\beta t)$, whose Laplace transformation is $1/(s-\beta)$.

Its z- transformation is specified by equation (2.4.12) as

$$H(z) = \text{Residue of} \left\{ \frac{1}{(s - \beta)(1 - e^{sT}z^{-1})} \right\} \text{ at } s = \beta$$

which evaluates as:

$$H(z) = 1 / (1 - e^{\beta T} z^{-1})$$

and this result is readily confirmed by means of the classical formula for a convergent geometric progression. A short compilation of Laplace and z- Transformation pairs is provided overleaf in Table 2.4.1 for the reader to gain experience in the application of equation (2.4.12). Although this tabulation and the partial fraction expansion technique enable the derivation of more complicated relationships, particularly comprehensive tables can be found in references (37) and (50).

Laplace transform	z- transform
$\exp(-skT)$	z^{-k}
$\frac{1}{s}$	$\frac{z}{z-1}$
$\frac{1}{s^2}$	$\frac{Tz}{(z-1)^2}$
$\frac{1}{s^2(s+a)}$	$\frac{(T/a)z}{(z-1)^2} - \frac{(1-e^{-aT})z}{a^2(z-1)(z-e^{-aT})}$
$\frac{1}{s^3}$	$\frac{Tz^2(z+1)}{2(z-1)^3}$

Table 2.4.1 Some Pairs of Laplace and z- Transformations

Because the factor exp(-sT) delays continuous data by exactly one sampling period, then its general effect is to change an arbitrary sequence x_0 ; x_1 ; x_2 .... into 0; x_0 ; x_1 ; x_2 .... It follows that if the z- transformation X(z) corresponds to the Laplace transformation X(s), then $z^{-1}X(z)$ corresponds to exp(-sT)X(s). Accordingly, the z- transformation of the D-A converter and plant model in equation (2.4.1) is evaluated in practice from equation (2.4.12) as:

$$H(z) = K(1 - z^{-1})\sum \text{Residues of} \left\{ \frac{G(s)}{s(1 - e^{sT}z^{-1})} \right\} \quad \text{at poles of } G(s)/s \qquad (2.4.14)$$

where the gain constant (K) has likely dimensions of mV or mA per least significant bit.

With regard now to the alternative evaluation of H(z) in equation (2.4.13), where the relevant first order poles are specified by equation (2.4.9) and (2.4.10) as:

$$p_n = \left[\log |z| - j(\Theta \pm 2\pi n)\right]/T \quad \text{for all integer } n \qquad (2.4.15)$$

the corresponding residues can be straightforwardly derived according to:

$$\left.\frac{\dfrac{d}{ds}\left[(s - p_n)H(s)\right]}{\dfrac{d}{ds}\left[1 - e^{sT}z^{-1}\right]}\right|_{s = p_n} = -\frac{1}{T} H(p_n) \quad \text{for all integer } n$$

so that:

$$H(z) = \frac{1}{T}\sum_{-\infty}^{\infty} H\left[\frac{1}{T}\log |z| - j\frac{1}{T}(\Theta + 2\pi n)\right] \qquad (2.4.16)$$

Setting:

$$z = \exp(sT)$$

in equations (2.4.5) and (2.4.16) produces by reference to equations (2.4.9) and (2.4.10) the more familiar form:

$$H^{*}(s) = \sum_{0}^{\infty} h_n e^{-snT} = \frac{1}{T}\sum_{-\infty}^{\infty} H(s - j2\pi n/T) \qquad (2.4.17)$$

which defines the *Pulse Laplace Transformation**. If the poles of the Laplace transformation H(s) lie strictly to the left of the imaginary axis in the s- plane, the above time domain series converges for:

$$s = j\omega \triangleq j2\pi f$$

to yield the single-sided *Pulse Fourier Transformation:*

---

*sometimes the terminology 'discrete' rather than 'pulse' is used

$$H^*(\omega) = \sum_{0}^{\infty} h_n e^{-j\omega nT} = \frac{1}{T}\sum_{-\infty}^{\infty} H(\omega - 2\pi n/T) \tag{2.4.18}$$

The conventional Fourier transformation of sampled continuous data is derived in Section 1.3 as:

$$X_p(\omega) = \sum_{-\infty}^{\infty} P_n' X(\omega - 2\pi n/T) \tag{2.4.19}$$

where from equation (1.3.10):

$$P_n' = \frac{\gamma}{T}\left[\frac{\sin(\pi n\gamma/T)}{\pi n\gamma/T}\right]\left[\frac{1}{1 + j2\pi n\tau/T}\right] \tag{2.4.20}$$

and

γ - the width of a sampling pulse of unit height

τ - time constant of sampling pulses

with

$\tau \ll \gamma \ll T$

Because the time intervals for data conversion and computer operations are so very much shorter than the sampling period for contemporary DDC systems, equation (2.4.19) is closely approximated over frequencies of practical interest by:

$$X_p(\omega) = \frac{\gamma}{T}\sum_{-\infty}^{\infty} X(\omega - 2\pi n/T) = \gamma X^*(\omega) \tag{2.4.21}$$

When binary coded computer pulses of width γ are entered into a staticisor register, the operation can be viewed conceptually in terms of equation (2.4.21) and the Laplace transfer function $K(1 - e^{-sT})/\gamma s$; because the pulse width factor must evidently cancel out. Thus, provided a D-A converter and staticisor are described by the Laplace transfer function:

$$H_0(s) = K \left[\frac{1 - e^{-sT}}{s} \right] \quad (2.4.22)$$

the conventional spectrum of their input signal from a digital controller can be taken as its Pulse Fourier transformation. This result is of value later in Chapter 5, where an analytical technique is developed for engineering the choice of the sampling frequency in a DCC system. Equation (2.4.22) characterises the dynamic behaviour of a so-called zero order hold (28, 37).

2.5 INTERSAMPLE BEHAVIOUR AND THE MODIFIED z-TRANSFORMATION

The fictitious sampling of the output signal from a DDC has been shown to enable its well-structured analysis in terms of pulse transfer functions. An immediate penalty, however, is that the control error is assessed at just the sampling instants so that the adequacy of the z- Transformation for design purposes is possibly questionable. Explicitly, what is the magnitude of the control error between the sampling instants? As described in Section 1.3, sampling frequencies must be high enough both to restrict irreversible baseband distortion, and to allow an adequate attenuation of the sideband spectra centred around multiples of the sampling frequency. In the time domain, baseband distortion creates a control error component with rates of change similar to the continuous input data, while imperfectly attenuated sidebands appear as much more rapidly varying *'ripple'* or *'chatter'*. Excessive ripple can cause the premature fatigue failure of a plant actuator as well as loss of control accuracy. The sampling frequency of a DDC system must therefore be designed high enough to achieve both a specified accuracy and operational life.

For illustrative purposes suppose that the continuous input data to a linearised DDC system is a sinusoid, whose frequency must

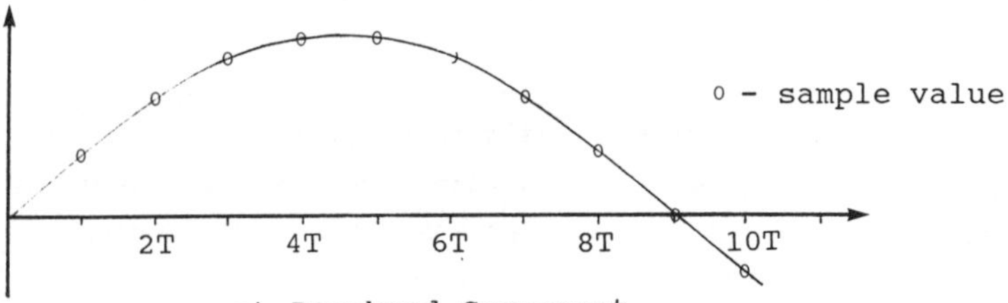

a) Baseband Component

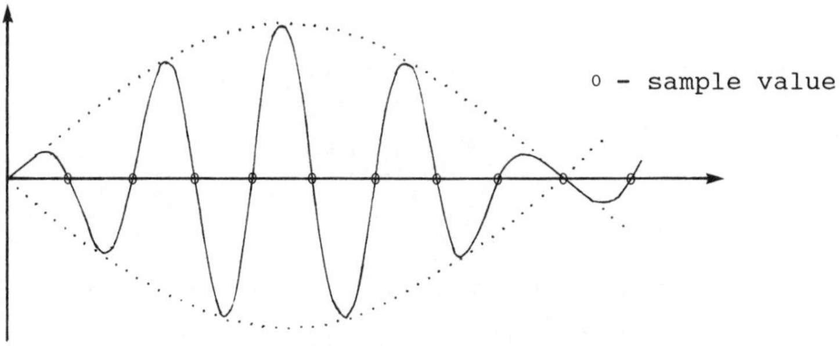

b) One of the Amplitude Modulated Components

Fig 2.5.1 Illustrating the Baseband and Ripple Components of the Control Error Signal

obviously be much less than the sampling frequency (1/T). It follows from the above discussion that the continuous control error signal contains the additive components shown in Fig 2.5.1. Although the amplitude of the baseband component can be properly assessed from sampled information acquired by means of the z-Transformation, this is seen not to be the case for the ripple components. The presently recommended design technique for DDC systems is based on the z- Transformation and a separate straightforward calculation of the steady-state *rms ripple power** using formulae developed in Chapter 5. To satisfy typical control accuracy requirements, it is generally the case

*rms ripple power - root mean square ripple power

that control error spectra are dominated by their basebands. Consequently, quite gross transient performance characteristics like speed of response and transient overshoot etc can be adequately measured from samples of the control error (z- Transformation analysis), while the usually more stringent steady-state performance can be adequately assessed in terms of error samples and the rms ripple power (51). Full details appear later in Chapters 5 and 7.

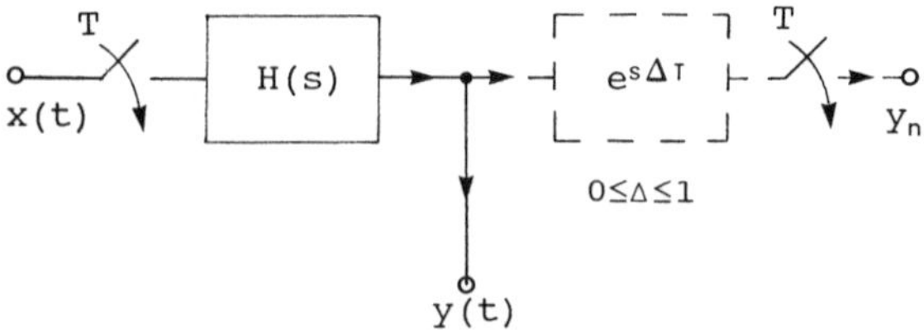

Fig 2.5.2 Illustrating the Derivation of the Modified z- Transformation

Early articles on the design of DDC systems advocated the *Modified z- Transformation* (52, 53) as a means of assessing intersample performance. As illustrated by Fig 2.5.2, the technique is based on fictitiously sampling the output signal from a plant after it has been advanced in time by ΔT with the Laplace transfer function $\exp(s\Delta T)$. Thus, the modified z- transformation corresponding to an arbitrary Laplace transfer function is defined similarly to equation (2.4.5) as:

$$H(z,\Delta) = \sum_{0}^{\infty} h(nT + \Delta T) z^{-n} \qquad \text{for } 0 \le \Delta \le 1 \qquad (2.5.1)$$

so that:

$$H(z) = H(z,\Delta)\Big|_{\Delta=0} \qquad (2.5.2)$$

Exactly the same procedure as that used to derive equations (2.4.4) and (2.4.8) shows that:

$$Y(z,\Delta) = H(z,\Delta)\ X(z) \tag{2.5.3}$$

where

$$H(z,\Delta) = \frac{1}{2\pi j}\int_{c-j\infty}^{c+j\infty} H(s)e^{s\Delta T}\left[\frac{1}{1 - e^{sT}z^{-1}}\right] ds \tag{2.5.4}$$

Inversion of a modified z- transformation likewise assumes the same form as equation (2.1.9):

$$y(\overline{n + \Delta T}) = \frac{1}{2\pi j}\int_{C} Y(z,\Delta)z^{n-1}dz \tag{2.5.5}$$

where the closed contour C encloses all the poles of $Y(z,\Delta)$. Intersample values of the continuous output and control error data are derived by inversion of modified z- transformations similar to equations (2.2.12) and (2.2.13):

$$Y(z,\Delta) = \left[\frac{\lambda D(z)G(z,\Delta)}{1 + \lambda D(z)G(z,\Delta)}\right] X(z) \tag{2.5.6}$$

$$E(z,\Delta) = \left[\frac{1}{1 + \lambda D(z)G(z,\Delta)}\right] X(z) \tag{2.5.7}$$

The complexity of equations (2.5.6) and (2.5.7) for practical systems militates against their analytical inversion, and a numerical inversion of the corresponding difference equations* for several fixed values of the parameter Δ provides the tractable alternative. Clearly the number of intermediate sequences to be used in an assessment remains an open question. For this reason,

*their coefficients are now periodic functions of time with period T

it is considered that the steady-state rms ripple power provides a relatively less ambiguous and more poignant measure of intersample error performance in a DDC system.

As illustrated by Fig 2.5.3, instabilities in the form of exponentially increasing sinuoids would remain undetected, if a DDC system were to be analysed by the z- Transformation alone. Indeed, it was originally argued that the possibility of the so-called *hidden oscillations* (53) made an analysis by the Modified z- Transformation imperative. However, their potential existence is patently obvious in the Laplace transfer function or the real frequency response function of a plant, where they would appear as resonances at around integral half-multiples of the sampling frequency. Competent control engineers would obviously avoid the excitation of such oscillatory modes by an apposite choice of sampling frequency. Accordingly, the Modified z-Transformation and hidden oscillations are viewed as topics of intellectual, rather than of practical, interest.

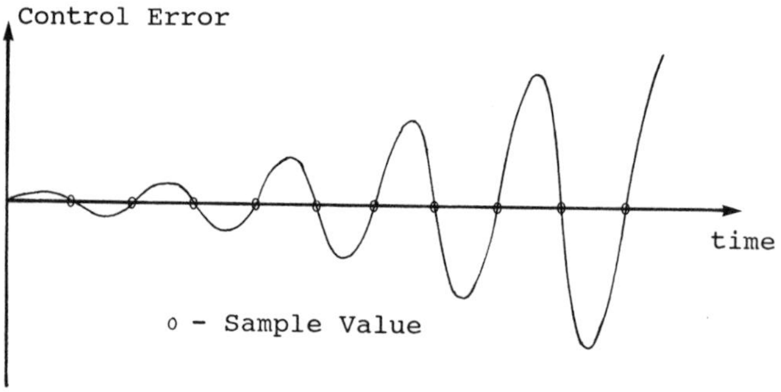

Fig 2.5.3 Hidden Oscillations

CHAPTER 3

Stability, Root Loci and Nyquist Diagrams

Und die Seele umbewacht
will in freien Flügen schweben. - H Hesse

3.1 STABILITY

The notion of dynamic stability is actually more esoteric than a cursory intuitive appraisal might suggest. To achieve greater intuitive appeal, the following introductory discussion is couched in terms of differential equations though the same concepts also relate to difference equations. ***Dynamic stability***, or its converse ***instability***, describes the nature of changes to the stationary (equilibrium) solution ($\hat{v}$) of the vector differential equation:

$$\dot{v} = f(v, x, t) \tag{3.1.1}$$

when its initial state vector (v_0) or its input vector (x) is perturbed. Control engineers generally refer to the non-linear function f as the 'system' and the stationary solution ($\hat{v}$) as the 'steady state'.

Fig 3.1.1 on the next page depicts a simple example in which a billiard ball rolls down a variable width groove in an inclined rigid surface under the action of a hypothetically variable gravitational force and conventional viscous damping. Motion down the centre of the groove represents the stationary solution ($\hat{v}$), whose stability is in question. If perturbations to the initial

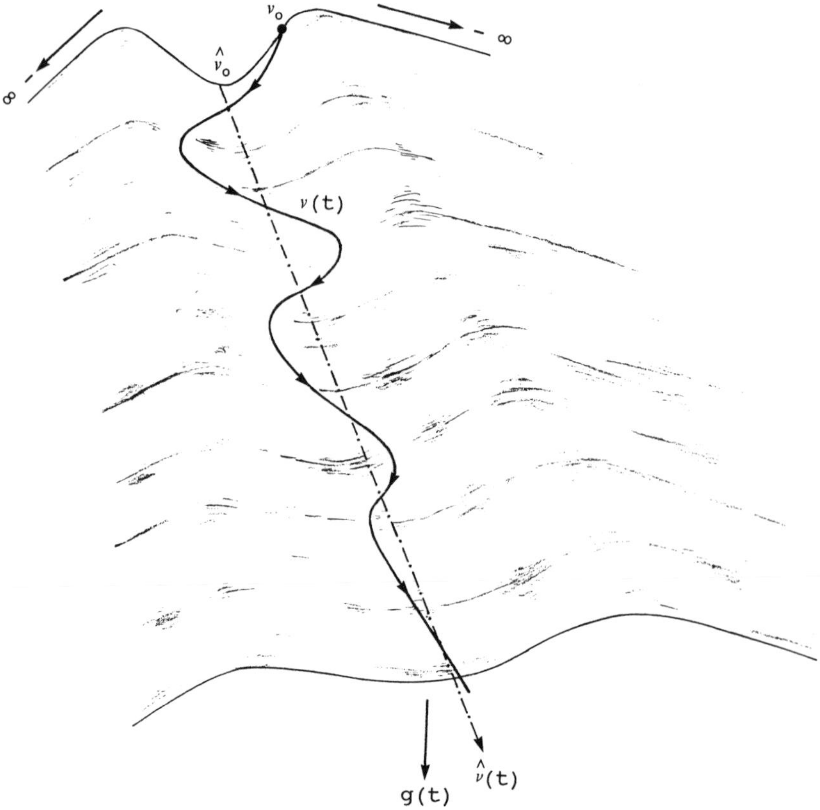

Fig 3.1.1 Illustrating Stability Concepts

displacement $v_0 - \hat{v}_0$ for example leave the ball sufficiently well within the groove, its subsequent motion can eventually converge in some manner to $\hat{v}$, so that a form of stability exists. On the other hand, sufficiently large displacements enable the subsequent trajectory to escape from the groove at some time, and the total divergence from $\hat{v}$ corresponds to instability. Dynamic systems can therefore be just *locally stable* or *globally stable*. A groove with monotonically steepening sides exemplifies the latter situation. By allowing the gravitational force (input function) to vary as in Fig 3.1.2a on the next page, a perturbed trajectory is encouraged to remain close to the stationary solution, while the opposite situation occurs in Fig 3.1.2b. A supermarket trolley-wheel, whose oscillations about its forward displacement decrease in amplitude with pushing-speed, provides an example of the familiar Van de Pol system (55) whose stability can be improved by an appropriate choice of input function.

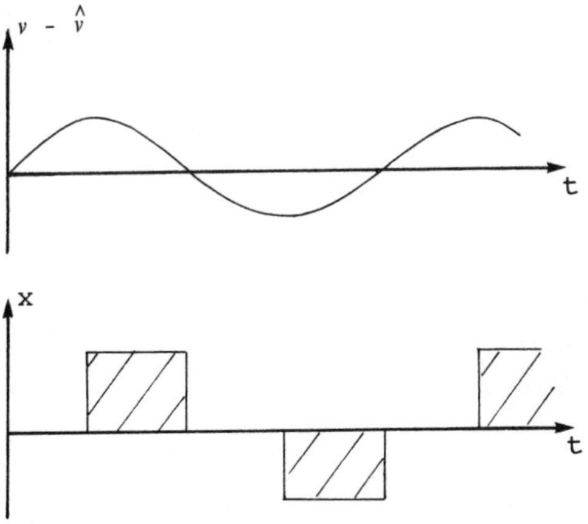

a) Stabilising Input Function

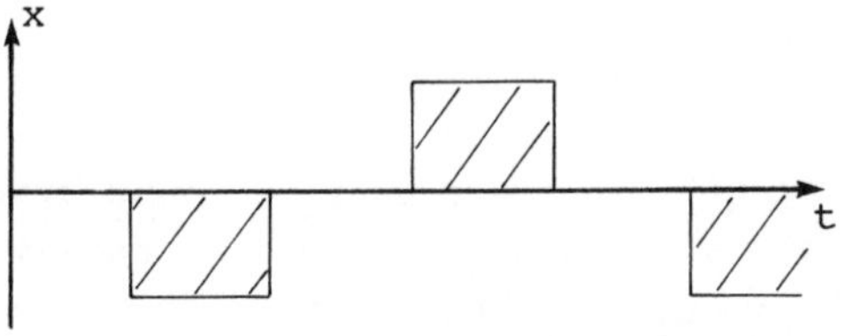

b) De-stabilising Input Function

Fig 3.1.2 The Influence of Input Function on the Dynamic Stability of a System

Mathematical investigations of stability must evidently embody a precise definition of the intuitive notion that some possible trajectories should not wander too far away from, or possibly converge to, the stationary solution. Many different forms of behaviour have been formally categorised and analysed (54, 55, 56, 57, 58). Although these studies provide fascinating insights into intellectually stimulating aspects of dynamics, the complexity of industrial processes (59) generally obstructs application of these techniques. Indeed, an objective of control engineering is not so much to investigate stability as to achieve it by the introduction of additional components or modifications to system topology. As described in Section 1.1, there is no real option for this purpose other than to deploy linearised time invariant plant models, and then vindicate the design by detailed non-linear simulations (59). Necessary and sufficient conditions for the system in equation (3.1.1) to be linear are that if:

$$\dot{\nu}_1 = f(\nu_1, x_1, t)$$

and (3.1.2)

$$\dot{\nu}_2 = f(\nu_2, x_2, t)$$

for arbitrary inputs x_1 and x_2 then:

$$\dot{\nu}_1 + \dot{\nu}_2 = f(\nu_1 + \nu_2, x_1 + x_2, t) \quad (3.1.3)$$

and also for any scalar value λ:

$$\lambda\dot{\nu} = f(\lambda\nu, \lambda x, t) \quad (3.1.4)$$

Equations (3.1.3) and (3.1.4) imply that a linear system is either globally stable or globally unstable. That is to say, a local behaviour pattern in a linear system obtains globally. The stringent convergence requirement now applied to a linear system is formally:

> A linear dynamical system is 'stable' if and only if any initial perturbation of a state vector induces a deviation from any stationary solution that remains arbitrarily small after some specifiable elapsed time.

(3.1.5)

A linear time-invariant discrete data system with a common sampling period (T) is described by a vector difference equation of the form:

$$\begin{aligned} \boldsymbol{v}_{n+1} &= A\ \boldsymbol{v}_n + B\ \boldsymbol{x}_n \\ \boldsymbol{y}_n &= C\ \boldsymbol{v}_n + D\ \boldsymbol{x}_n \end{aligned} \qquad (3.1.6)$$

where

A, B, C, D - matrices with constant real entries

$\boldsymbol{v}$ - state vector ; $\boldsymbol{x}$ - input vector

$\boldsymbol{y}$ - observed output vector

and the subscript n denotes the value of its associated vector at time nT. This concise formulation characterises both single- and multi-input systems. The state vector sequence ($\boldsymbol{v}$) is simply the coordinates with respect to a set of reference vectors for expressing the solution of equation (3.1.6). Like our familiar 3-dimensional world, there are many possible sets of reference vectors (bases) in which the solution can be specified, but that of the canonical realisation in equation (2.3.5) is as good as any for investigating the stability of a single-input system. If the state vector sequence in this representation of the pulse transfer function:

$$H(z) = \left\{ \sum_{k=o}^{K} a_k\, z^{K-k} \right\} \Big/ \left\{ 1 + \sum_{k=1}^{K} b_k\, z^{K-k} \right\} = A(z)/B(z) \qquad (3.1.7)$$

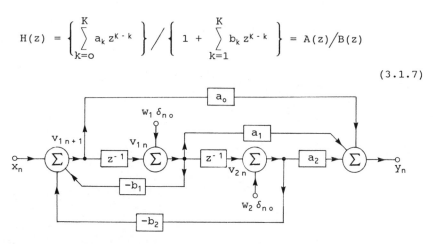

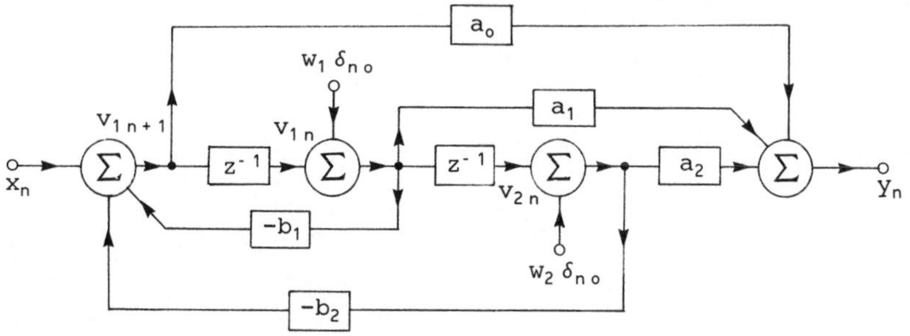

Fig 3.1.3 Perturbation of State Vector Components for a Second-Order System

is initially perturbed by the vector sequence $w\delta_{no}$, as illustrated by the second-order example in Fig 3.1.3, the corresponding temporal equations are*:

$$\left.\begin{aligned} v_{1\,n+1} &= \sum_{k=1}^{K} (-b_k)\, v_{k\,n} + x_0 + \sum_{k=1}^{K} (-b_k)\, w_k\, \delta_{no} \\ v_{k+1\,n+1} &= v_{k\,n} + w_k\, \delta_{no} \quad \text{for} \quad 2 \le k \le K \\ y_n &= \sum_{k=1}^{K} (a_k - a_0 b_k)\, v_{k\,n} + a_0 x_n + \sum_{k=1}^{K} (a_k - a_0 b_k)\, w_k\, \delta_{no} \end{aligned}\right\} \tag{3.1.8}$$

Although the above equations are readily cast into the matrix format of equation (3.1.6), the recursive property of the state vector components enables the terser format above. The stationary solution of equation (3.1.8) for any scalar input $\{x_n\}$ is derived by nulling the perturbation vector $w\delta_{no}$, and the corresponding state vector and scalar response sequences are now denoted by $\hat{v}$ and $\hat{y}$ respectively. Defining the deviation quantities:

$$\psi = v - \hat{v} \quad \text{and} \quad \theta = y - \hat{y} \tag{3.1.9}$$

their z-transformations are derived from equation (3.1.8) as:

$$z\Psi_1(z) = \sum_{k=1}^{K} (-b_k)\Psi_k(z) + \sum_{k=1}^{K} (-b_k)\, w_k \tag{3.1.10}$$

$$\Psi_k(z) = z^{-k+1}\Psi_1(z) + \sum_{i=1}^{K} w_i\, z^{-k+i} \quad \text{for } 2 \le k \le K \tag{3.1.11}$$

---

*recall that $\delta_{no} = 1$ for $n = 0$; otherwise it is zero.

$$\Theta(z) = \sum_{k=1}^{K} (a_k - a_o b_k)\,\Psi_k(z) + \sum_{k=1}^{K} (a_k - a_o b_k) w_k \qquad (3.1.12)$$

Substituting equation (3.1.11) into equation (3.1.10) yields:

$$\Psi_1(z) = P(z)/B(z) \qquad (3.1.13)$$

where P(z) is a polynomial of degree N - 1 in z with coefficients that are functions of $\{b_k\}$ and $\{w_k\}$. Equations (3.1.11) and (3.1.12) establish that the convergence of the sequences $\{\theta\}$ and $\{\psi_{k\geq 2}\}$ to zero (or their unbounded divergence) as n tends to infinity is dependent only* on the asymptotic behaviour of $\{\psi_1\}$. Consequently, the z-transformation $\Psi_1(z)$ contains all the information on the stability or otherwise of the linear discrete data system H(z). In Section 2.1, it is shown that a sequence converges to zero if and only if the poles of its z-transformation have magnitudes strictly less than unity. Now if:

$$\lim_{n\to\infty} \psi_{1n} = 0$$

then by definition, the magnitude of ψ_1 remains arbitrarily small after a definable elapsed time. Hence in terms of statement (3.1.5):

> A linear discrete data system with a rational pulse transfer function is stable if and only if all its poles have magnitudes strictly less than unity.
>
> (3.1.14)

The unit circle in the z-plane is therefore a stability boundary for linear time invariant discrete data systems, and Section 3.3 confirms its totally analogous role to the imaginary axis of the s- plane in the analysis of linear continuous data systems.

---

*note that z^{-k} just delays the corresponding temporal sequence by kT

It would be patently irksome if the engineering of adequate stability in a DDC system necessitated repeated factorisation of its overall pulse transfer function in order to establish its pole locations. More significantly, little or no insight into the choice of appropriate compensating elements would be available in such circumstances. Fortunately the Root Loci (60) and Nyquist Diagram (61) methods, which alleviate these difficulties so effectively for the design of continuous data control systems, also largely apply to DDC systems. Basic descriptions of these techniques as applied to DDC systems complete this chapter, while their inclusion as part of a comprehensive engineering design procedure follows later in Chapter 7.

3.2 ROOT LOCI

A control system is broadly required to follow its stationary (steady-state) response for any particular input with small enough deviations after a prescribed time. *Adequate stability*, rather than just mere stability, is the practical requirement. Section 3.1 and the Inversion Integral (2.1.9) show that the transient deviation from a stationary solution is determined by the poles of a pulse transfer function*. Accordingly, the designer must manipulate the poles of a linearised system model so as to achieve the required transient performance. In this context, a recurring problem is the choice of a scalar gain constant (λ) so that the poles of a minor or overall feedback system take up suitable locations. These poles are generally the algebraic roots of a *characteristic equation* with the form:

$$1 + \lambda F(z) = 0 \qquad (3.2.1)$$

where

*The following discussion applies equally well to continuous data systems, for which replace 'pulse transfer function' by 'Laplace transfer function'.

λ - adjustable gain parameter

$F(z)$ - a pulse transfer function

The Root Locus Diagram (60) outlined below maps the roots of equation (3.2.1) in the z-plane as a continuous function* of the adjustable gain parameter. Pole positions are afterwards related to the significant features of their sometimes oscillatory contribution to a temporal response, thereby enabling a straightforward choice of scalar gain. Finally in this section, important distinctions are identified between s- and z- plane applications of the technique.

The construction and engineering applications of Root Loci are thoroughly described for example by Truxal (61), and a brief outline is considered to be sufficient here. An elegant feature of the technique is that fractorisations are required for just the 'open loop' poles and zeros of F(z). Accordingly, equation (3.2.1) is first written as:

$$F(z) = \prod_{1}^{K} (z - z_k) \Big/ \prod_{1}^{K} (z - p_k) = (1/\lambda) \exp\left\{ \pm j(2N + 1)\pi \right\} \tag{3.2.2}$$

from which the various constructional algorithms are easily deduced. When the scalar gain parameter (λ) approaches zero, the right-hand-side of equation (3.2.2) tends to infinity, so that a locus always starts from the poles $\{p_k\}$ of F(z). Likewise, when the gain becomes arbitrarily large, the right-hand-side of equation (3.2.2) approaches zero, so that a locus always terminates on the zeros $\{z_k\}$ of F(z). Because the gain parameter can always be set to the required magnitude, the phase criterion:

$$\text{Arg } F(z) = \sum_{1}^{1} \text{Arg}(z - z_k) - \sum_{1}^{K} \text{Arg}(z - p_k) = \pm (2N + 1)\pi \tag{3.2.3}$$

*The roots of a finite polynomial are continuous functions of its parameters (10)

effectively determines a locus. As illustrated by Fig 3.2.1, equation (3.2.3) requires that the phase angles of vectors from the open loop poles and zeros to a point on a Locus sum algebraically to an integral odd multiple of ± 180°.

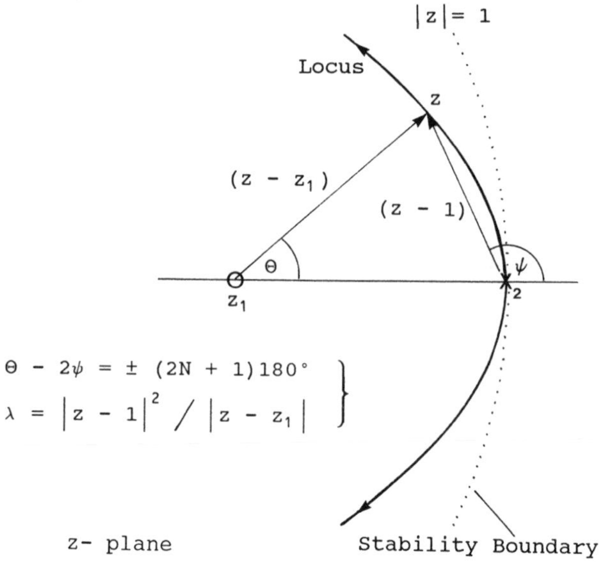

Fig 3.2.1 Illustrating the Phase Angle Criterion and the Calibration of the Root Locus for $F(z) = (z - z_1) / (z - 1)^2$

Although the construction of a locus might appear to be somewhat difficult at first sight, simple rules (61) facilitate its construction. For instance, the number of open loop poles always exceeds the number of zeros in practice, and the excess is termed the ***rank*** (R) of F(z). Some parts of a locus must therefore terminate at infinity and in doing so become asymptotic to the straight lines:

$$\text{Arg}(z - z_i) = \Theta_i \tag{3.2.4}$$

whose intercept (z_i) and angles (Θ_i) with the positive real axis are readily derived as follows. At arbitrarily large values of

$|z|$, equation (3.2.2) requires that points on a locus satisfy:

$$-\lambda = z^R \approx (z - z_i)^R \tag{3.2.5}$$

so that:

$$\theta_i = \pm (2N + 1)\pi/R \tag{3.2.6}$$

Furthermore, straightforward algebraic division of equation (3.2.2) yields:

$$-\lambda = z^R + \left[\sum_1^K p_k - \sum_1^K z_k \right] z^{R-1} + \ldots\ldots \tag{3.2.7}$$

and by equating coefficients of z^{R-1} in equations (3.2.5) and (3.2.7) the real intercept of the asymptotes is derived as:

$$z_i = \left[\sum_1^K p_k - \sum_1^K z_k \right] / R \tag{3.2.8}$$

Once the construction of a locus is complete in terms of the phase angle criterion of equation (3.2.3), corresponding scalar gains can be superimposed from equation (3.2.2) as:

$$\lambda = \left\{ \prod_1^K |z - p_k| \right\} / \left\{ \prod_1^K |z - z_k| \right\} \tag{3.2.9}$$

Suitable locations for the closed loop poles lying on a locus still remain to be decided, and in this respect the more familiar continuous systems are examined first of all.

If a real pole or complex pole pair of a Laplace transfer function lies much closer to the imaginary axis than other poles and zeros, then the Inversion Integral (1.2.27) shows that this pole (pair) provides the largest and slowest contribution (residue) to the transient deviation about any stationary solution. Such so-called ***dominant poles*** occur surprisingly frequently in practice. Consequently, control engineers often study first of all the dynamics of the first- and second-order continuous data systems:

$$K_1(s) = 1/(\tau s + 1) \tag{3.2.10}$$

$$K_2(s) = \omega_0^2/(s^2 + 2\zeta\omega_0 s + \omega_0^2) \tag{3.2.11}$$

in order to relate the position and time-domain behaviour of dominant poles. This correspondence can be deduced from the response (y) of these basic systems to the unit step function:

$$\left.\begin{aligned} U(t) &= 1 \quad \text{for } t > 0 \\ &= 0 \quad \text{otherwise} \end{aligned}\right\} \tag{3.2.12}$$

Because the first-order system $K_1(s)$ with a simple pole at $-1/\tau$ elicits the strictly monotonic response:

$$y(t) = 1 - \exp(-t/\tau) \tag{3.2.13}$$

the location and time-domain behaviour of a real dominant pole are easily related. The second-order system $K_2(s)$ elicits the oscillatory response:

$$y(t) = 1 - \left[\sqrt{1 - \zeta^2}\right]^{-1} . \exp(-\zeta\omega_0 t) . \sin(\omega_n t + \phi) \tag{3.2.14}$$

where

ζ - damping factor ; ω_0 - characteristic frequency

$$\phi = \tan^{-1}\left\{\sqrt{1 - \zeta^2}/\zeta\right\} \; ; \; \omega_n = \omega_0\sqrt{1 - \zeta^2} \tag{3.2.15}$$

With reference to Fig 3.2.2a on the next page, the ***transient overshoot*** (M_T) and ***response time*** ($\tilde{t}$) serve to characterise the transient deviations about a stationary solution that result from a dominant complex pole pair. Differentiating equation (3.2.14) yields these transient performance parameters as:

$$M_T = \exp\left(-\pi\zeta\Big/\sqrt{1-\zeta^2}\right) \quad ; \quad \tilde{t} = \pi/\omega_n$$

Designers generally seek a transient overshoot of between 5 and 16% which correspond to damping factors of 0.7 to 0.5 respectively. As depicted in Fig 3.2.2b the cosine of the angle subtended by a complex pole at the origin equals the damping factor of the pair. Consequently, normal practice is to select a gain constant (λ) so that the dominant poles of a linearised model lie in a sector with a half angle of between 45 and 60°. However, the presence of nearby closed loop zeros* reduces the response time whilst increasing the transient overshoot, and appropriate allowances should be made in such circumstances (61). It is now shown that the situation with DDC systems is far less convenient.

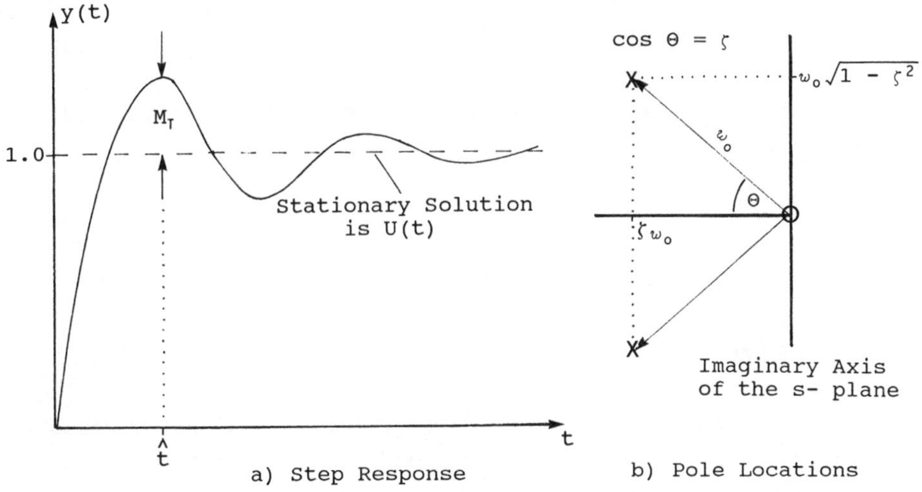

Fig 3.2.2 Step Response and Pole Locations for the Classical Second-Order Laplace Transfer Function

The second-order discrete data system that forms the dual of $K_2(s)$ in equation (3.2.11) is given by:

---

*In Fig 2.2.3, these zeros are those of D(z)G(z)

$$K_2(z) = \left|1 - re^{j\Theta}\right|^2 \Big/ (z - re^{j\Theta})(z - re^{-j\Theta}) \qquad (3.2.16)$$

where

$$0 < r < 1 \quad \text{and} \quad 0 < \Theta < \pi/2$$

Its complex conjugate poles are located at $re^{\pm j\Theta}$ in the z- plane, and its response $\{y_k\}$ to the unit step function evaluates from the Inversion Integral (2.1.9) as:

$$y_n = 1 + r^{n-1} \left\{ \left|1 + re^{j\Theta}\right|^2 \Big/ \sin\Theta \right\} \sin(n\Theta - \phi) \qquad (3.2.17)$$

$$\phi = \text{Arg}\ (1 + re^{j\Theta}) \qquad (3.2.18)$$

By means of the discrete turning point criterion:

$$y_{n+1} - y_n = 0 \qquad (3.2.19)$$

the response time ($\tilde{n}T$) and transient overshoot (M_T) are derived respectively as:

$$\tilde{n}T = \pi T/\Theta \qquad (3.2.20)$$

and

$$M_T = r^{\pi/\Theta} \qquad (3.2.21)$$

Equation (3.2.21) shows that loci of constant transient overshoot are far more complicated for a dominant pole pair in the z- plane than for the s- plane. Consequently, manually constructed root loci for DDC systems are less easily interpreted than those for continuous data systems. Moreover, even though the spirals of constant transient overshoot can be readily superimposed on any z- plane root locus plot by personal computer software, personal experience is that in some cases less insight is available as regards appropriate forms of controller, or the influence of neighbouring poles or zeros. It is particularly informative in this respect to consider the root locus design of a series* digital controller to achieve a response time of around 3 seconds for the plant and D-A converter system:

$$G(s) = \frac{(1 - e^{-sT})}{s^3 (s + 1)} \quad \text{with} \quad T = 0.1 \text{ second} \qquad (3.2.22)$$

Indeed, when confronted with the above linearised model of a practical system, the author could effect suitable compensation only through the Nyquist Diagram procedure in Chapter 7. Accordingly, the Root Locus Diagram is not recommended as a means of precisely engineering the compensator parameters for a DDC system.

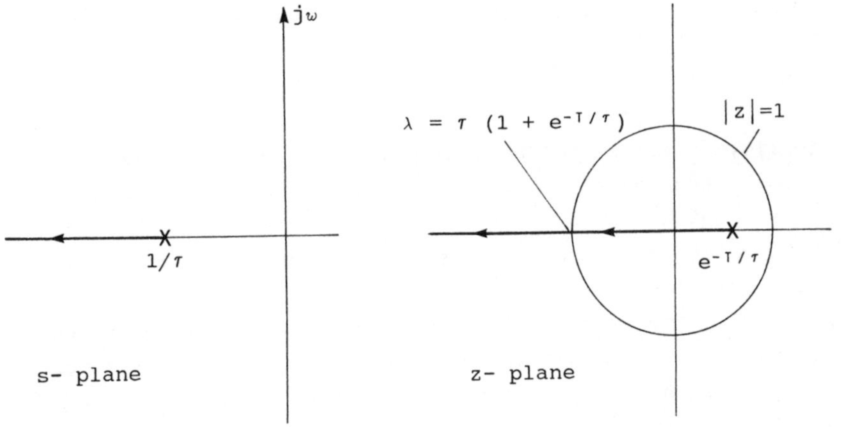

a) Root Locus of $\lambda/(\tau s + 1)$ b) Root Locus of $\lambda\tau^{-1}z/(z - e^{-T/\tau})$

Fig 3.2.3 Root Loci for First Order Plants with Gain

It should be noted that the sampling process always aggravates to some extent the stability of a closed loop control system. To illustrate the point, consider a unity feedback system that is constructed around the first-order Laplace transfer function $K_1(s)$ in equation (3.2.10) and an arbitrary scalar gain parameter (λ).

---

* as the controller in Fig 2.2.3

The root locus in Fig 3.2.3a establishes that the continuous data feedback system remains stable for all values of the gain parameter. However, as shown in Fig 3.2.3b, the introduction of simple error sampling with any period (T) renders the overall system unstable for

$$\lambda \geq \tau \left\{ 1 + \exp(- T/\tau) \right\} \tag{3.2.23}$$

In essence, the degradation in stability stems from the fundamental precept of control engineering:

> Any time delay in effecting appropriate control action encourages instability

Equation (3.2.23) reveals that such time delays are not absolute, but must be related to the response time of the system under consideration.

Finally, it should not be construed from the preceding discussion that a step function always provides a suitable test input for establishing the transient characteristics of control systems which have possibly other than largely dominant poles. Because the energy content* of a step function decays away rapidly with frequency, poles representing high frequency modes of temporal behaviour might not be excited strongly enough to be readily discernible against the inevitable background noise. Also, the long-term offset created by a step input can be detrimental to plant operation or efficiency. For these reasons, pseudo random codes (63, 64) or pulse inputs (59) are sometimes preferable alternatives to a step function input.

---

*Strictly, the energy spectrum of a unit step function U(t) does not exist - see Section 1.2. However, exp(-bt).U(t) tends to U(t) as b tends to zero and the Fourier Transformation of exp(-bt).U(t) is $1/(j\omega + b)$, so that in this limiting sense the energy spectrum of U(t) has meaning

3.3 REAL FREQUENCY RESPONSE

As described in Chapter 2, the response of a linear discrete data system to an arbitrary input can be specified in terms of z-transformations by:

$$Y(z) = H(z)X(z) \tag{3.3.1}$$

where:

$Y(z)$ - output ; $X(z)$ - input

$H(z)$ - pulse transfer function of the system

The actual output sequence is derived from the Inversion Integral (2.1.9) as:

$$y(n) = \frac{1}{2\pi j}\int_C H(z)X(z)z^{n-1}dz \tag{3.3.2}$$

where the closed contour (C) of integration encloses all the poles of $H(z)X(z)$. By Cauchy's Theory of Residues (10, 12) equation (3.3.2) evaluates in general as:

$$y(n) = y_H(n) + \hat{y}(n) \tag{3.3.3}$$

where:

$y_H(n)$ - contribution from the poles of $H(z)$ only

$\hat{y}(n)$ - contribution from the poles of $X(z)$ only

If the system $H(z)$ is stable with all its poles strictly inside the unit circle, the deviations $\{y_H(n)\}$ from the stationary response $\{\hat{y}(n)\}$ are shown in Section 3.1 to become eventually arbitrarily small. For the particular input sequence:

$$x_n = \exp(j\omega nT) \tag{3.3.4}$$

for which:

$$X(z) = z \Big/ \left(z - e^{j\omega T}\right) \tag{3.3.5}$$

the stationary (steady state) response of a stable system is therefore:

$$\hat{y}_n = \text{Residue of } \left\{ H(z) z^n / (z - e^{j\omega T}) \right\} \text{ at } z = e^{j\omega T}$$

or explicitly:

$$\hat{y}_n = H^*(\omega) e^{j\omega nT} \tag{3.3.6}$$

where:

$$H^*(\omega) \triangleq H(z) \Big|_{z=e^{j\omega T}} = \sum_{0}^{\infty} h_n \; e^{-j\omega nT} \tag{3.3.7}$$

According to equation (2.4.18), $H^*(\omega)$ is the Pulse Fourier transformation of the sequence $\{h_n\}$ and it is periodic with period $2\pi/T$ rad/s. It is frequently convenient to express this complex valued function in polar coordinate form as:

$$H^*(\omega) = \left|H^*(\omega)\right| \underline{/H^*(\omega)} \tag{3.3.8}$$

and because terms in the sequence $\{h_n\}$ are real, equation (3.3.7) implies that:

$$H^*(-\omega) = \left|H^*(\omega)\right| \underline{/-H^*(\omega)} \tag{3.3.9}$$

That is to say, $H^*(-\omega)$ is the complex conjugate of $H^*(\omega)$. By expanding the sinusoidal input sequence:

$$x_n = \sin(\omega nT) \qquad (3.3.10)$$

as:

$$x_n = \left[e^{j\omega nT} - e^{-j\omega nT} \right] / 2j \qquad (3.3.11)$$

it follows by the principle of linear superposition from equations (3.3.6) and (3.3.9) that the corresponding stationary response is:

$$\hat{y}_n = \left| H^*(\omega) \right| \sin\left[\omega nT + \underline{/H^*(\omega)} \right] \qquad (3.3.12)$$

Fig (3.3.1) illustrates equations (3.3.10) and (3.3.12), which show that the function $H^*(\omega)$ determines the relative amplitudes and phase of the envelopes that contain the sinusoidal input and output sequences in the steady state. In this respect, and in others to be identified later, the function $H^*(\omega)$ is analogous to the familiar real frequency response function of a linear continuous data system. Accordingly, it is termed the *pulse real frequency response function* and, furthermore, the unit circle in the z- plane is seen from equation (3.3.7) to play the same role as the imaginary axis in the s- plane to the Laplace Transformation.

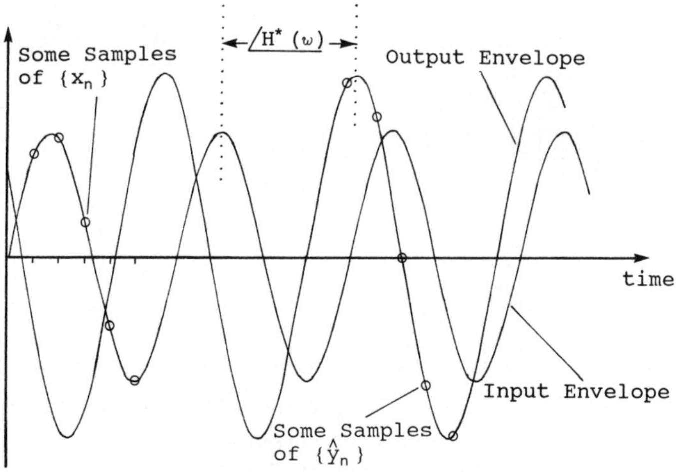

Fig 3.3.1 Illustrating a Pulse Real Frequency Response Function

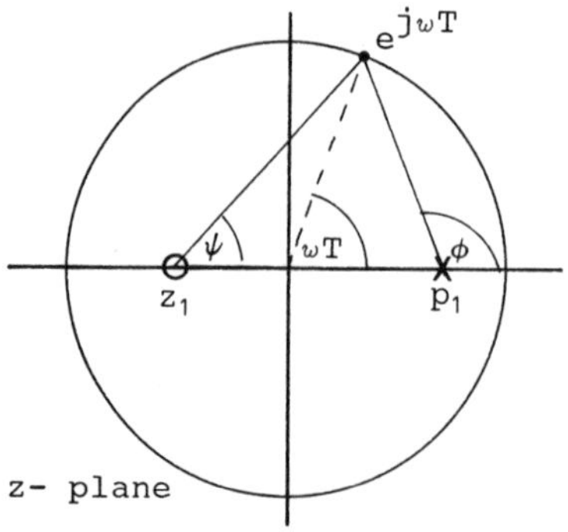

Given for example: $H(z) = (z - z_1)/(z - p_1)$

then: $H^*(\omega) = (L_z/L_p)\underline{/\psi - \phi}$

where:

L_z - length of vector from z_1 to $\exp(j\omega T)$

L_p - length of vector from p_1 to $\exp(j\omega T)$

Fig 3.3.2 Geometrical Interpretation of a Real Frequency Response Function in terms of Pole-Zero Locations

Valuable engineering insight is acquired by interpreting the pulse real frequency response function of a discrete data system in terms of the pole-zero locations of its transfer function. The pulse real frequency response corresponding to:

$$H(z) = \prod_{k=1}^{K} (z - z_k) \Big/ \prod_{k=1}^{K} (z - p_k) \qquad (3.3.13)$$

at an angular frequency ω is specified in polar coordinates by:

$$\left|H^*(\omega)\right| = \prod_{k=1}^{K} \left|e^{j\omega T} - z_k\right| \Big/ \prod_{k=1}^{K} \left|e^{j\omega T} - p_k\right|$$

$$\text{Arg } H^*(\omega) = \sum_{1}^{K} \underline{/(e^{j\omega T} - z_k)} - \sum_{1}^{K} \underline{/(e^{j\omega T} - p_k)} \qquad (3.3.14)$$

which is represented graphically in Fig 3.3.2. A pole sufficiently close to the unit circle is seen to create an amplifying resonance at an angular frequency (ω_p) that is approximately related to its angular displacement (ϕ) in the z-plane by:

$$\omega_p = \phi/T \tag{3.3.15}$$

Likewise, similarly placed zeros produce attenuating notches. If all poles of H(z) and X(z) lie strictly inside the unit circle, it follows from Section 2.4 that the response is completely specified by the Pulse Fourier transformation:

$$Y^*(\omega) = H^*(\omega)X^*(\omega) \tag{3.3.16}$$

Appendix A3.3 demonstrates that $\left|X^*(\omega)\right|^2$ can be regarded as the 'energy' spectrum of its sequence $\{x_n\}$, so that a pulse real frequency response function actually filters an input signal on a similar physical basis to a continuous data system. Thus, because a DDC system is usually required to respond near perfectly to the low frequency components of its input, equation (3.3.16) establishes that the overall real frequency response function ideally should be unity around zero frequency*. Also because an amplifying resonance relatively accentuates the corresponding frequency components of an input, the temporal response contains oscillations in the same frequency range. In broad terms therefore, the overall pulse real frequency response function of a DDC system should not contain strong resonances. However, engineering design requires a much more precise relationship between time and frequency domain behaviour as addressed next in Section 3.4. Although the arguments there are for the most part intuitively appealing or rigorous, limited references are made to material presented later on. If a reader has any doubts after an initial reading, further retrospective consideration should be given after actually applying the techniques in Chapters 5 and 7.

3.4 TIME AND FREQUENCY DOMAIN BEHAVIOUR

Engineering judgement or actual measurements usually indicate that the continuous input data to a DDC system has significant spectral components only up to some angular frequency ω_B. Consequently, its overall pulse real frequency response function

---

* sometimes referred to as DC for obvious reasons

$K^*(\omega)$ should be approximately unity over the frequency range $[-\omega_B \; ; \; \omega_B]$*. However, various forms of inertia coerce the response of engineering systems towards zero with increasing frequency so that the compromise often sought is for:

$$\left| K^*(\omega_B) \right|^2 = 1/2 \qquad (3.4.1)$$

That is, the overall system bandwidth should broadly equate with signal bandwidth. The common sense in the rule-of-thumb that 'Around ten points are required to graph a sinuosoid' is self-evident. This forms a basis for the criterion that the bandwidth required from a DDC system must not exceed one-tenth the angular sampling frequency. Mathematically, the intuitively reasonable requirement is that:

$$\omega_B T \leq \pi/5 \qquad (= 0.63) \qquad (3.4.2)$$

Early publications on the choice of sampling frequency (69, 70) propose similar laxly formulated criteria. If the spectrum of a continuous data input were to be strictly bandlimited between $\pm\omega_B$, then application of the analysis in Section 1.3 shows that it could be recovered perfectly from samples taken at intervals of π/ω_B using the idealised low pass filter in equation (1.3.5). On the next page Fig 3.4.1 illustrates this result, which was discovered by Nyquist in 1928 (71) and which sets the ultimate constraint of:

$$\omega_B T \leq \pi \qquad (3.4.3)$$

As previously discussed, a DDC system cannot operate in non-real time nor can its real frequency response reduce abruptly to zero. Furthermore, input spectra are almost certainly not bandlimited in practice so that equation (3.4.2) defines a more realistic upper-bound. Indeed, experience in applying the analysis in Chapter 5

---

*Unless appreciable interference is superimposed, but this represents a detailed refinement of the present argument

suggests that even smaller system bandwidth to sampling frequency ratios are necessary to meet contemporary performance specifications. Similarly small ratios are also found to be necessary in Chapter 7 to engineer satisfactory digital phase advance type controllers. For all these diverse reasons, it is concluded that equation (3.4.2) defines the practical upper-bound on bandwidth to sampling frequency ratio for DDC systems.

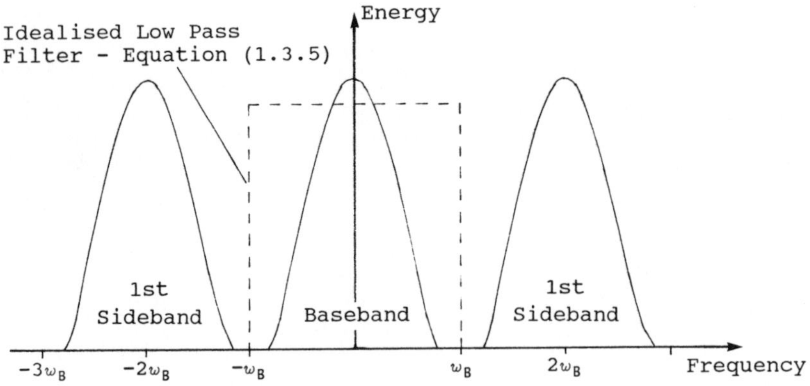

Fig 3.4.1 Illustrating Nyquist's Sampling Theorem for strictly Bandlimited Signals

As described in Section 3.2, the first- and second order discrete data systems:

$$K_1(z) = (1 - a)/(z - a) \quad \text{with} \quad 0 < a < 1 \tag{3.4.4}$$

and:

$$K_2(z) = |1 - re^{j\Theta}|/(z - re^{j\Theta})(z - re^{-j\Theta}) \tag{3.4.5}$$

with

$$0 < r < 1 \quad ; \quad 0 < \Theta < \pi/2$$

often closely approximate the dominant dynamic characterics of linearised systems. Useful design information can therefore be acquired by comparing their step and real frequency responses. The step function response of the first order discrete data system is evidently:

$$Y(z) = (1 - a)z/(z - 1)(z - a) \tag{3.4.6}$$

which by the Inversion Integral (2.1.9) translates into the monotonic time domain sequence:

$$y_n = 1 - a^n \tag{3.4.7}$$

If the normalised response time ($\tilde{n}$) is somewhat arbitrarily defined as:

$$0.5 = 1 - a^{\tilde{n}} \tag{3.4.8}$$

then:

$$\tilde{n} = - \log 2/\log a \tag{3.4.9}$$

The bandwidth of the first order system is derived from equations (3.4.1) and (3.4.4) as:

$$\omega_B T = \cos^{-1}\left\{(8a - 3a^2 - 3)/2a\right\} \tag{3.4.10}$$

Numerical solutions of equations (3.4.9) and (3.4.10) are compiled in Table 3.4.1 for a range of pole locations (a). Within the practical constraint imposed by equation (3.4.2), bandwidth and response time of the first order model are seen to be related for design purposes by:

$$\tilde{n}T = 1.2/\omega_B \tag{3.4.11}$$

a	0.6	0.7	0.8	0.9	0.95	0.99
$\omega_B T$	0.927	0.632	0.390	0.183	0.0889	0.0174
$\tilde{n}T\omega_B$	1.26	1.23	1.21	1.20	1.20	1.20

Table 3.4.1 Bandwidths and Response Times for the First Order Discrete Data System

The magnitude of the pulse real frequency response for the second order system in equation (3.4.5) can be written as:

$$\left| K_2^* (\omega) \right| = F_1 (0) \Big/ \sqrt{F_1 (\omega) F_2 (\omega)} \qquad (3.4.12)$$

where:

$$\left. \begin{aligned} F_1 (\omega) &= 1 + r^2 - 2r \cos(\omega T - \Theta) \\ F_2 (\omega) &= 1 + r^2 - 2r \cos(\omega T + \Theta) \end{aligned} \right\} \qquad (3.4.13)$$

Differentiation of the product $F_1 F_2$ establishes that an amplifying resonance occurs in $K^* (\omega)$ at the angular frequency:

$$\omega_p T = \cos^{-1} \left\{ \left(\frac{1 + r^2}{2r} \right) \cos \Theta \right\} \qquad (3.4.14)$$

with the corresponding *peak magnification factor* specified by:

$$M_p = F_1 (0) \Big/ \sqrt{F_1 (\omega_p) F_2 (\omega_p)} \qquad (3.4.15)$$

Its transient overshoot (M_T) and response time ($\tilde{n}T$), which characterise the maximum deviation from a stationary response, are defined in equations (3.2.20) and (3.2.21) as:

$$M_T = r^{\pi/\Theta} \qquad (3.4.16)$$

and

$$\tilde{n}T = T\pi/\Theta \tag{3.4.17}$$

For the second order continuous data system $K_2(s)$ in equation (3.2.11), the corresponding results are (61):

$$M_p = \left[2\varsigma\sqrt{1 - \varsigma^2}\,\right]^{-1} \tag{3.4.18}$$

and

$$M_T = \exp(-\pi\varsigma/\sqrt{1 - \varsigma^2}) \quad \text{at} \quad \omega_p\tilde{t} = \pi\sqrt{(1 - 2\varsigma^2)/(1 - \varsigma^2)} \tag{3.4.19}$$

which are functions only of the damping factor (ς). Unlike therefore the discrete data system for which

$$M_p = M_p(r,\Theta) \quad \text{and} \quad M_T = M_T(r,\Theta) \tag{3.4.20}$$

these parameters for the continuous data system are uniquely related as in Table 3.4.2. Despite the lack of such a precise one to one correspondence for the discrete data system, an adequate design relationship can be established by first identifying practical constraints on its pole locations.

M_p	1.05	1.13	1.23	1.35	1.50
M_T	0.10	0.15	0.20	0.25	0.30
$\omega_p\tilde{t}$	2.14	2.51	2.70	2.82	2.90

Table 3.4.2 Peak Magnification and Transient Parameters for the Second Order Continuous Data System - a unique correspondence

The pulse real frequency response function $K_2^*(\omega)$ in Fig 3.4.2 on the next page illustrates that:

$$\omega_p \leq \omega_B \tag{3.4.21}$$

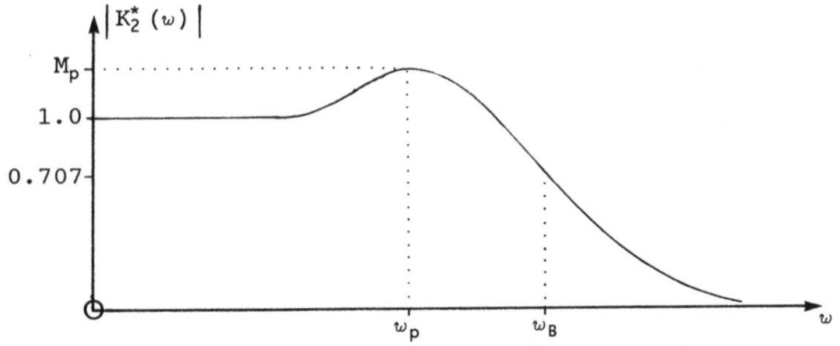

Fig 3.4.2 Pulse Real Frequency Response of the Second Order Discrete Data System $K_2(z)$ in Equation (3.4.5)

and so by equations (3.3.15) and (3.4.2):

$$\Theta/\pi \le 0.2 \qquad (3.4.22)$$

Thus a necessary condition for an adequate suppression of the spectral sidebands by a DDC system is that its dominant complex pole-pair are located within a sector of width $\pm 0.2\pi$ radians or $\pm$ 36°. In fact this constraint is actually too liberal, because equation (3.4.17) shows that when

$$\Theta/\pi = 0.2$$

the maximum transient overshoot (M_T) occurs on just the fifth output sample. The functional dependence of the transient overshoot (M_T) and the resonant frequency - response time product ($\omega_p \tilde{n}T$) on the peak magnification (M_p) at any fixed value of Θ/π can be readily computed using the numerical scheme in Fig 3.4.3 on page 106. Corresponding results in Table 3.4.3 on the next page reveal that, within the practical constraint imposed by equation (3.4.22), these parameters match for design purposes those of the continuous data system in Table 3.4.2. Finally, it should be noted that control systems are not generally engineered to have precise transient overshoots and rise times; rather these parameters are quite broadly specified as for example:

$$M_T \le 0.15 \quad \text{and} \quad \tilde{n}T \le 3 \text{ seconds}$$

M_p	1.07	1.16	1.26	1.40	1.54	
M_T	0.10	0.15	0.20	0.25	0.30	$\theta/\pi = 0.2$; $\tilde{n} = 5$
$\omega_p \tilde{n} T$	2.30	2.60	2.76	2.86	2.94	

M_p	1.05	1.14	1.24	1.36	1.51	
M_T	0.10	0.15	0.20	0.25	0.30	$\theta/\pi = 0.10$; $\tilde{n} = 10$
$\omega_p \tilde{n} T$	2.18	2.53	2.71	2.83	2.90	

M_p	1.05	1.13	1.23	1.36	1.50	
M_T	0.10	0.15	0.20	0.25	0.30	$\theta/\pi = 0.05$; $\tilde{n} = 20$
$\omega_p \tilde{n} T$	2.15	2.51	2.70	2.82	2.90	

M_p	1.05	1.13	1.23	1.35	1.50	
M_T	0.10	0.15	0.20	0.25	0.30	$\theta/\pi = 0.02$; $\tilde{n} = 50$
$\omega_p \tilde{n} T$	2.14	2.51	2.70	2.82	2.90	

M_p	1.05	1.13	1.23	1.35	1.50	
M_T	0.10	0.15	0.20	0.25	0.30	$\theta/\pi = 0.01$; $\tilde{n} = 100$
$\omega_p \tilde{n} T$	2.14	2.51	2.70	2.82	2.90	

Table 3.4.3 Peak Magnification and Transient Parameters for the Second Order Discrete Data System for poles with various angular coordinates (θ)

The close affinity between practical DDC system design concepts and continuous data systems is further re-enforced in the next Section 3.5, which is concerned with Nyquist Diagrams.

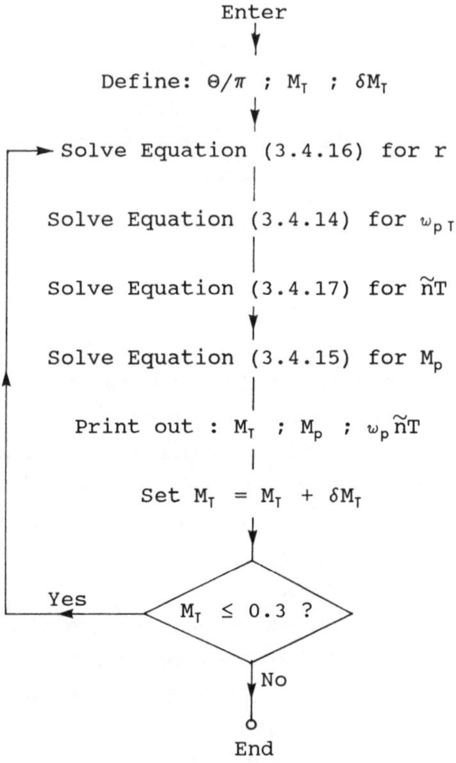

Fig 3.4.3 Numerical Scheme to calculate M_T and $\omega_p \tilde{n}T$ as a function of M_p for a Pole-Pair with constant angular coordinate Θ

3.5 NYQUIST DIAGRAMS

A control engineer generally manipulates the poles and zeros of a linearised system model in order to achieve a specified transient performance within the constraints imposed by mechanical construction and operational environment. This objective can often be achieved by introducing: series compensators as in Fig 2.2.3, networks or components in the feedback path (72), minor feedback loops (73, 74), divided reset (75) to compensate for backlash in gearing (76) and resilience in driveshafts (72), or feedforward (77, 78). In practice, these techniques are applied jointly or severally according to specific traditional experience or operational constraints. Within the broad framework of design, a recurring problem is the choice of scalar gain constant (λ) to position the poles of a minor or overall feedback loop so as to endow adequate stability. These poles are actually the zeros of a function:

$$1 + \lambda F(z) \qquad (3.5.1)$$

which forms the denominator of a pulse transfer function. The Root Locus approach to this problem is shown in Section 3.2 to be far less attractive for DDC than for continuous data systems. However, the established close relationship between time and frequency domain behaviour now provides both the basis and motivation for developing a Nyquist Diagram solution for DDC systems. Later in Chapter 7, the use of Nyquist Diagrams is extended to devise suitable series compensating elements which modify the function F(z). Furthermore, because the effects of design uncertainties (79) and different operating conditions (59) can also be assessed by the technique, it can be concluded that the Nyquist Diagram is far more versatile than just an algorithm for locating the zeros of $1 + \lambda F(z)$ with variations in λ.

Many texts on complex variable theory derive what might well be designated the '*encirclement theorem*'

$$\int_{\mathcal{C}} \frac{G'(z)}{G(z)}\, dz = \int_{G(\mathcal{C})} \frac{du}{u} = 2\pi j(Z - P)\,\mathrm{sgn}\, \mathcal{C} \qquad (3.5.2)$$

where

G - meromorphic function ; G' - differential coefficient of G

$\mathcal{C}$ - closed contour that actually crosses neither poles nor zeros of G

$u = G(z)$

Z - number of zeros of G inside C

P - number of poles of G inside C

and

$$\left.\begin{aligned} \operatorname{sgn} C &= 1 \text{ if } C \text{ encircles its interior points anticlockwise} \\ &= -1 \text{ if } C \text{ encircles its interior points clockwise} \end{aligned}\right\} \tag{3.5.3}$$

For present purposes, the function G is identified with $1 + \lambda F$ whose zeros are the poles of the closed loop system under investigation. The closed contour C becomes that shown in Fig 3.5.1, and it encloses in the negative (clockwise) sense all zeros of $1 + \lambda F$ lying strictly outside the unit circle. Linearised control system models frequently possess integrating

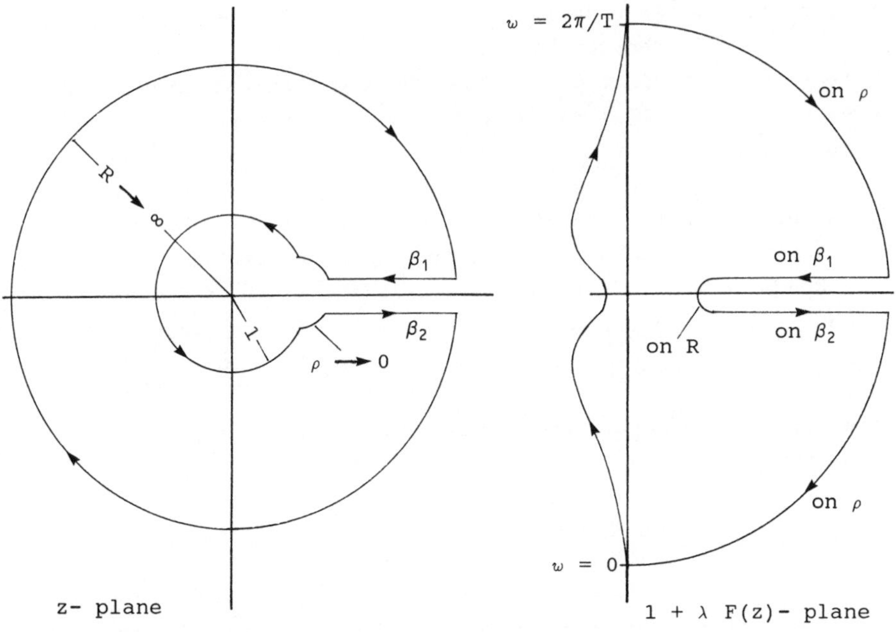

Fig 3.5.1 Contours associated with Nyquist's Criterion

elements which endow $1 + \lambda F$ with poles at $1\angle 0$. To satisfy conditions of the encirclement theorem, the contour C contains an arbitrary small detour of radius ρ about this point. Also no poles or zeros of $1 + \lambda F$ occur anywhere else on the unit circle in practical control systems except for a few special values of the scalar gain constant. Indeed, in this respect, the object is to ensure that all these zeros lie sufficiently well inside the unit circle to ensure an adequately damped transient response. Finally, the parallel segments β_1 and β_2 lie infinitesimally close to the real axis, so that for effectively all practical situations* the contour C satisfies conditions of the encirclement theorem whilst encircling any destabilising zeros. The integrand u^{-1} in equation (3.5.2) has a first order pole at the origin of the $1 + \lambda F$- plane, so that each complete anti-clockwise encirclement of the origin by $1 + \lambda F(C)$ contributes $2\pi j$ to its residue. Consequently, a necessary and sufficient condition for system stability is that $1 + \lambda F(C)$ encircles the origin P times in the positive anti-clockwise sense. However, for the reasons about to be given, it is more convenient to map C into the F- plane. As illustrated in Fig 3.5.2 on the next page, this alternative enables Nyquist's Criterion to be cited as:

> A pulse transfer function with denominator $1 + \lambda F(z)$ is stable if and only if the contour $\lambda F(C)$ encircles the *critical point* (-1, 0) in the F- plane a total of P times anti-clockwise

(3.5.4)

where $\lambda F(C)$ is evidently the contour $F(C)$ radially amplified by the scalar λ.

Although the construction of a Nyquist Diagram $F(C)$ appears a

---

*Diffusion processes for example create irrational transfer functions. These require ad hoc modifications to the contour C, which must lie on one sheet of the appropriate Riemann surface for equation (3.5.2) to be valid (81)

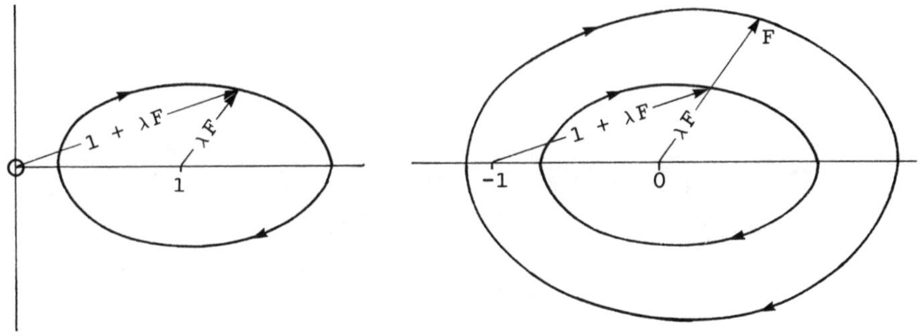

Fig 3.5.2 Transferring Contours between the 1 + λF- and F- planes

somewhat intricate problem at first sight, considerable simplification is readily achieved. By the Initial Value Theorem (2.1.6):

$$f_o = \lim_{z \to \infty} F(z) \tag{3.5.5}$$

so that the infinite circle in the z -plane reduces to the easily determined point f_o in the F- plane. Apart from the arbitrary small circular detour about $1\angle 0$, the remainder of the unit circle maps into the real frequency response function $|F^*(\omega)|\angle F^*(\omega)$ which is either calculable or measurable by experiment. The parallel paths β_1 and β_2 along the real axis are largely irrelevant, but the infinitesimal circular detour around $1\angle 0$ can present problems to the unwary. For instance, consider a unity feedback DDC system whose overall pulse transfer function is:

$$K(z) = \frac{F(z)}{1 + F(z)} \tag{3.5.6}$$

If the open loop pulse transfer function is:

$$F(z) = 1/z(z - 1) \tag{3.5.7}$$

then because 1 + F has no poles within the contour *C*, system stability requires that its Nyquist diagram does not encircle the

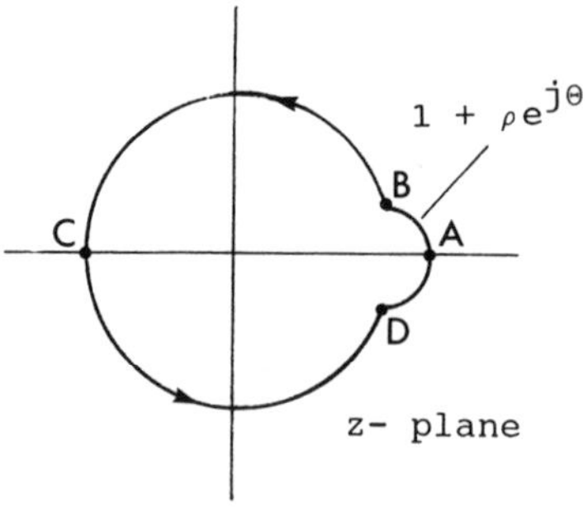

a) Part of the *C* Contour

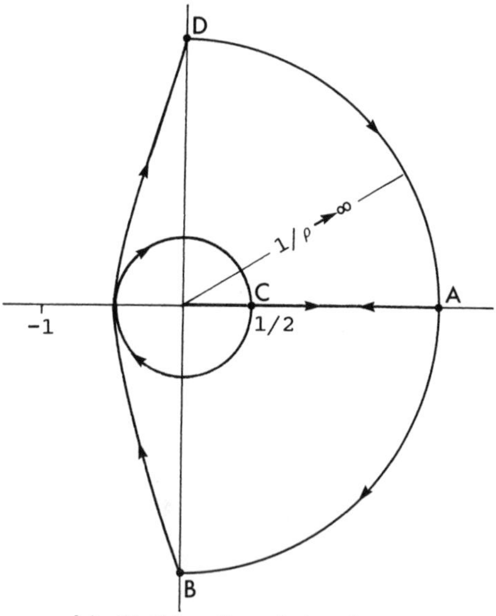

b) Stable Nyquist Diagram
for F in Equation (3.5.7)

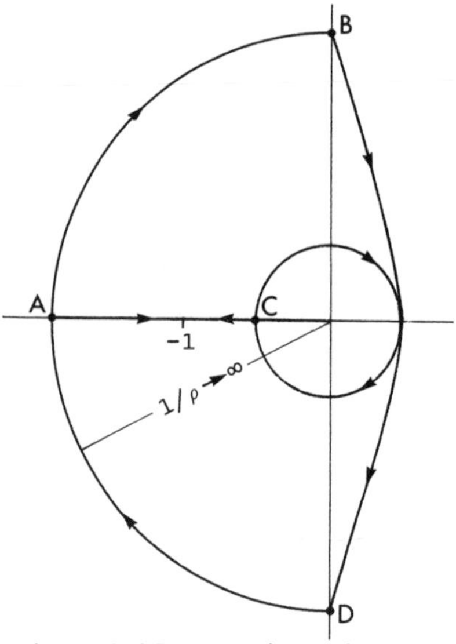

c) Unstable Nyquist Diagram
for F in Equation (3.5.8)

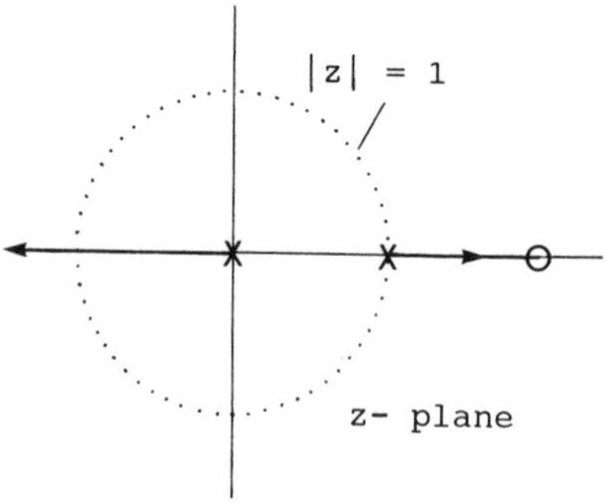

d) Root Locus of Equation (3.5.8)

Fig 3.5.3 Examples of Nyquist Diagrams which denote Stable and Unstable Closed Loop Control depending on the mapped Infinitesimal Circular Detour around $1\angle 0$

critical point (-1, 0). A sketch of F(*C*) in Fig (3.5.3b) displays no such encirclements so that the DDC system is stable. On the other hand, although the mapped real frequency response (B- C- D in Fig 3.5.3a) of the open loop pulse transfer function:

$$F(z) = (z - 2)/3z(z - 1) \qquad (3.5.8)$$

lies 'encouragingly' to the right of (-1, 0) in Fig (3.5.3c), the mapped circular detour (D- A- B) now causes the complete locus to encircle the critical point once clockwise. Because no poles of F lie within *C*, then according to equation (3.5.2) the closed loop system is unstable with one of its two poles outside the unit circle. An easily constructed Root locus in Fig 3.5.3d confirms the instability. These examples underline the importance of mapping the complete contour *C*, or alternatively confirming stability by other means. In general, interplay between a Root locus and Nyquist diagram promotes understanding of plant dynamics, though the latter is unequivocally recommended for DDC system design.

Equation (3.2.22) specifies a linearised model for an industrial plant whose operational environment dictated its stabilisation by a series controller. The design procedure in Chapter 7 provides a suitable controller with pulse transfer function:

$$D(z) = \frac{10(z - 0.975)(z - 0.9)}{z(z - 0.8 - j0.3)(z - 0.8 + j0.3)} \qquad (3.5.9)$$

A sketch of the corresponding Nyquist diagram in Fig 3.5.4b exhibits typically more tortuous behaviour then the academic examples in Fig 3.5.3. There is an obvious requirement for a systematic procedure to decide encirclements or otherwise of the critical point. One approach involves labelling and sub-dividing the z- and F- plane contours as in Fig 3.5.4 on the next page. Mathematics (10) aligns with intuition to define the number of encirclements as the change in argument by multiples of 360° of a radius vector from the critical point after it completes one full

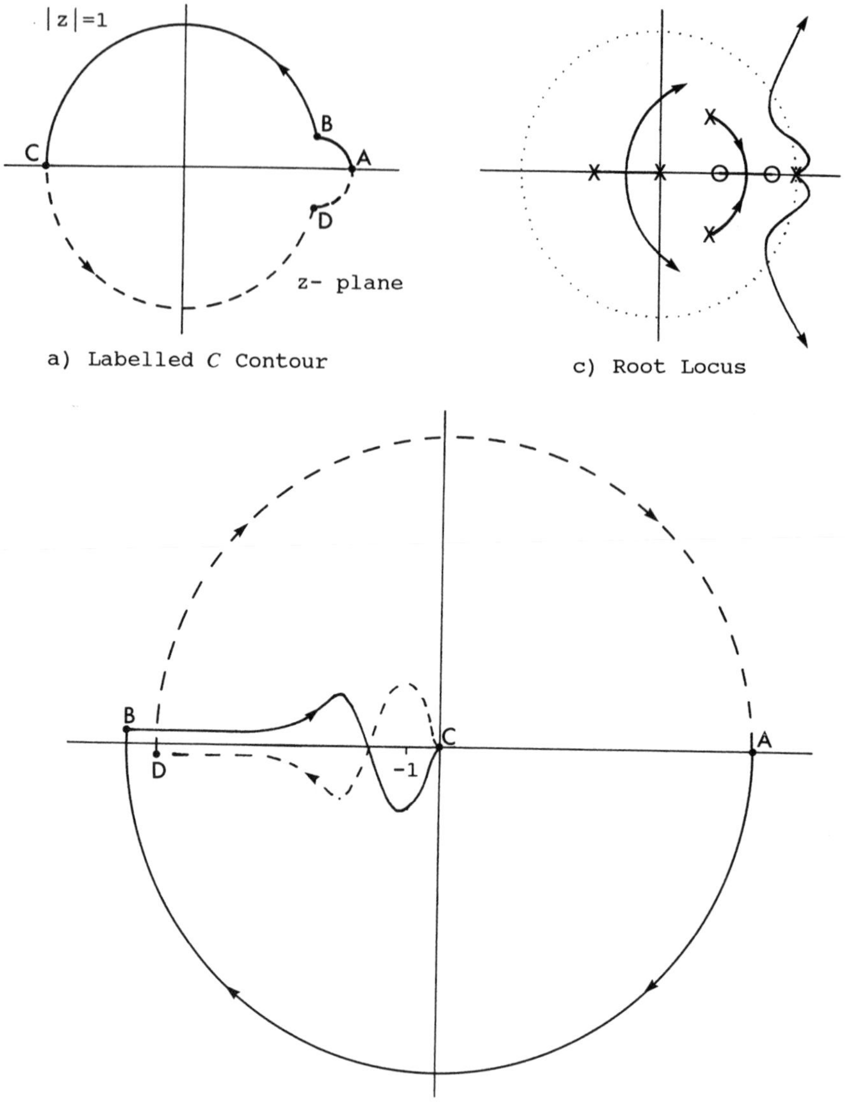

a) Labelled *C* Contour

c) Root Locus

b) Nyquist Diagram

Fig 3.5.4 Illustrating Conditional Stability and a Systematic Procedure for determining Encirclements

sweep of the contour in its reference direction. Table 3.5.1 quantifies the changes in argument of this radius vector as it sweeps over the labelled paths in Fig 3.5.4b. Because these argument changes are cumulative and because this particular open loop system has no poles outside the contour ABCD in the z- plane, it follows that the zero encirclements of the Nyquist diagram imply that the closed loop system is stable. Even after several years' experience and acquired confidence, it is still advisable to confirm a closed loop stability deduction by plotting the corresponding Root locus as in Fig 3.5.4c.

A - B	-180°
B - C	+180°
C - D	+180°
D - A	-180°
Total Change in Argument =	0°

Table 3.5.1 Regarding Encirclements for the Nyquist Diagram in Fig 3.5.4b

If the scalar gain constant of the controller in equation (3.5.9) is reduced sufficiently, the above procedure confirms that the shrunken Nyquist diagram encircles the critical point twice in a clockwise direction. There are therefore two closed loop poles (zeros of 1 + DG) outside the unit circle. Control systems which become unstable by a reduction in controller gain constant are designated *conditionally stable*. Some processes in the nuclear and electrical supply industries (59) for example are required to be *fail-safe*, so that their conditionally stable control would not meet with operational approval.

Frequency values (rad/s) are readily superimposed on a Nyquist diagram during its construction. By measuring off vectors from the origin and critical point to these identified points on a locus, corresponding values of the overall real frequency response can be computed for the unity* feedback system:

$$K^*(\omega) = \frac{\lambda F^*(\omega)}{1 + \lambda F^*(\omega)} \qquad (3.5.10)$$

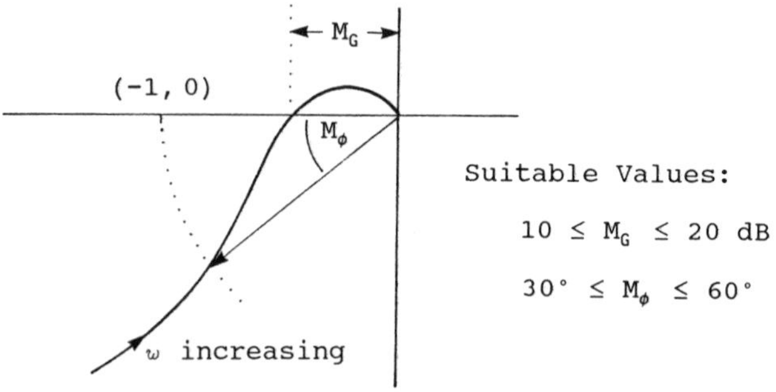

Fig 3.5.5 Illustrating the Gain-Phase Margin Criterion

Close approximations to the resonant frequency (ω_p) and the peak magnification (M_p) can be derived by evaluating equation (3.5.10) at a few points around where the locus lies closest to the critical point. Granted that a DDC system has dominant poles and a realistic sampling frequency, then desirable behaviour of its Nyquist diagram can be identified in terms of the relationship between peak magnification and transient overshoot (M_T) in Table 3.4.3. Although the familiar *M-Circles* (80) can assist in this respect, the *Gain-Phase Margin Criterion* (5) illustrated in Fig 3.5.5 is preferred because it provides some accommodation for variations in system parameters due to uncertainties in physical data, ageing or marginally different operating conditions (59). If a Nyquist diagram exhibits two resonant frequencies, or equivalently two separate approaches towards the (-1, 0) point, a slowly decaying 'tail' is produced in its temporal step response

*When a feedback path contains additional components (eg a tachometer etc), the numerator of equation (3.5.10) becomes evidently different from λF. However, the following principles are still relevant

as a result of the lower resonant frequency. Under these circumstances, it is advisable to compute the time domain response after each re-shaping of the locus in order to derive the necessary insight for further modifications and compromise.

3.6 INVERSE NYQUIST DIAGRAMS

If the function G in equation (3.5.2) is now identified with the inverse of an overall pulse transfer function K(z) and if the contour C is the z- plane contour in Fig 3.5.1, then there are $-(Z' - P')$ anticlockwise encirclements of the origin by $K^{-1}(C)$

where:

Z' - number of zeros of $K^{-1}(z)$ within C

P' - number of poles of $K^{-1}(z)$ within C

Because the zeros (poles) of K^{-1} are the poles (zeros) of K, then necessary and sufficient conditions for system stability are that the *Inverse Nyquist diagram* $K^{-1}(C)$ encircles the origin of the K^{-1}- plane a total of P' times anticlockwise. For the feedback system shown in Fig 3.6.1 on the next page, the overall pulse transfer function is:

$$K(z) = \left[1 + F(z)B(z) \right]^{-1} F(z) \tag{3.6.1}$$

and because its inverse is evidently:

$$K^{-1}(z) = F^{-1}(z) + B(z) \tag{3.6.2}$$

then

$$K^{-1}(C) = F^{-1}(C) + B(C) \tag{3.6.3}$$

In the special case of a unity feedback system (B = 1), encirclements by the inverse locus $K^{-1}(C)$ can be derived as shown in Fig 3.6.2 from just the open loop locus $F^{-1}(C)$ by taking the (-1, 0) point in the F^{-1}- plane as the origin of the K^{-1} -plane. With a non-unity feedback system, the two loci $F^{-1}(C)$ and $B(C)$ are added vectorially with the origin remaining at (0,0) in the combined F^{-1} and B- plane. Prior to the widespread availability of digital computers, vector operations were implemented graphically and the addition of $F^{-1}(C)$, $B(C)$ in this format was much easier* than the multiplication of $F(C)$ and $B(C)$ required in equation (3.6.1). Another historic advantage claimed for the Inverse Nyquist Diagram technique is the simpler construction of M-Circles (73), but this became obsolescent with the advent of the Gain-Phase Margin Criterion. However, the real advantage of the Inverse Nyquist Diagram method lies in its generalisation to multi-input multi-output systems (82). In this context, F and B in equations (3.6.1) and (3.6.2) generalise to pulse transfer function matrices. The less convoluted form of the latter enables a far simpler mathematical analysis, and the development of efficient computer software (83) makes the *Inverse Nyquist Array* a successful practical design technique. Further discussion lies outside the scope of this specialised text and lucid descriptions are provided in references (84) and (85). More details concerning the frequency domain design of specifically single-input DDC systemsby the conventional Nyquist Diagram procedure are given in Chapter 7.

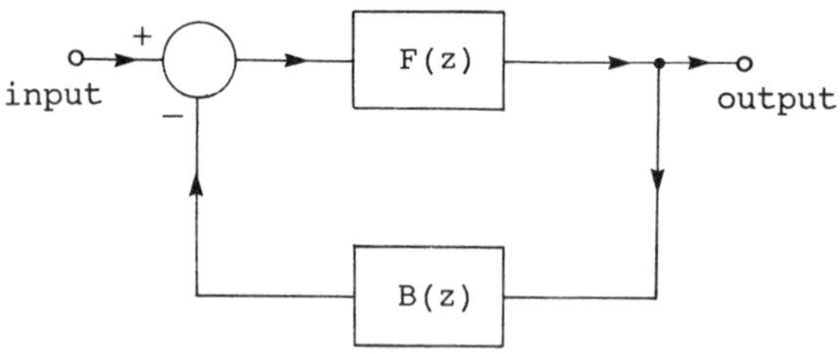

Fig 3.6.1 A Non-unity Feedback Control System

---

*hence the motivation for the later use of Bode plots

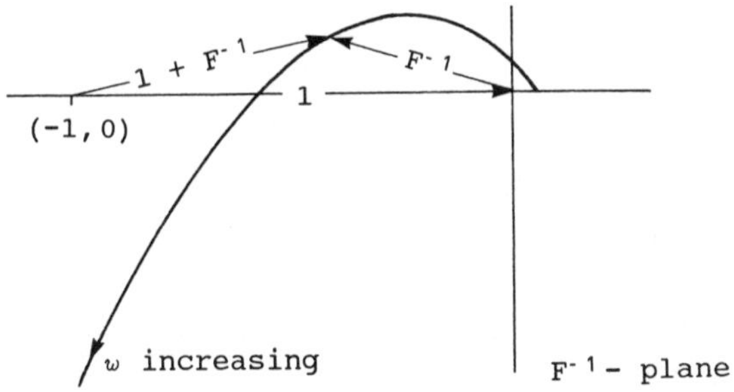

Fig 3.6.2 Relating to the Inverse Nyquist Diagram for a Unity Feedback System

3.7 THE BILINEAR TRANSFORMATION

In order to achieve a near maximum utilisation of available bandwidth or a marked reduction of the interference on received data (5, 65, 66), communications engineers have devised over the years a wealth of effective synthesis techniques and empirical designs for continuous data filters. However, these filters cannot generally be directly transcribed into digital designs by simply effecting their z- transformation. As revealed by equation (2.4.16), distortion of the required baseband function $H(\omega)$ by sidebands at $\pm 2\pi n/T$ renders:

$$H^*(\omega) \neq H(\omega) \quad \text{for} \quad -\pi/T \leq \omega \leq \pi/T \qquad (3.7.1)$$

For high performance wideband filters with very sharp cut-off rates etc, this difference can be highly significant (67). Such aliasing or folding (67) can be reduced by restricting the frequency range of $H(\omega)$ with certain spectral window functions (67). Alternatively, a digital filter can synthesised from a continuous data parent by the bilinear transformation:

$$sT/2 = (z - 1)/(z + 1) \qquad (3.7.2)$$

which constitutes a one to one correspondence between the z- plane and the horizontal strip in the s- plane:

$$-\pi/T \leq \mathbf{I}(s) \leq \pi/T \tag{3.7.3}$$

Defining the coordinates:

$$s = \sigma + j\omega \quad \text{and} \quad z = \exp(\sigma' T + j\omega' T) \tag{3.7.4}$$

then the relationship between the real frequency variables ω and ω' is derived from equation (3.7.2) as:

$$\omega T/2 = \tan(\omega' T/2) \tag{3.7.5}$$

The so-called *break frequencies*, where pass and stop bands start or finish, are the principal parameters for filter synthesis by the bilinear transformation. Firstly, a suitable set of break frequencies $\{\omega'_m\}$ is chosen for the required digital filter with its sampling period meeting the now obvious constraint of:

$$\omega'_m << \pi/T \quad \text{for all } m \tag{3.7.6}$$

Next, these break frequencies are *prewarped* (45, 67) according to equation (3.7.5) to yield the corresponding set $\{\omega_m\}$. A continuous data filter of the required type and order with break frequencies $\{\omega_m\}$ is then realised from basic low pass Laplace transfer functions by applying standard conversion formulae (65). Finally, application of the bilinear transformation in equation (3.7.3) produces a pulse transfer function with the desired break frequencies $\{\omega'_m\}$ characterising its real frequency response. However, because of some inevitable spectral folding, the synthesis cannot precisely match the pass and stop band behaviour of a continuous data parent. For instance, if a wide band digital differentiator is realised from the parent sT/2, then it follows from equation (3.7.5) that the progeny:

$$H(z) = (2/T)(z - 1)/(z + 1) \tag{3.7.7}$$

has the required linearly rising frequency response just over the limited range:

$$|\omega'| < 0.2\pi/T \tag{3.7.8}$$

In principle, the bilinear transformation can be applied also to the design of controllers for DDC systems. For this purpose, the pulse transfer function of a linearised plant model is first mapped into the so-called *w- plane* by the inverse of equation (3.7.2):

$$z = (1 + w)/(1 - w) \tag{3.7.9}$$

where:

$$w \triangleq sT/2 \tag{3.7.10}$$

A conventional design procedure is then used to create a suitable controller D(w), which is finally transformed into the required digital controller D(z) by setting:

$$w = (z - 1)/(z + 1) \tag{3.7.11}$$

After the original paper by Johnson et al (68), several reports have confirmed the effectiveness of the technique. Here, however, design in terms of pulse real frequency response functions is strongly favoured because:

- no additional algebraic manipulation is involved

- as shown in Section 3.4, totally reliable relationships exist in practice between behaviour in the frequency and time domains

- as shown in Sections 3.5 and 3.6, the familiar Nyquist Diagram techniques apply

- behaviour in the pass and stop bands is not precisely transferred between the w- and z- planes by the bilinear transformation; so some uncertainty exists.

CHAPTER 4

Stochastic Input Signal Analysis

...... under the sun ......
time and chance happeneth to them all. - Ecclesiastes

4.1 RELEVANT ASPECTS OF MATHEMATICAL PROBABILITY

The fluctuations in an electric current form one example of the many naturally occurring forms of stochastic process. If these fluctuations are recorded against time for a number of apparently identical devices, an obvious feature is their indeterminate amplitude. However, even a cursory examination suggests that certain amplitudes are more likely than others. This prompts a study of such stochastic time domain functions by the mathematical theory of probability. The basic concept underlying mathematical probability is that of relating an event which might occur to the complete collection of events that could occur (86). An *ensemble* consists of the set of all possible events from nominally identical sources. For the electronic noise example cited above, its ensemble consists of the fluctuation records taken from an indefinitely large number of nominally identical devices tested under nominally the same conditions. A stochastic process is defined by its ensemble; any member of which represents a particular *realisation*. An ensemble enables the creation of a set of functions which specify the probability that any finite set of amplitudes $\{x(t_i)\,|\,i \le m\}$ from a particular realisation lie within arbitrary limits. For example, the probability that a realisation $x(t)$ of the ensemble $\{X\}$ has successive amplitude samples satisfying:

$$\left.\begin{array}{l} x_1 \le x(t_1) \le x_1 + \delta x_1 \\ x_2 \le x(t_2) \le x_2 + \delta x_2 \\ \cdots\cdots\cdots\cdots\cdots\cdots \\ x_m \le x(t_m) \le x_m + \delta x_m \end{array}\right\} \quad (4.1.1)$$

is expressed in terms of an mth order probability density function as (87, 88):

$$P_X(x_1, x_2, \ldots x_m)\,dx_1\,dx_2 \ldots dx_m \quad (4.1.2)$$

By definition *probability density functions* satisfy:

$$\oint_{-\infty}^{\infty} P_X(x_1, x_2, \ldots x_m)\,dx_1\,dx_2 \ldots dx_m = 1 \quad (4.1.3)$$

and

$$P_X(x_1, x_2, \ldots x_m) \ge 0 \quad (4.1.4)$$

Equations (4.1.1) and (4.1.2) implicitly imply that the behaviour of a realisation before t_m influences its behaviour at that instant. Such dependence naturally occurs because practical systems have inherent stray capacitance or inertia etc which create realisations with finite spectral bandwidths, and therefore restricted rates of change. It is concluded that truly random* signals cannot exist, and consequently signals requiring a statistical characterisation are sometimes described as *stochastic* (89). Signals not requiring a statistical characterisation, like the sinusoid in equation (3.3.10), are termed *deterministic*.

The set of probability distributions that characterise a stochastic process generally depend on the time instants that the amplitudes of a realisation are in question. If every translation in time of an ensemble leaves its complete set of probability distributions unaltered, the process is described as *strict sense*

---

*an uncontrolled or unguarded state (90)

stationary (89). Probability distributions for strict sense stationary processes are therefore dependent on time differences rather than the actual time instants themselves. Strict sense stationarity is over-restrictive and unnecessary for most, if not all, engineering purposes. It evolves as sufficient that just the first- and second-order probability distributions are invariant of time translation. Such stochastic processes are termed *wide sense stationary* (88, 89). Under certain idealised conditions, a complete set of probability distributions can be derived theoretically (88), and electrical Shot Noise (91) is one such process. However, this detailed analysis is out of the question in most instances. Moreover, even if it were feasible, there remains the formidable problem of determining the probability functions that characterise the response of a system. Less complete information must therefore usually suffice for practical systems, whose response to stochastic inputs can be assessed only in terms of certain statistical parameters. These statistical parameters must necessarily possess special attributes. In the first place, they must be feasibly calculable either theoretically from a mathematical model or experimentally from an analysis of empirical data. Secondly, knowledge of their values at the input of a system should generally allow a tractable evaluation of the corresponding output values. Finally, they must sufficiently characterise their parental probability density functions as to enable an assessment of system behaviour. These requirements are met for many practical purposes by the mean and autocorrelation function of a process. The *mean* is defined by:

$$\overline{x(t_1)} = \int_{-\infty}^{\infty} x_1 \; P_X(x_1)\,dx_1 \tag{4.1.5}$$

and the *autocorrelation function* by:

$$\phi_X(t_1, t_2) = \overline{x(t_1)\; x(t_2)} = \iint_{-\infty}^{\infty} x_1 x_2 \; P_X(x_1, x_2)\,dx_1\,dx_2 \tag{4.1.6}$$

If a stochastic process is wide sense stationary, the above

equations simplify to:

$$\overline{x} = \int_{-\infty}^{\infty} x \; P_X(x)\,dx \tag{4.1.7}$$

and:

$$\phi_X(t_2 - t_1) = \iint_{-\infty}^{\infty} x_1 x_2 \; P_X(x_1, x_2)\,dx_1\,dx_2 \tag{4.1.8}$$

so that now:

$$\phi_X(t_1 - t_2) = \phi_X(t_2 - t_1) \tag{4.1.9}$$

Stationary *Normal or Gaussian* processes receive considerable attention because the autocorrelation function determines their complete set of probability functions (88, 89). Probability distributions or related statistical parameters for the response of linear (92) and certain non-linear systems (88, 92) to a Normally distributed input are in consequence relatively easy to derive. By virtue of the *Central Limit Theorem* (87), Normal processes broadly result from the linear superposition of a large number of separate effects. It is therefore not surprising that the convolution operation of a linear system produces a Normal output process in response to a Normal input (92). However, any other input into a linear or non-linear system generally produces an output with a radically different probability distribution from itself (88, 92).

An ensemble is a philosophical concept whose existence is assumed a priori to enable a precise formulation of a statistical problem. Measurements across an ensemble are evidently impossible in practice, where the available data for a process is contained in finite length recordings of a few realisations. The *Birkhoff Ergodic Hypothesis* (86, 88, 89) provides the bridge between theory and experiment by equating in a probabilistic sense the

time domain properties of a single realisation to the statistical properties of its ensemble. In essence, ergodicity implies that if one 'representative' member of an ensemble is observed for long enough, all salient features of its ensemble would occur.
Defining*:

$$<x> = \lim_{\varsigma \to \infty} \frac{1}{2\varsigma} \int_{-\varsigma}^{\varsigma} x(t)dt \tag{4.1.10}$$

and

$$<x(t)x(t+\tau)> = \lim_{\varsigma \to \infty} \frac{1}{2\varsigma} \int_{-\varsigma}^{\varsigma} x(t)x(t+\tau)dt \tag{4.1.11}$$

then with probability one for a 'representative' member of an ergodic ensemble:

$$\overline{x} = <x> \tag{4.1.12}$$

and

$$\phi_x(\tau) = \overline{x(t)x(t+\tau)} = <x(t)x(t+\tau)> \tag{4.1.13}$$

A *representative* member of an ensemble is by definition a realisation with a non-zero probability of occurrence, so that the distinction to be drawn is between the possible and the probable. An example of a non-representative realisation would be a coin tossing experiment in which an indefinite number of obverse sides occurred. Although precise mathematical conditions can be formulated for a process to be ergodic (89), it is largely impossible to establish that they are actually satisfied in practice. Often there is a strong intuitive basis for assuming ergodicity, while sometimes the hypothesis is adopted with just a pious hope that the end will eventually justify the means.

---

*For stochastic sequences, the following integrations are replaced by summations

Stationarity is a necessary, but not sufficient, condition for a process to be ergodic. For example, a Normally distributed ensemble of constant functions is stationary but patently non-ergodic (93).

An autocorrelation function relates to just a single ensemble, but sometimes the objective is to establish a causal relationship between two ensembles {X} and {Y}. In these circumstances, the cross correlation function defined by:

$$\phi_{XY}(t_1, t_2) = \overline{x(t_1)y(t_2)} = \iint_{-\infty}^{\infty} xy\ P_{XY}(x, y)\,dx\,dy \tag{4.1.14}$$

finds application (96). If the ensembles are jointly wide sense stationary, the above equation gives*:

$$\phi_{XY}(\tau) = \overline{x(t)y(t + \tau)} = \phi_{YX}(-\tau) \tag{4.1.15}$$

Furthermore if the ensembles are jointly ergodic, then with probability one:

$$\phi_{XY}(\tau) = \lim_{\varsigma \to \infty} \frac{1}{2\varsigma} \int_{-\varsigma}^{\varsigma} x(t)y(t + \tau)\,dt \tag{4.1.16}$$

Largely to ease nomenclature, this brief introduction has concerned time functions representing continuous data. Non-stationary discrete stochastic processes also arise naturally, as for example the development of bacterial infections in humans and animals (94). However, in the present context, attention is directed at stochastic sequences that have been created artificially by the periodic sampling of wide sense stationary continuous data. More complete descriptions of the general theory of stochastic processes can be found in the cited references.

---

*

$$\overline{x(t)y(t + \tau)} = \overline{x(t - \tau)y(t)} = \overline{y(t)x(t - \tau)}$$

4.2 A GEOMETRICAL INSIGHT INTO CORRELATION ANALYSIS

A correlation function is formulated in Section 4.1 as a mathematical abstraction, but here a geometrical interpretation is developed to promote a deeper understanding. Suppose a variate u defines the amplitude of a realisation at some specified time instant, and the variate v is a similar quantity from the same realisation or from another of a different ensemble. Plotting experimental values of v against u from a large number of measurements creates a *scatter diagram* (87) like that in Fig 4.2.1. Any tendency for v to be functionally dependent on u manifests itself as a greater density of points along a particular locus. This latent curve or functional relationship is termed the *regression* (87) of v on u, and the derivation of such relationships is a prime concern in some branches of experimental science*. Although a complicated functional dependence may actually obtain, the simple linear regression:

$$v' = a + bu \qquad (4.2.1)$$

is now proposed, in which the coefficients {a, b} are to be

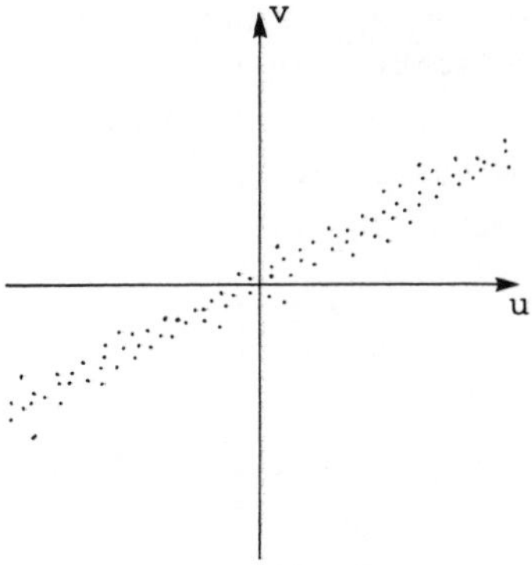

Fig 4.2.1 A Scatter Diagram

---

*eg heat transfer correlations for chemical and mechanical engineering design

selected so as to minimise the mean square error quantity:

$$\overline{\epsilon^2} = \overline{(v - v')^2} \tag{4.2.2}$$

Adopting for ease of nomenclature the normalised variates:

$$U = u - \overline{u} \quad \text{and} \quad V = v - \overline{v} \tag{4.2.3}$$

it can be shown (87, 88) that the optimum straight line is:

$$V' = \left(\overline{UV} / \overline{U^2} \right) . U \tag{4.2.4}$$

with a corresponding mean square error of:

$$\overline{\epsilon^2_{min}} = \overline{V^2} \left\{ 1 - \left(\overline{UV} \right)^2 / \overline{U^2}\ \overline{V^2} \right\} \tag{4.2.5}$$

Because this error quantity is necessarily positive, it follows that:

$$(\overline{UV})^2 \leq \overline{U^2} . \overline{V^2} \tag{4.2.6}$$

When equality obtains in the above expression the mean square error, and therefore the scatter in Fig 4.2.1, is zero. Under these conditions, a perfectly linear causal dependence exists between U and V. On the other hand, when:

$$\overline{UV} = \overline{U} . \overline{V} \tag{4.2.7}$$

the scatter diagram is widely, but not necessarily uniformly, dispersed and the variates are described as *linearly independent* or *uncorrelated*.

Variates are described as *statistically independent* if and only if any joint probability density function equates with the product of corresponding first-order distributions. In particular, if the normalised variates U and V are statistically independent, their correlation as defined by equation (4.1.6) or (4.1.14) reduces to:

$$\overline{UV} = \left[\int_{-\infty}^{\infty} U\; P_U(U)\,dU\right]\left[\int_{-\infty}^{\infty} V\; P_V(V)\,dV\right] \qquad (4.2.8)$$

so that:

$$\overline{UV} = 0 \qquad (4.2.9)$$

Two statistically independent normalised variates are therefore uncorrelated, and because they are not causally related, points on their scatter diagram are completely randomly dispersed over the U-V plane.

In general, inferences based on the *correlation coefficient*:

$$\rho(U,V) = \frac{UV}{\sqrt{U^2\;V^2}} \qquad (4.2.10)$$

approaching unity or zero should be viewed circumspectly; or even sceptically. For example, increased infantile survivability in the Soviet Union correlates with exposure to atomic radiation from the damaged Chernobyl nuclear power station (95). The exposure is evidently beneficial to neither pregnant mothers nor their children, and the true dependence is on the subsequently intensified medical care of its local population. Similar to digital computer simulations, an unphysical basis (garbage in) for a correlation can yield quite ridiculous conclusions (garbage out). As another example, consider a discrete normalised variate (U) which takes all integer values between ±8 inclusive with equal probability. If U is processed by a square law device to create another variate V, their cross correlation is:

$$\overline{UV} = \sum_{-8}^{8} k.k^2\,(1/17) = 0 \qquad (4.2.11)$$

for all positive integer values of p and q.

It might be judged advisable from the above discussion to avoid correlation techniques, or at least leave their interpretation to the cognoscente. However, quite unambiguous and easily interpreted relationships are subsequently shown to exist for linearised systems with stochastic inputs.

4.3 POWER SPECTRA

An additive combination of deterministic and stochastic signals often forms the input to a linearised control system model. Because linearity implies that the output is the sum of its separate responses to each component, all subsequent discussion tacitly considers stochastic processes that do not contain any deterministic components like a non-zero mean etc. Consequently, if $\{X(t)\}$ devotes a wide sense stationary process (ensemble) in continuous time, its autocorrelation function $\phi_X(\tau)$ can be assumed to approach zero as the temporal separation (τ) becomes increasingly large. Although such asymptotic behaviour is necessary for the existence of the Fourier Transformation pair:

$$\left.\begin{aligned} \Phi_X(\omega) &= \int_{-\infty}^{\infty} \phi_X(\tau)\, e^{-j\omega\tau} d\tau \\ \phi_X(\tau) &= \frac{1}{2\pi}\int_{-\infty}^{\infty} \Phi_X(\omega)\, e^{j\omega\tau} d\omega \end{aligned}\right\} \qquad (4.3.1)$$

it is not sufficient. As described in Section 1.2, an auto-correlation function must actually decay to zero fast enough to be both integrable and absolutely square integrable. Viewed from the time domain, there is little or no physical evidence to suggest a general compliance with these conditions. Once again, however, deeper insight is achieved through the frequency domain. Noting that $\phi_X(o)$ represents the mean square value, or the statistically expected* power in the case of an electrical signal, the function $\Phi_X(\omega)$ is appropriately termed the *power spectrum* of the stochastic process $\{X(t)\}$. Because inertia or stray capacitance etc of practical sources constrains power spectral components to reduce asymptotically at a minimum rate of 20 dB/decade, the existence of the Fourier Integral pair (4.3.1) can therefore be intuitively inferred in practice. This last statement evidently presupposes a relationship between the power spectrum of a stochastic process and the real frequency response function of a linear system. Section 4.7 formally establishes this functional dependence, and in addition relates the cross correlation between input and output to a system's real frequency response function. Accordingly, it is also reasonable here to assume the existence of:

$$\left.\begin{aligned} \Phi_{XY}(\omega) &= \int_{-\infty}^{\infty} \phi_{XY}(\tau)\, e^{-j\omega\tau} d\tau \\ \phi_{XY}(\tau) &= \frac{1}{2\pi} \int_{-\infty}^{\infty} \Phi_{XY}(\omega)\, e^{j\omega\tau} d\omega \end{aligned}\right\} \qquad (4.3.2)$$

Electronic instruments determine the power spectrum of a stochastic process by a numerical Fourier Transformation (97) of its correlation function** as estimated from a periodically sampled, finite length, recording of a single realisation. Because it is usually impossible to establish rigorously the

---

*mean and (statistical) expectation are synonyms in this text

**If a spectrum is derived from the periodogram $\left|X(\omega,\Lambda)\right|^2/\Lambda$ where:

$$X(\omega,\Lambda) = \int_0^{\Lambda} x(t)\,\exp(-j\omega t)\,dt$$

the variance of this estimate about the true (mean) value $\Phi_X(\omega)$ does not necessarily converge to zero as the measurement duration (Λ) increases - see page 106 of reference (88)

ergodicity of an ensemble, such results are regarded as representative purely on intuitive grounds.

The two-sided Laplace Transformation pair:

$$\left.\begin{aligned} \Phi(s) &= \int_{-\infty}^{\infty} \phi(\tau)\, e^{-s\tau} d\tau \\ \phi(\tau) &= \frac{1}{2\pi j} \int_{-j\infty}^{j\infty} \Phi(s)\, e^{s\tau} ds \end{aligned}\right\} \quad (4.3.3)$$

is obtained from equation (4.3.1) or (4.3.2) by setting:

$$s = j\omega$$

At first sight there might appear to be an ambiguity in the inversion of such a spectrum function by the method of residues (10, 12). Poles inside the right-half s- plane can evidently represent either decreasing functions in negative time or increasing functions in positive time. However, the uncertainty is easily resolved because a correlation function must eventually

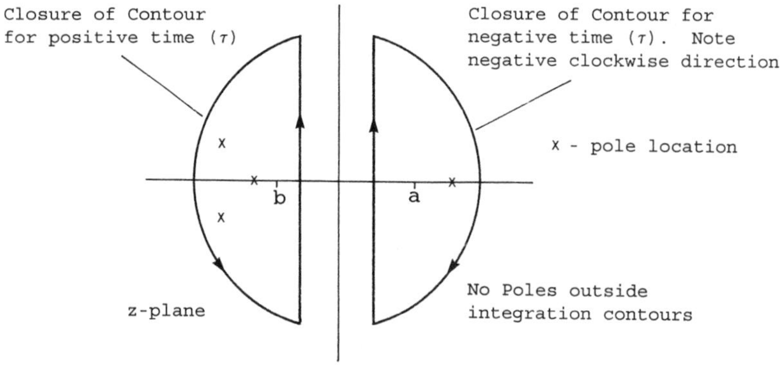

Fig 4.3.1 Pole Locations for a typical Power Spectrum and the Inversion Contours

decay towards zero as the magnitude of the temporal separation (τ) increases. Consequently, those poles lying strictly in the right-half s- plane define the negative time behaviour (decreasing); while the remaining poles lying strictly in the left-half s- plane define the positive time behaviour (decreasing) of the correlation function. Because correlation functions are real valued, it follows from Section 1.2 that the poles of a spectrum function are real or complex conjugate pairs of entities. Fig 4.3.1 exemplifies such pole positions and the appropriate infinite semicircles for closing the line contour [$-j\infty$, $j\infty$]. In that the depicted poles are located asymmetrically about the imaginary axis, it follows that:

$$\phi(-\tau) \neq \phi(\tau)$$

so that a cross correlation function is involved here. Except for the physically non-existent realisation:

$$\Phi_x(s) = A \quad \text{a constant} \tag{4.3.4}$$

which is termed *white noise** (88, 89, 93), a power spectrum always has at least one pole in each half of the s- plane, so that it converges asymptotically to zero at least as fast as $1/s^2$. According to the discussion in Appendix B2.4, its integral around either infinite semicircle then converges to zero even for zero time. Consequently:

$$\phi(o) = \frac{1}{2\pi j} \int_{-j\infty}^{j\infty} \Phi(s)\,ds \tag{4.3.5}$$

when the contour is closed by a left- or a right-infinite semicircle. Equation (4.3.5) is of course latent in the assumed existence of the Fourier transformation $\Phi(\omega)$.

---

*Like daylight, all frequencies are present. However, it's a poor analogy, as the Planck-Boltzmann equation(98) shows that different frequencies have different intensities

As an illustrative example of the above concepts consider:

$$\Phi(\omega) = a^2/(\omega^2 + a^2) \tag{4.3.6}$$

Setting:

$$s = j\omega$$

the corresponding two-sided Laplace transformation is obtained as:

$$\Phi(s) = a^2/(a^2 - s^2) \tag{4.3.7}$$

whose simple poles at ±a are symmetrically located about the imaginary axis. The negative time behaviour of the corresponding correlation function is determined by the residue at s = a as:

$$\phi(\tau) = (a/2)\ \exp(a\tau) \qquad \text{for } \tau \leq 0$$

and the positive time behaviour by the residue at s = -a as:

$$\phi(\tau) = (a/2)\ \exp(-a\tau) \qquad \text{for } \tau \geq 0$$

Combining the above results yields:

$$\phi(\tau) = (a/2)\ \exp(-a|\tau|) \tag{4.3.8}$$

which is the autocorrelation function of a so-called ***Markovian process*** (89, 93).

4.4 PULSE POWER SPECTRA

If wide sense stationary continuous data is sampled periodically, the ensemble of sample pulses is evidently wide sense stationary when viewed as sequences*. Furthermore, if the

---

*each sample pulse has its amplitude indexed by the corresponding integral multiple of T. For example x(nT) becomes x_n

correlation function of the continuous data is $\phi(\tau)$, the *correlation sequence* of the discrete data is clearly $\phi(kT)$ where k denotes a positive or negative integer. Because the break frequencies of a power spectrum $\Phi(\omega)$ reflect the pole locations of $\Phi(s)$, the previous general discussion implies that continuous wide sense stationary data can be characterised in practice by correlation functions that are a sum of finitely many two-sided exponentially decaying components. It is therefore reasonable to define the *pulse power spectrum* of such a derived correlation sequence as the *two-sided z- transformation*:

$$\Phi(z) = \lim_{N \to \infty} \sum_{-N}^{N} \phi_k z^{-k} \tag{4.4.1}$$

because it converges at least on the unit circle (Γ) specified by:

$$|z| = 1$$

To distinguish between discrete and continuous data processes, the real frequency behaviour of a pulse power spectrum is denoted by:

$$\Phi^*(\omega) = \sum_{-\infty}^{\infty} \phi_k \, e^{-j\omega kT} \tag{4.4.2}$$

Round-off errors associated with arithmetic multiplication in digital computers are an example of an intrinsically discrete stochastic process. As described later in Chapter 6, its auto-correlation sequence for fixed-point rounded multiplication is reasonably approximated by:

$$\phi_k = (q^2/12)\,\delta_{k0} \tag{4.4.3}$$

and for fixed-point unrounded multiplication a similar analysis establishes:

$$\phi_k = (q^2/3)\,\delta_{k0} \tag{4.4.4}$$

where as defined in equation (1.4.2):

q - width of quantisation or 1 least significant bit

and

$$\left.\begin{aligned} \delta_{k0} &= 1 \quad \text{for} \quad k = 0 \\ &= 0 \quad \text{otherwise} \end{aligned}\right\} \tag{4.4.5}$$

A discrete data process whose autocorrelation sequence is a singleton like equation (4.4.3) or (4.4.4) is loosely termed 'random', or because its pulse spectrum function takes the form:

$$\Phi^*(\omega) = \phi_0 \tag{4.4.6}$$

the description '*white noise*' is also applied (28). Efficient numerical algorithms for generating essentially white noise sequences are provided in references (99, 100). Amongst other important practical applications, white noise sequences can be used to create discrete data processes with arbitrary pulse spectra by means of the *shaping filter* technique detailed in Section 4.8.

Excluding the special case of white noise, a correlation sequence generally consists in practice of two-sided exponentially decaying components*. Consequently, its subsequence $\{\phi_k \mid k \geq 0\}$ has the one-sided z- transformation:

$$F_+ = \sum_{0}^{\infty} \phi_k z^{-k} \tag{4.4.7}$$

whose poles exist strictly within the unit circle as real or as complex conjugate pairs of entities. Furthermore, the subseries

---

*a DC bias or non-zero mean can be treated separately as described at the beginning of Section 4.3

of functions $\{\phi_k z^{-k} | k \geq 0\}$ is normally summable <u>outside</u> any circle of radius $R_+ < 1$, which lies outside all the poles of F_+. Likewise, the subsequence $\{\phi_k | k \leq 0\}$ has the one-sided z-transformation:

$$F_- = \sum_{-\infty}^{0} \phi_k z^{-k} \tag{4.4.8}$$

whose poles by means of the simple change of variable:

$$z \rightarrow 1/z \tag{4.4.9}$$

are seen to exist strictly outside the unit circle as real or complex conjugate pairs of entities. Furthermore, the subseries of functions $\{\phi_k z^{-k} | k \leq 0\}$ is normally summable <u>inside</u> any circle of radius $R_- > 1$, which lies inside all the poles of F_-. Hence a pulse spectrum function as defined in equation (4.4.1) is normally summable in the region:

$$R_+ < |z| < R_- \tag{4.4.10}$$

Because the integral of the sum of a normally summable series equates with the sum of its separately integrated terms (10), it follows that for any positive or negative integer n:

$$\frac{1}{2\pi j} \int_\Gamma \Phi(z) z^{n-1} dz = \sum_{-\infty}^{\infty} \phi_k \frac{1}{2\pi j} \int_\Gamma z^{n-1-k} dz$$

Applying equation (2.1.8) yields the *two-sided Inversion Integral*:

$$\phi_n = \frac{1}{2\pi j} \int_\Gamma \Phi(z) z^{n-1} dz \tag{4.4.11}$$

It is patently clear from the derivation of equation (4.4.11) that the positive time behaviour of $\{\phi_n\}$ corresponds to the sum of the residues at poles of $\Phi(z) z^{n-1}$ lying within the anticlockwise contour formed by the unit circle (Γ). On the other hand, it might seem alien that the negative time behaviour is governed by the residues at poles lying 'outside' the unit circle, which is now the clockwise contour of integration. Most simply*, the paradox is resolved by stereographically projecting (46, 102) the z- plane on to a sphere whose 'South Pole' lies at the origin. The transformation in equation (4.4.9) can be regarded as shifting the viewpoint to its 'North Pole', so that the mapped unit circle now encloses the other poles in the negative mathematical sense (ie clockwise).

For the particular case of an autocorrelation function, equation (4.1.9) asserts that:

$$\phi(-k) = \phi(k)$$

so that from equation (4.4.1) the corresponding pulse power spectrum satisfies:

$$\Phi(z) = \Phi(z^{-1}) \tag{4.4.12}$$

Consequently, the zeros of $\Phi(z)$ must lie actually on the unit circle or exist as mutually reciprocal pairs. By defining:

$$\Psi(z) = 1/\Phi(z)$$

its singularities (in practice its poles) are seen to exist also as mutually reciprocal pairs**. Pole locations with their positive and negative time connotations as well as regions of normal summability are illustrated in Fig 4.4.1.

To exemplify the above analysis, consider the autocorrelation

---

*Readers who demand ultimate rigour regarding the 'inside' and 'outside' of a contour are referred to reference (101)

**Note a pole actually on the unit circle would not contribute a decaying temporal component, so that their existence can be denied in the present circumstances

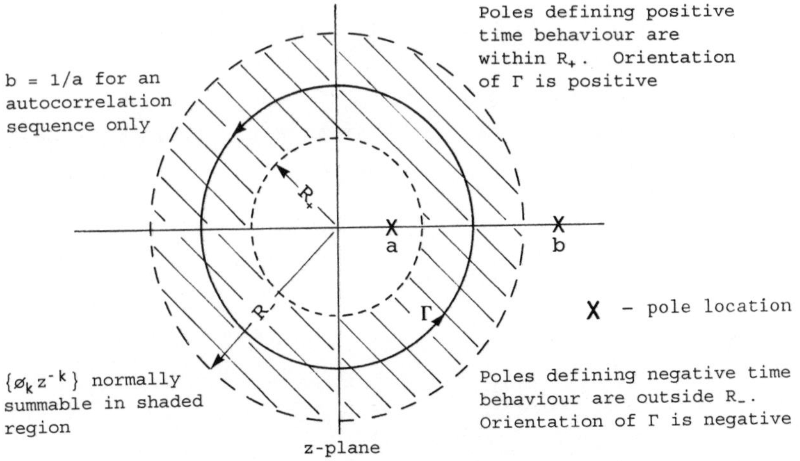

Fig 4.4.1 Pole Locations etc for a simple Pulse Power Spectrum

sequence:

$$\phi_k = (a/2)\ \exp(-a|k|) \tag{4.4.13}$$

indexed by positive and negative values of the integer k. In the region:

$$\exp(-a) < |z| < \exp(a)$$

the series of functions $\{\exp(-a|k|)z^{-k}\}$ is normally summable, and the pulse power spectrum as defined by equation (4.4.1) is:

$$\Phi(z) = \frac{a}{2}\left\{\sum_{-\infty}^{0} e^{ak}z^{-k} + \sum_{0}^{\infty} e^{-ak}z^{-k} - 1\right\}$$

which simplifies to:

$$\Phi(z) = \frac{az(e^{-a} - e^{a})}{2(z - e^{-a})(z - e^{a})} \tag{4.4.14}$$

Equation (4.4.14) clearly exhibits the reciprocal property for the poles of an autocorrelation sequence. Because the pulse spectrum function has a single zero on the unit circle, equation (4.4.12) is also satisfied. For the inversion of equation (4.4.14) by (4.4.11), the simple pole at e^{-a} determines the positive time behaviour of its correlation sequence as:

$$\phi_n = \left\{ \frac{a(e^{-a} - e^{a})z^n}{2(z - e^{a})} \right\}_{z = e^{-a}} \qquad \text{for } n \geq 0$$

or

$$\phi_n = (a/2)\exp(-an) \qquad \text{for } n \geq 0 \qquad (4.4.15)$$

Because the contour of integration is in the negative mathematical sense (clockwise) for those poles 'outside' the unit circle which determine the negative time behaviour of its correlation sequence, then:

$$\phi_n = -\left\{ \frac{a(e^{-a} - e^{a})z^n}{2(z - e^{-a})} \right\}_{z = e^{a}} \qquad \text{for } n \leq 0$$

or

$$\phi_n = (a/2)\exp(an) \qquad \text{for } n \leq 0 \qquad (4.4.16)$$

Combining equations (4.4.15) and (4.4.16) yields the original equation (4.4.13), so demonstrating the consistency of the previously derived mathematical procedures.

4.5 THE PULSE SPECTRUM OF A SAMPLED CONTINUOUS DATA PROCESS

If a wide sense stationary process with autocorrelation function $\phi_x(\tau)$ is periodically sampled with period T, the sample amplitudes viewed in discrete time are evidently characterised by the autocorrelation sequence $\phi_x(kT)$ with k a positive or negative integer. Generally in practice a correlation function can be adequately approximated by a finite sum of two-sided exponentially decaying terms. It is therefore reasonable to define the pulse power spectrum of the derived stochastic sequence as in equation (4.4.1):

$$\Phi_x(z) = \lim_{N\to\infty} \sum_{-N}^{N} \phi_x(kT)\, z^{-k} \tag{4.5.1}$$

where from equation (4.3.3):

$$\phi_x(\tau) = \frac{1}{2\pi j} \int_{-j\infty}^{j\infty} \Phi_x(s)\; e^{s\tau} ds \tag{4.5.2}$$

By virtue of equation (4.1.9), equation (4.5.1) can be expanded as:

$$\Phi_x(z) = \lim_{N\to\infty} \left\{ \phi_x(o) + \sum_{N}^{1} \phi_x(kT)\, z^{k} + \sum_{1}^{N} \phi_x(kT)\, z^{-k} \right\} \tag{4.5.3}$$

in which the positive time sequence $\{\phi_x(kT) \mid k \geq 0\}$ is derived from equation (4.5.2) by closing the line contour $[-j\infty, j\infty]$ with the infinite left-semicircle. Expressions are now derived for:

$$L_1 = \lim_{N\to\infty} \sum_{N}^{1} \phi_x(kT)\, z^{k}$$

$$L_2 = \lim_{N\to\infty} \sum_{1}^{N} \phi_x(kT)\, z^{-k} \tag{4.5.4}$$

so that the required spectrum function can then be expressed as (10, 46):

$$\Phi_x(z) = \phi_x(o) + L_1 + L_2 \tag{4.5.5}$$

The two-sided exponentially decaying components of a practical autocorrelation function correspond in general to a power spectrum whose poles lie outside the infinite strip:

$$-a < \mathbb{R}(s) < a \quad \text{for some } a > 0 \tag{4.5.6}$$

Because displacements of a closed integration contour over analytic regions* of a complex valued integrand leave the integral unaffected (10, 12), the limits under consideration can be written as:

$$L_1 = \lim_{N \to \infty} \sum_{N}^{1} \left[\frac{1}{2\pi j} \int_{b-j\infty}^{b+j\infty} \Phi_x(s) \; (e^{sT} z)^k \, ds \right] \tag{4.5.7}$$

and

$$L_2 = \lim_{N \to \infty} \sum_{1}^{N} \left[\frac{1}{2\pi j} \int_{b-j\infty}^{b+j\infty} \Phi_x(s) \; (e^{sT} z^{-1})^k \, ds \right] \tag{4.5.8}$$

where from the inequality (4.5.6) there is freedom to choose:

$$-a < b < 0 \tag{4.5.9}$$

This slight modification of the integration contour, which is depicted in Fig 4.5.1, simplifies a later analysis of the asymptotic behaviour of the integrand around the sector S_L of the infinite semi-circle.

Apart from the trivial idealised case of white noise, power spectrum functions usually behave as:

$$\lim_{s \to \infty} \left| s^2 \, \Phi_x(s) \right| = d \quad \text{a constant} \tag{4.5.10}$$

It is shown in Appendix A2.4 that the series of functions $\{\Phi_x(s)(e^{sT} z^{-1})^k\}$ in equation (4.5.8) is normally summable with respect to the Laplace variable provided:

---

* containing no singularities such as poles

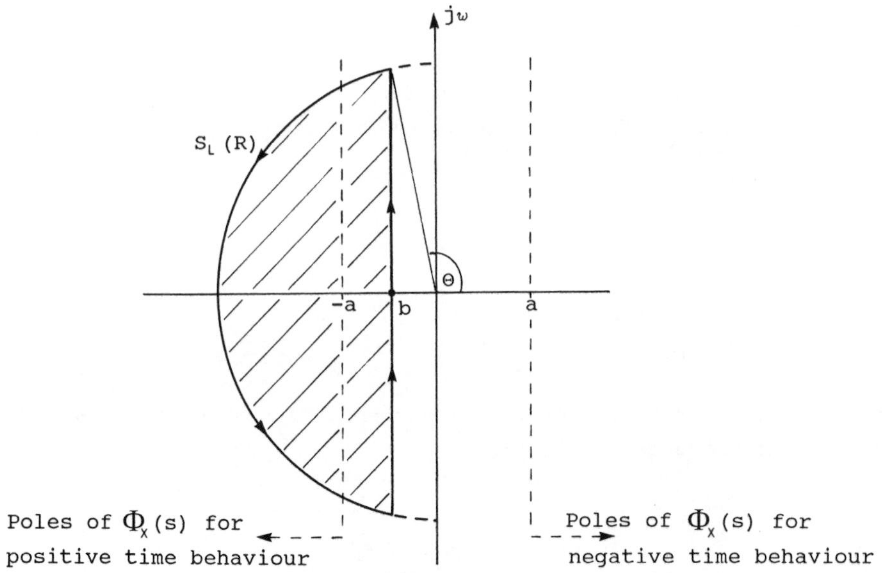

Fig 4.5.1 Illustrating the Closed Contour of Integration for Equations (4.5.7), (4.5.8) and (4.5.19)

$$|z| > \exp(bT) \tag{4.5.11}$$

Likewise, the series of functions $\{ \Phi_x(s)(e^{sT}z)^k \}$ in equation (4.5.7) is normally summable provided that:

$$|z| < \exp(-bT) \tag{4.5.12}$$

The two series are therefore both normally summable within the annulus:

$$\exp(bT) < |z| < \exp(-bT) \tag{4.5.13}$$

within which their limits are unaffected by any rearrangement of their terms, and within which the order of summation and integration can be interchanged (10). Accordingly the limits under consideration can be expressed as:

$$L_1 = \frac{1}{2\pi j} \int_{b-j\infty}^{b+j\infty} \Phi_x(s) \left\{ \lim_{N \to \infty} \sum_{1}^{N} (e^{sT} z)^k \right\} ds \tag{4.5.14}$$

and

$$L_2 = \frac{1}{2\pi j} \int_{b-j\infty}^{b+j\infty} \Phi_x(s) \left\{ \lim_{N \to \infty} \sum_{1}^{N} (e^{sT} z^{-1})^k \right\} ds \tag{4.5.15}$$

The inequality (4.5.13) ensures convergence of both of the above geometrical progressions to yield:

$$L_1 = \frac{1}{2\pi j} \int_{b-j\infty}^{b+j\infty} \Phi_x(s) \left\{ \frac{ds}{1 - e^{sT} z} \right\} \tag{4.5.16}$$

$$L_2 = \frac{1}{2\pi j} \int_{b-j\infty}^{b+j\infty} \Phi_x(s) \left\{ \frac{ds}{1 - e^{sT} z^{-1}} \right\} \tag{4.5.17}$$

Hence the required pulse spectrum function is derived from equation (4.5.5) as:

$$\Phi_x(z) = \frac{1}{2\pi j} \int_{b-j\infty}^{b+j\infty} \Phi_x(s)\ F(s,z) ds \tag{4.5.18}$$

where

$$F(s,z) = \left(1 - e^{2sT}\right) \Big/ \left(1 - e^{sT} z\right)\left(1 - e^{sT} z^{-1}\right) \tag{4.5.19}$$

A sector $S_L(R)$ of a finite left-semicircle, like that shown in Fig 4.5.1, corresponds to the open set of points:

$$u = Re^{j\Theta} \quad \text{for} \quad \pi/2 < \Theta < 3\pi/2 \tag{4.5.20}$$

On this locus the exponential term in equation (4.5.19) behaves as:

$$\exp(uT) = \exp(-RT\left|\cos\Theta\right|).\exp(jRT\sin\Theta) \quad 4.5.21)$$

Because the arguments $\pi/2$ and $3\pi/2$ have been deliberately excluded:

$$\left|\cos \Theta\right| \neq 0 \quad (4.5.22)$$

and it follows that within the bounds imposed by equation (4.5.13):

$$\lim_{R\to\infty} \left(1 - e^{2uT}\right) = 1 \quad ; \quad \lim_{R\to\infty} \left(1 - e^{uT}z\right) = 1$$

and (4.5.23)

$$\lim_{R\to\infty} \left(1 - e^{uT}z^{-1}\right) = 1$$

so that (10, 46):

$$\lim_{R\to\infty} F(s,z) = 1 \qquad \text{for} \quad \pi/2 < \Theta < 3\pi/2 \quad (4.5.24)$$

The above result and equation (4.5.10) imply that the integrand of equation (4.5.18) converges to zero as $1/R^2$ over the circular sector $S_L(R)$. Consequently, the integral itself converges to zero as $1/R$ over $S_L(R)$ giving:

$$\frac{1}{2\pi j} \int_{S_L} \Phi_x(s) \; F(s,z)ds = 0 \quad (4.5.25)$$

where by definition:

$$S_L = \lim_{R\to\infty} S_L(R)$$

Hence by Cauchy's theory of residues (10, 12) it follows that:

$$\Phi_x(z) = \sum \text{Residues of } \Phi_x(s)\,F(s,z) \text{ at the poles of } \Phi_x(s) \text{ inside the left-half s- plane} \tag{4.5.26}$$

By way of an example, the pulse power spectrum corresponding to:

$$\Phi_x(s) = a^2/(a^2 - s^2) = a^2/(a - s)(a + s) \tag{4.5.27}$$

is now determined using equation (4.5.26) rather than by the direct summation of its correlation sequence:

$$\phi_x(kT) = (a/2)\ \exp(-a|kT|)$$

as effected in Section (4.4). Because this particular power spectrum has just one relevant simple pole at s = -a, equation (4.5.26) is readily evaluated as:

$$\Phi_x(z) = \frac{a}{2}\left(\frac{1 - e^{-2aT}}{\left(1 - e^{-aT}z\right)\left(1 - e^{-aT}z^{-1}\right)}\right) \tag{4.5.28}$$

which after slight manipulation becomes identical to equation (4.4.14). Even for such a simple case the contour integral technique facilitates the computation, while with more complicated spectra the relative economy of effort is even greater.

For the purpose of developing an alternative expansion to equation(4.5.18), it is first noted that the even property of an autocorrelation function enables the factorisation:

$$\Phi_x(s) = \Psi(s) + \Psi(-s) \tag{4.5.29}$$

where $\Psi(s)$ has real or complex conjugate pairs of poles inside the left-half s- plane only. Because the function $\Psi(-s)$ has no poles within the contour [b-j∞, b+j∞] when closed by S_L, equation (4.5.18) can be written as:

$$\Phi_x(z) = \frac{1}{2\pi j} \int_{b-j\infty}^{b+j\infty} \Psi(s)\ F(s,z)ds \qquad (4.5.30)$$

The function F(s,z) possesses simple poles located strictly to the right of the line [$-j\infty$, $j\infty$] at:

$$p_n = \pm\ (\log z + j2\pi n)/T \qquad \text{for } \pm \text{ integer } n \qquad (4.5.31)$$

A stereographic projection (46, 102) reduces the infinite sector S_L to a mere point, and whether the closed contour of integration encloses the poles of $\Psi(s)$ or F(s,z) is seen to be debatable in this context. A similar question is whether the North or South Pole of the earth is enclosed by the equator orientated in the West to East direction? Clearly, the equatorial contour encloses both Poles, but in a different rotational sense. Viewed from the North Pole, it is enclosed in the positive anticlockwise direction. On the other hand, the South Pole sees itself as enclosed in the negative clockwise direction. Accordingly, in the case of equation (4.5.30) it follows that:

$$\Phi_x(z) = -\sum \text{Residues of } \Psi(s).F(s,z) \text{ at the poles of } F(s,z) \qquad (4.5.32)$$

These residues at the pole locations specified by equation (4.5.31) are given by:

$$\lim_{s \to p_n} (s - p_n)\ \Psi(s)\ F(s,z) \qquad \text{for all } \pm \text{ integer } n \qquad (4.5.33)$$

Because the above limit behaves indeterminantly (0/0), it can be evaluated as the quotient of:

$$\lim_{s \to p_n} \frac{d}{ds} \left[(s - p_n)\ \Psi(s)(1 - e^{2sT}) \right] = \Psi(p_n) \left[(1 - \exp(2p_n T) \right]$$

and

$$\lim_{s \to p_n} \frac{d}{ds} \left[(1 - e^{sT} z)(1 - e^{sT} z^{-1}) \right]$$

$$= -T \exp(p_n T) . \left[z + z^{-1} - 2 \exp(p_n T) \right]$$

Noting that:

$$\exp(p_n T) = \exp(\pm \log z) = z^{\pm 1}$$

then equation (4.5.32) becomes:

$$\Phi_x(z) = \sum_{-\infty}^{\infty} \frac{1}{T} \left\{ \Psi(\log z + j2\pi n/T) + \Psi(- \log z + j2\pi n/T) \right\}$$

and from equation (4.5.29):

$$\Phi_x(z) = \frac{1}{T} \sum_{-\infty}^{\infty} \Phi_x(\log z + j2\pi n/T) \tag{4.5.34}$$

It is convenient to express the above as a Pulse Laplace Transformation by setting*:

$$z = e^{sT} \qquad \text{or} \qquad sT = \log z$$

to yield:

---

*The Principal Value is used here so that log z is indeed a function (single valued) in the strict mathematical sense (46). In this particular context, the z- plane is a Riemann surface with a countably infinite number of sheets and a Branch Cut along the positive real axis (101, 114). Though the contour of integration is thereby distorted, the end-result is still equation (4.5.35)

$$\Phi_x^*(s) = \frac{1}{T}\sum_{-\infty}^{\infty} \Phi_x(s + j2\pi n/T) = \frac{1}{T}\sum_{-\infty}^{\infty} \Phi_x(s - j2\pi n/T) \quad (4.5.35)$$

Equation (4.5.35) declares the now familiar theme; the periodic sampling of continuous data creates an infinite number of sidebands which are formed by shifting the unsampled spectrum up and down the real frequency axis by integral multiples of 1/T Hz.

The next Section 4.6 establishes the spectral response of both continuous and discrete linear systems to wide sense stationary inputs. Such attention to continuous data systems might seem incongruous in a text about DDC systems, but the motivation is indeed sound. Firstly, by demonstrating the similarities between the Laplace- and z- Transformations, a bridge is provided for the interchange of design experience. Secondly, it should be recalled (and even cast in tablets of stone) that the performance of a DDC system must be judged generally in the context of continuous time. Here the strategy advocated for this purpose invokes two disparate, yet coupled, procedures. By means of the z- Transformation, a satisfactory performance is engineered at the sampling instants. Laplace Transformation techniques are then deployed to ensure that intersample deviations are adequately constrained by the choice of a suitable sampling frequency. In this framework, the output samples generated by a digital controller must be viewed as a function in continuous time rather than the number sequence implicit in equation (4.5.35). Analytical procedures for both continuous and discrete data systems are therefore required for the design of DDC systems.

4.6 INPUT-OUTPUT RELATIONSHIPS

Following the switching-on of a linear system, an analysis based on the inversion integral has shown that the output (y) contains contributions from both its set of natural oscillatory modes (poles) and its input signal (x). In the case of a stable system, the natural modes of oscillation decay to insignificant amplitudes after sufficient time has elapsed, and then the response becomes

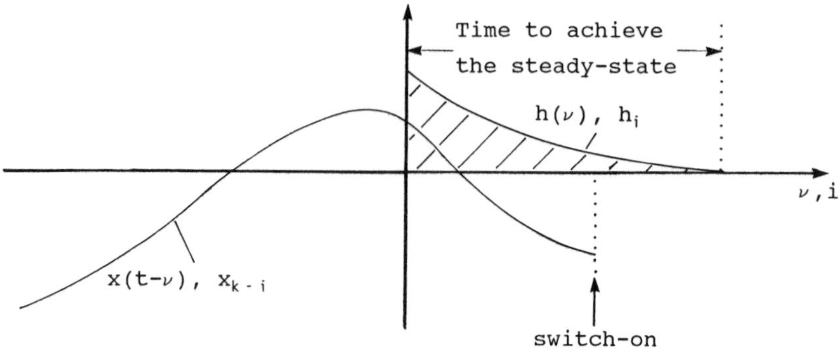

Fig 4.6.1 The Weighting Function (or sequence) and Steady State Behaviour

effectively determined by the input signal alone. That is to say, the system achieves its steady state or stationary response. An alternative, yet equivalent viewpoint, is provided by the convolution integral (1.2.31):

$$y(t) = \int_{0^+}^{t} h(\nu)x(t - \nu)d\nu \qquad (4.6.1)$$

in which the weighting or impulse response function h(t) of a system satisfies the causality constraint:

$$h(t) = 0 \quad \text{for} \quad t \leq 0 \qquad (4.6.2)$$

and for present purposes

$$x(t) = 0 \quad \text{for} \quad t \leq 0 \qquad (4.6.3)$$

As illustrated by Fig 4.6.1, the steady-state condition exists in a linear system when the weighting function has virtually 'forgotten' the switching-on instant, and the input signal for positive time appears over its 'effective width'. Under these conditions, the output response is given for all practical purposes by equation (4.6.1) as:

$$y(t) = \int_{0^+}^{\infty} h(\nu)x(t - \nu)d\nu \tag{4.6.4}$$

Subsequent analysis in this text relates exclusively to the steady state behaviour of stable linear systems with wide sense stationary inputs, and in this way follows techniques developed by Wiener (86). More recent work by Kalman (103, 104) extends statistically optimum filtering to include the transient period of linear systems. Because his time-varying filters involve a particularly undemanding numerical computation (compared to the esoteric analytical spectral factorisation procedure in the Wiener approach), recent developments in μ- processors have enabled a diversity of practical applications (105, 106, 107).

If the input (x) to a linear continuous data system is a realisation of a wide sense stationary ensemble {X}, the auto-correlation function of its steady state response (y) is derived from equations (4.1.6) and (4.6.4) as:

$$\phi_y(t, t + \tau) = \iint_{-\infty}^{\infty} \left\{ \iint_{0^+}^{\infty} h(\eta)h(\nu)\; x_1 x_2 \; d\eta d\nu \right\} P_X(x_1, x_2)dx_1 dx_2 \tag{4.6.5}$$

where for ease of nomenclature:

$$x_1 = x(t - \eta) \quad \text{and} \quad x_2 = x(t + \tau - \nu)$$

If the input x(t) is an amplitude modulated pulse train, the integrand of equation (4.6.5) is continuous over the region of integration in practice*, but this region is unbounded. Consequently, the familiar elementary conditions (108, 109) for rearranging an iterated integral are not satisfied and a more powerful theorem is required. Towards this end observe that an input signal is always bounded in reality so that for values of

---

* refer to the discussion after equation (1.3.9)

the independent variables:

$$\left|h(\eta)h(\nu)\ x_1 x_2\ P_X(x_1, x_2)\right| \le B^2 \left|h(\eta)\right| \left|h(\nu)\right| P_X(x_1, x_2)$$

Granted system stability so that h(t) is absolutely integrable, it then follows from equation (4.1.3) that the integrand of equation (4.6.5) is also absolutely integrable over its infinite multidimensional domain. Hence the conditions of the generalised Fubini theorem (15) are satisfied, and the iterated integral can be evaluated in any order. In particular:

$$\phi_Y(t, t+\tau) = \iint_{0^+}^{\infty} h(\eta)h(\nu) \left\{ \iint_{-\infty}^{\infty} x_1 x_2 P_X(x_1, x_2) dx_1 dx_2 \right\} d\eta d\nu$$

which readily simplifies to:

$$\phi_Y(\tau) = \iint_{0^+}^{\infty} h(\eta)h(\nu)\ \phi_X(\tau - \nu + \eta) d\eta d\nu \qquad (4.6.6)$$

Equation (4.6.6) establishes that a wide sense stationary input to a linear time invariant continuous data system elicits a wide sense stationary response. The corresponding power spectral density function is derived from equation (4.3.1) as:

$$\Phi_Y(\omega) = \int_{-\infty}^{\infty} \left\{ \iint_{-\infty}^{\infty} h(\eta)h(\nu)\ \phi_X(\tau - \nu + \eta) d\eta d\nu \right\} e^{-j\omega\tau} d\tau$$

A similar argument to that above justifies rearranging the order of integration to:

$$\Phi_Y(\omega) = \iint_{0^+}^{\infty} h(\eta)h(\nu) \left\{ \int_{-\infty}^{\infty} \phi_X(\tau - \nu + \eta) e^{-j\omega\tau} d\tau \right\} d\eta d\nu \qquad (4.6.7)$$

By effecting the transformation:

$$\eta \rightarrow \eta \quad ; \quad \nu \rightarrow \nu \quad ; \quad \tau \rightarrow \tau - \nu + \eta$$

whose Jacobian determinant is unity, equation (4.6.7) reduces to:

$$\Phi_Y(\omega) = \left(\int_{0^+}^{\infty} h(\eta) e^{j\omega\eta} d\eta \right) \left(\int_{0^+}^{\infty} h(\nu) e^{-j\omega\nu} d\nu \right) \Phi_X(\omega)$$

which further reduces to:

$$\Phi_Y(\omega) = \left| H(\omega) \right|^2 \Phi_X(\omega) \tag{4.6.8}$$

or alternatively in terms of two-sided Laplace transformations:

$$\Phi_Y(s) = H(-s) H(s) \Phi_X(s) \tag{4.6.9}$$

Equation (4.6.8) confirms the previously made assertion that practical power spectra reduce in intensity at a minimum rate of 20 dB/decade due to intrinsic inertia or stray capacitance.

If the input to a linear continuous data system is a realisation of a wide sense stationary process, the cross correlation function with its steady state output is obtained from equations (4.1.4) and (4.6.4) as:

$$\phi_{XY}(t, t+\tau) = \iint_{-\infty}^{\infty} \left(\int_{0^+}^{\infty} h(\nu) x_1 x_2 d\nu \right) P_{XY}(x_1, x_2) dx_1 dx_2$$

where for ease of nomenclature

$$x_1 = x(t) \quad \text{and} \quad x_2 = x(t + \tau - \nu)$$

For a stable system, rearrangement of the order of integration is justified as above to yield:

$$\phi_{XY}(t, t+\tau) = \int_{0^+}^{\infty} h(\nu) \left(\iint_{-\infty}^{\infty} x_1 x_2 P_{XY}(x_1, x_2) dx_1 dx_2 \right) d\nu$$

which reduces to:

$$\phi_{XY}(\tau) = \int_{0^+}^{\infty} h(\nu)\ \phi_X(\tau - \nu)d\nu \qquad (4.6.10)$$

According to equation (4.3.2), the corresponding cross power spectral density function is defined by:

$$\Phi_{XY}(s) = \int_{-\infty}^{\infty}\int_{0^+}^{\infty} h(\nu)\phi_X(\tau - \nu)e^{-s\tau}d\nu d\tau$$

After a simple transformation of variables and transposition of the order of integration, the above equation simplifies to:

$$\Phi_{XY}(s) = \left[\int_{0^+}^{\infty} h(\nu)e^{-s\nu}d\nu\right]\Phi_X(s)$$

or

$$\Phi_{XY}(s) = H(s)\,\Phi_X(s) \qquad (4.6.11)$$

Equation (4.6.11) provides an obvious basis for measuring the real frequency response function of a linear continuous data system in terms of particular statistical properties of its input and output signals. A detailed comparison with the more usual deterministic technique is provided later in Section 4.7. Next, the analogues of equations (4.6.9) and (4.6.11) for discrete data systems are derived, and these are found to confirm yet again the close similarities between Laplace and z- Transformation analyses.

Linear superpostion implies that the response $\{y_k\}$ of a linear discrete data system to an input sequence $\{x_k\}$ necessarily takes the form of equation (2.4.2):

$$y_k = \sum_{0}^{k} h_i\, x_{k-i} \qquad (4.6.12)$$

where

$$H(z) = \sum_{0}^{\infty} h_i z^{-i} \tag{4.6.13}$$

denotes the particular pulse transfer function involved. Under the steady state conditions illustrated by Fig 4.6.1, equation (4.6.12) becomes:

$$y_k = \sum_{0}^{\infty} h_i x_{k-i} \tag{4.6.14}$$

If the input is a realisation of a wide sense stationary sequence ensemble, the autocorrelation sequence of this steady state response is specified similarly to equation (4.6.5) by:

$$\phi_Y(k,\ k+n) = \iint_{-\infty}^{\infty} \left(\sum_{0}^{\infty} \sum_{0}^{\infty} h_i h_m \ x_1 x_2 \right) P_X(x_1, x_2)\, dx_1\, dx_2 \tag{4.6.15}$$

where for notational convenience:

$$x_1 = x_{k-i} \quad \text{and} \quad x_2 = x_{k+n-m} \tag{4.6.16}$$

In practice, input signals and their joint probability density functions are bounded, so there exists a positive real number (B) such that for all values of the independent variables:

$$\left| h_i h_m \ x_1 x_2 \ P_X(x_1, x_2) \right| \le B \left| h_i \right| \left| h_m \right|$$

Granted a stable system, none of its poles lie outside a disc of some radius $r < 1$ in the z- plane. As proved in Appendix A4.6, the series $\{h_i\}$ is then absolutely convergent so that:

$$\left| h_i h_m \ x_1 x_2 \ P_X(x_1, x_2) \right| \le B \left(\sum_{0}^{\infty} \left| h_i \right| \right)^2$$

Thus the series of functions $\{h_i h_m \; x_1 x_2 \; P_X(x_1, x_2)\}$ is normally summable with respect to the variables x_1 and x_2, and its integration and summation commute to yield:

$$\phi_Y(k, \; k + n) = \sum_{0}^{\infty} \sum_{0}^{\infty} h_i h_m \iint_{-\infty}^{\infty} x_1 x_2 P_X(x_1, x_2) dx_1 dx_2$$

or

$$\phi_Y(n) = \sum_{0}^{\infty} \sum_{0}^{\infty} h_i h_m \; \phi_X(n - m + i) \qquad (4.6.17)$$

Equation (4.6.17) establishes that a wide sense stationary input to a linear time invariant discrete data system elicits a wide sense stationary steady-state response. The corresponding pulse spectral density function is defined by equation (4.4.1) as:

$$\Phi_Y(z) = \lim_{N \to \infty} \sum_{-N}^{N} \sum_{0}^{\infty} \sum_{0}^{\infty} h_i h_m \phi_X(n - m + i) z^{-n} \qquad (4.6.18)$$

Appendix A4.6 establishes that the series $\{h_m z^{-m}\}$ is normally summable outside a disc of some radius $r < 1$ in the z- plane. By means of the transformation:

$$z^{-1} \to z$$

the same result demonstrates that the series $\{h_i z^i\}$ is normally summable inside a disc of radius $1/r > 1$ in the z- plane. If a pulse spectrum results from a periodically sampled auto-correlation function having a two-sided Laplace transformation, equation (4.5.13) reveals that $\{\phi_X(k) z^{-k}\}$ is normally summable in an annulus containing the unit circle. Thus the three series share a common region of normal summability typified by:

$$R_I \leq |z| \leq R_o$$

with:

$R_I < 1$ and $R_0 > 1$

In this region the sum of any finite subseries of $\{h_i h_m \phi_X(n - m + i) z^{-n}\}$ is evidently bounded by:

$$\left(\sum_{0}^{\infty} |h_i| R_0^i\right)\left(\sum_{0}^{\infty} |h_m| R_I^{-m}\right)\left(\sum_{0}^{\infty} |\phi_X(k)| (R_0^k + R_I^{-k})\right)$$

so the series is normally summable (10, 46). Consequently, its sum can be evaluated in any order (10, 38) and in particular by summing first over:

$k = n - m + i$

equation (4.6.18) becomes:

$$\Phi_Y(z) = H(z^{-1})H(z)\Phi_X(z) \qquad (4.6.19)$$

If the input to a linear discrete data system is a wide sense stationary sequence, the cross correlation function of its steady state output is obtained from equations (4.1.14) and (4.6.14) as:

$$\phi_{XY}(k, k + n) = \iint_{-\infty}^{\infty} \left(\sum_{0}^{\infty} h_i \; x_1 x_2\right) P_{XY}(x_1, x_2) dx_1 dx_2 \qquad (4.6.20)$$

where for ease of nomenclature:

$x_1 = x_k$ and $x_2 = x_{k+n-i}$

For a stable system, similar considerations as for equation (4.6.15) justify interchanging the order of integration and summation to yield:

$$\phi_{XY}(n) = \sum_{0}^{\infty} h_i \quad \phi_X(n - i) \tag{4.6.21}$$

and the corresponding pulse spectral density function is defined by equation (4.4.1) as:

$$\Phi_{XY}(z) = \sum_{-\infty}^{\infty} \sum_{0}^{\infty} h_i \, \phi_X(n - i) z^{-n} \tag{4.6.22}$$

As with regard to equation (4.6.18), the sum of any finite subseries of $\left\{ \left| h_i \, \phi_X(n-i) z^{-n} \right| \right\}$ is bounded by:

$$\left(\sum_{0}^{\infty} \left| h_i \right| R_0^i \right) \left(\sum_{0}^{\infty} \left| \phi_X(k) \right| (R_0^k + R_I^{-k}) \right)$$

so the series is normally summable. Its sum can be effected therefore in any order, and in particular by summing first over:

$$k = n - i$$

equation (4.6.22) becomes

$$\Phi_{XY}(z) = H(z) \, \Phi_X(z) \tag{4.6.23}$$

Comparisons of equation (4.6.19) with (4.6.9) and equation (4.6.23) with (4.6.11) reveal again the close similarity between Laplace and z- Transformation analyses. This resemblance is even more striking when expressed in terms of real frequency functions:

$$\Phi_Y(\omega) = \left| H(\omega) \right|^2 \Phi_X(\omega) \quad ; \quad \Phi_Y^*(\omega) = \left| H^*(\omega) \right|^2 \Phi_X^*(\omega) \tag{4.6.24}$$

and

$$\Phi_{XY}(\omega) = H(\omega) \, \Phi_X(\omega) \quad ; \quad \Phi_{XY}^*(\omega) = H^*(\omega) \, \Phi_X^*(\omega) \tag{4.6.25}$$

Some particularly useful developments of these stochastic input-output relationships are described in the following Section 4.7.

4.7 MULTIPLE UNCORRELATED INPUTS

A stable linear system is excited at two separate inputs by the uncorrelated wide sense stationary signals u and x. Equation (4.6.4) defines the steady state response of a continuous data system to each of these inputs applied separately as:

$$v(t) = \int_{0^+}^{\infty} g(\nu)u(t-\nu)d\nu \quad \text{and} \quad y(t) = \int_{0^+}^{\infty} h(\nu)x(t-\nu)d\nu \tag{4.7.1}$$

where g and h denote the pertinent weighting functions. The corresponding results for a discrete data system are given by equation (4.6.14) as:

$$v_k = \sum_{0}^{\infty} g_i u_{k-i} \quad \text{and} \quad y_k = \sum_{0}^{\infty} h_i x_{k-i} \tag{4.7.2}$$

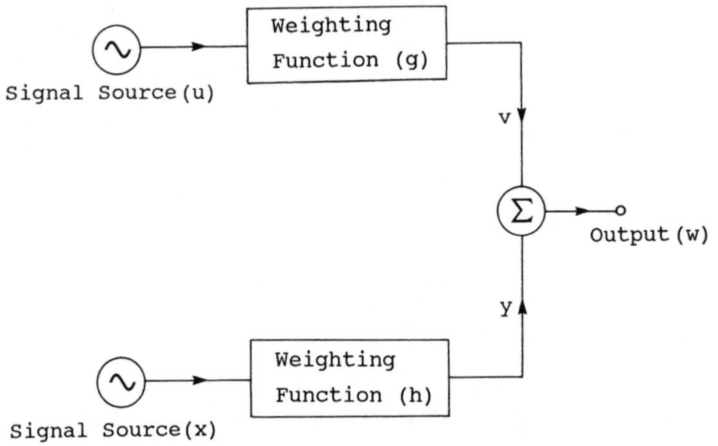

Fig 4.7.1 A Linear System with Two Separate Inputs

By virtue of its linearity, the response of the system when both sources operate simulataneously is:

$$w(t) = v(t) + y(t) \qquad \text{or} \qquad w_k = v_k + y_k \tag{4.7.3}$$

which is illustrated in terms of transfer functions by Fig 4.7.1. It follows that the autocorrelation function of this combined response is:

$$\left.\begin{aligned} \phi_W(t,\ t+\tau) &= \phi_V(\tau) + \phi_{VY}(t,\ t+\tau) + \phi_{YV}(t,\ t+\tau) + \phi_Y(\tau) \\ \phi_W(k,\ k+n) &= \phi_V(n) + \phi_{VY}(k,\ k+n) + \phi_{YV}(k,\ k+n) + \phi_Y(n) \end{aligned}\right\} \tag{4.7.4}$$

where:

ϕ_V - autocorrelation function (sequence) of the response to the input u alone

ϕ_Y - autocorrelation function (sequence) of the response to the input x alone (4.7.5)

The cross correlation terms are obtained from equations (4.7.1) or (4.7.2) for example as:

$$\phi_{VY}(t,\ t+\tau) = \overline{\int_{0^+}^{\infty}\!\!\int g(\nu)h(\eta)u(t-\nu)x(t+\tau-\eta)d\nu d\eta} \tag{4.7.6}$$

or

$$\phi_{VY}(k,\ k+n) = \overline{\sum_{0}^{\infty}\sum g_i h_m \ u_{k-i} \ x_{k+n-m}} \tag{4.7.7}$$

where the 'bar' notation denotes the integration that effects the ensemble averaging process. Similar convergence considerations to those involved in the derivation of equations (4.6.6) and (4.6.17) enable an interchange of the order of integration or the order of integration and summation to yield:

$$\left.\begin{aligned} \phi_{VY}(\tau) &= \iint_{0^+}^{\infty} g(\nu)h(\eta)\ \phi_{ux}(\tau - \eta + \nu)d\nu d\eta \\ \phi_{VY}(n) &= \sum_{0}^{\infty}\sum g_i h_m\ \phi_{ux}(n - m + i) \end{aligned}\right\} \qquad (4.7.8)$$

Because the signal sources are uncorrelated with zero means*:

$$\phi_{UX}(\tau) \equiv 0 \text{ for all } \tau \quad ; \quad \phi_{UX}(n) \equiv 0 \text{ for all } n \qquad (4.7.9)$$

so that equations (4.7.8) reduce to:

$$\phi_{VY} = \mathbf{0} \qquad (4.7.10)$$

From equation (4.1.9):

$$\phi_{YV}(\tau) = \phi_{VY}(-\tau) \quad ; \quad \phi_{YV}(n) = \phi_{VY}(-n) \qquad (4.7.11)$$

and consequently equations (4.7.4) simplify to:

$$\phi_W = \phi_V + \phi_Y \qquad (4.7.12)$$

which expressed in terms of spectral density functions becomes:

$$\Phi_W = \Phi_V + \Phi_Y \qquad (4.7.13)$$

Noting the definitions (4.7.5) and substituting equation (4.6.9) or (4.6.19) yields:

$$\left.\begin{aligned} \Phi_W(s) &= G(-s)G(s)\,\Phi_U(s) + H(-s)H(s)\,\Phi_X(s) \\ \Phi_W(z) &= G(z^{-1})G(z)\,\Phi_U(z) + H(z^{-1})H(z)\,\Phi_X(z) \end{aligned}\right\} \qquad (4.7.14)$$

---

*deterministic components can be linearly superimposed after a separate calculation

Thus the spectral response of a linear system to two uncorrelated signal sources is the sum of the spectral responses to each input operating individually.

Mathematical induction readily enables an extension of equations (4.7.14) to an arbitrary finite number of uncorrelated signal sources. Suppose there are N+1 such sources denoted by $\{x_p | p \leq N+1\}$ with corresponding separate responses $\{y_p | p \leq N+1\}$ and transfer functions $\{H_p | p \leq N+1\}$. Equation (4.7.12) is assumed true for N uncorrelated sources. Writing:

$$w = \left(\sum_{p=1}^{N} y_p \right) + y_{N+1} \qquad (4.7.15)$$

it follows as in equation (4.7.4) that:

$$\phi_W = \phi_\Sigma + \phi_{\Sigma . Y N+1} + \phi_{Y N+1 . \Sigma} + \phi_{Y N+1} \qquad (4.7.16)$$

where

Σ - the ensemble containing $\sum_{p=1}^{N} y_p$

The cross correlation $\phi_{\Sigma . Y N+1}$ is derived from equation (4.7.6) or (4.7.7) as:

$$\phi_{\Sigma . Y(N+1)}(\tau) = \sum_{1}^{N} \iint_{0^+}^{\infty} h_p(\nu) h_{N+1}(\eta) \overline{x_p(t-\nu) x_{N+1}(t+\tau-\eta)} d\nu d\eta$$

or:

$$\phi_{\Sigma . Y N+1}(n) = \sum_{1}^{N} \sum_{0}^{\infty} \sum h_p(i) h_{N+1}(m) \overline{x_p(k-i) x_{N+1}(k+n-m)}$$

(4.7.17)

and because the stochastic sources are uncorrelated, these cross correlations are null so that equation (4.7.16) reduces to:

$$\phi_W = \phi_\Sigma + \phi_{Y\,N+1}$$

Applying the inductions hypothesis yields:

$$\phi_W = \sum_{p=1}^{N+1} \phi_{Y\,p}$$

so that equation (4.7.14) is true for an arbitrary finite number of uncorrelated sources. Explicitly:

$$\left.\begin{aligned}\Phi_W(s) &= \sum_{p=1}^{N} H_p(-s)\; H_p(s)\,\Phi_{X\,p} \\ \Phi_W(z) &= \sum_{p=1}^{N} H_p(z^{-1})\; H_p(z)\,\Phi_{X\,p}\end{aligned}\right\} \qquad (4.7.18)$$

The z- transform relationships in equation (4.7.18) find an important application in Chapter 6 with regard to quantifying the total rms noise generated by the uncorrelated sources of round-off error in the realisation of a digital controller or filter.

With the same nomenclature as above define:

$$w = \sum_{p=1}^{N} y_p$$

then similar considerations to those that yielded equation (4.7.17) give for example:

$$\left.\begin{aligned}\phi_{X\,1.W}(\tau) &= \sum_{1}^{N} \int_{o^+}^{\infty} h_p(\nu)\,\overline{x_1(t)\,x_p(t+\tau-\nu)}\,d\nu \\ \phi_{X\,1.W}(n) &= \sum_{1}^{N} \sum_{o}^{\infty} h_p(i)\,\overline{x_1(k)\,x_p(k+n-i)}\end{aligned}\right\} \qquad (4.7.19)$$

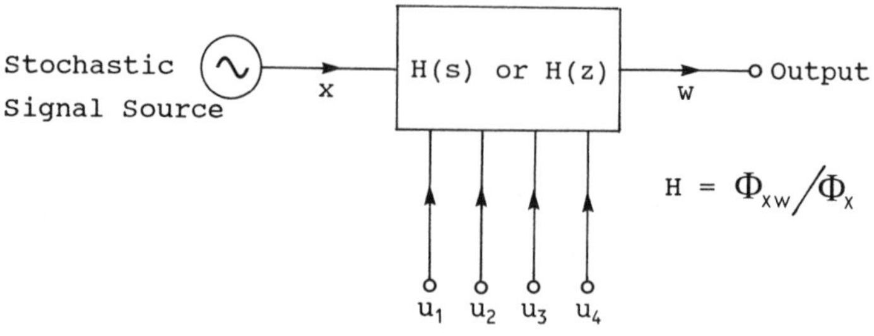

Fig 4.7.2 A Stochastic Frequency Response Measurement

Because the signal sources are mutually uncorrelated with zero means, the above equations simplify to:

$$\left.\begin{aligned} \phi_{x1.w}(\tau) &= \int_{0^+}^{\infty} h_1(\nu)\,\phi_{x1}(\tau - \nu)\,d\nu \\ \phi_{x1.w}(n) &= \sum_{0}^{\infty} h_1(i)\,\phi_{x1}(n - i) \end{aligned}\right\} \qquad (4.7.20)$$

which as in Section 4.6 provide the spectral relationships:

$$\left.\begin{aligned} \Phi_{x1.w}(s) &= H_1(s)\,\Phi_{x1}(s) \\ \Phi_{x1.w}(z) &= H_1(z)\,\Phi_{x1}(z) \end{aligned}\right\} \qquad (4.7.21)$$

Equations (4.7.21) enable the stochastic measurement of a real frequency response function as illustrated by Fig 4.7.2. The procedure is evidently simplest when it is known a priori that the excitation (x) has an effectively white spectrum.

For practical purposes it must be assumed that the excitation

(x) and intrinsic noise sources (u_p) belong to mutually ergodic ensembles, so that the required cross correlation for example is given with probability one by:

$$\left.\begin{aligned} \phi_{xw}(\tau) &= \lim_{\eta\to\infty} \frac{1}{\eta}\int_0^{\infty} x(t)\; w(t+\tau)dt \\ \phi_{xw}(n) &= \lim_{N\to\infty} \frac{1}{N}\sum_{0}^{N} x_k w_{k+n} \end{aligned}\right\} \qquad (4.7.22)$$

and the origin (t=0 or k=0) marks the beginning of an essentially steady state recording. Measurement accuracy inevitably suffers because the correlation functions or sequences must be estimated from just finite length recordings. A computed real frequency function therefore contains errors due to the imperfect rejection of plant noise ($\phi_{uw}\neq 0$) and the imperfect recovery of the required correlations ϕ_x and ϕ_{xw}. Usually these finite sampling errors can be made arbitrarily small by using long enough recordings, provided that a spectrum is derived from an initial estimate of its correlation function and <u>not</u> from its periodogram (88)*. However, significant residual errors can still remain if the non-linearities in an actual system create large enough intermodulation products between the excitation (x) and a noise source (u). For instance, a product somewhat simplistically proportional to x u^2 correlates with the input excitation to make the integrand of an equation like (4.7.19) non-zero.

Frequency response function measurements using deterministic input signals involve the excitation of a system with sinusoids or single pulses (59) whose amplitudes, though much larger than ambient noise levels, are small enough to preserve essentially its linearity about an operating point. Because the input spectrum of a single pulse can be made to contain all significant frequencies at an adequate instantaneous power level, a single test advantageously provides the required frequency response function.

---

*refer to the footnote below equation (4.3.2)

However, the necessity that the induced perturbations stand out clearly above plant noise levels can render deterministic measurements unacceptable in some cases. For example, the corresponding reactivity changes in experimental nuclear reactors could erroneously initiate an emergency trip (63), while in other cases the degradation in product quality might preclude such a test (113). There are therefore good reasons for alternative frequency response measurements that demand much lower levels of plant excitation. The stochastic technique in Fig 4.7.2 achieves this objective, but at the expense of an increased measurement time. An effectively white noise source, that provides a largely uniform input spectrum over the significant frequency range of a system, evidently forms the stochastic equivalent of a deterministic pulse generator.

4.8 SHAPING FILTERS AND MEAN SQUARE VALUE CALCULATIONS

If a power spectrum function is the ratio of two real polynomials* in s or z, its poles and zeros are necessarily real or complex conjugate pairs of entities (10, 12, 29). Furthermore, because an autocorrelation function is an even function of its temporal separation variable, it follows directly from the definition of its power spectrum that:

$$\Phi_x(s) = \Phi_x(-s) \quad \text{and} \quad \Phi_x(z) = \Phi_x(z^{-1}) \tag{4.8.1}$$

Consequently, these power spectra can always be factored as:

$$\Phi_x(s) = \chi(s)\,\chi(-s) \quad \text{and} \quad \Phi_x(z) = \chi(z)\,\chi(z^{-1}) \tag{4.8.2}$$

where

$\chi(s)$ has real or complex pairs of poles and zeros in the left-half s- plane only

and

---

*a rational function

$\chi(z)$ has real or complex pairs of poles and zeros inside the unit circle

Noting that:

$$\lim_{\tau \to \infty} \phi_x(\tau) = 0 \qquad \text{and} \qquad \lim_{n \to \infty} \phi_x(n) = 0 \qquad (4.8.3)$$

then the discussions in Sections 4.3 and 4.4 imply that the poles of $\chi(s)$ and $\chi(z)$ (which govern behaviour for positive time) must lie strictly within the left-half s- plane or the unit circle respectively. Also because stray capacitance or inertia etc are intrinsic aspects of practical systems, equation (4.7.9) implies that continuous data processes have power spectra which converge to zero asymptotically faster than $1/s^2$. These observations imply that $\chi(s)$ corresponds to the transfer function of a stable lumped parameter continuous data system, and that $\chi(z)$ corresponds to the pulse transfer function of a stable discrete data system. Operational amplifier circuits which realise the general biquadratic 'building block':

$$(a_2 s^2 + a_1 s + a_0)/(s^2 + b_1 s + b_0)$$

are available (110) to synthesise Laplace transfer functions like $\chi(s)$. The various programming techniques described in Section 2.3 enable the synthesis of pulse transfer functions like $\chi(z)$. If a white noise process with power spectrum:

$$\Phi_N(s) = A \qquad \text{or} \qquad \Phi_N(z) = A \qquad (4.8.4)$$

forms the input to a so-called *shaping filter* (111) with transfer function:

$$\chi(s)/\sqrt{A} \qquad \text{or} \qquad \chi(z)/\sqrt{A} \qquad (4.8.5)$$

then it follows from equations (4.7.9) and (4.7.19) that the respective output spectrum is $\Phi_x(s)$ or $\Phi_x(z)$. Equations (4.8.2), (4.8.4) and (4.8.5) represent the essential elements for

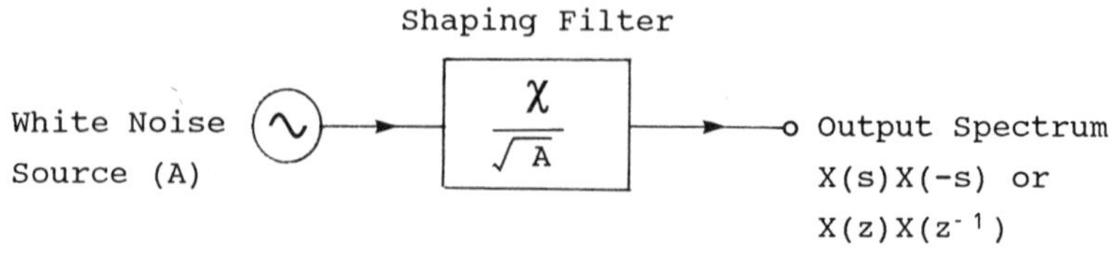

Fig 4.8.1 Synthesis of Rational Power Spectrum Function

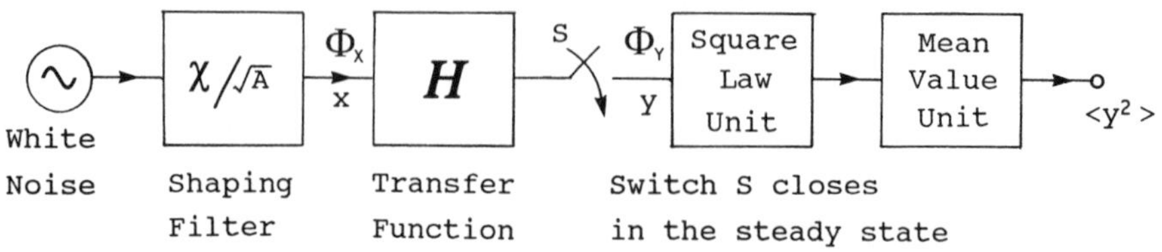

Fig 4.8.2 A Stochastic Measurement of an Ensemble Mean Square Value

sythesising in the manner of Fig 4.8.1 an arbitrary rational power spectrum using a white noise source (99, 100, 112, 113). In practice, continuous data with a perfectly white spectrum cannot exist due to the now obvious constraint of a finite bandwidth (power), while statistical sampling errors on necessarily finite recordings detract from the certainty that a discrete process is truly white. Nevertheless, provided the 'white noise' generator maintains a uniform spectral intensity over a wide enough bandwidth in relation to $\chi(s)$ or $\chi(z)$, an adequate approximation to the required power spectrum is achieved.

Because a shaping filter and white noise source can match any particular rational input spectrum (Φ_x), the combination elicits an identical spectral response from a linearised system model.

Provided the white noise generator creates a realisation of an ergodic ensemble, the ensemble mean square value of a steady state output variable (y) could be measured as in Fig 4.8.2 where:

$$\left.\begin{aligned} \langle y^2 \rangle &= \lim_{\Lambda \to \infty} \frac{1}{\Lambda} \int_0^{\Lambda} y^2(t)\,dt \\ \langle y^2 \rangle &= \lim_{N \to \infty} \frac{1}{N+1} \sum_0^{N} y_2(i) \end{aligned}\right\} \qquad (4.8.6)$$

Unfortunately the accuracy of such estimates suffers from: errors induced by the necessity of finite length recordings, the time wasted until a system achieves its steady state and the relatively slow convergence of statistical measurements (93, 115). A more efficient deterministic calculation is now described which uses the above spectral factorisation and equation (4.6.9) or (4.6.19). In these terms, the inversion integral (4.3.3) or (4.4.11) specifies the required ensemble mean square value as:

$$\phi_Y(o) = \frac{1}{2\pi j} \int_{-j\infty}^{j\infty} H(-s)H(s)\chi(s)\chi(-s)\,ds$$

or (4.8.7)

$$\phi_Y(o) = \frac{1}{2\pi j} \int_{\Gamma} H(z^{-1})H(z)\chi(z)\chi(z^{-1})z^{-1}\,dz$$

where

Γ - unit circle in the z- plane
χ - spectral factor for Φ_x

If g is a weighting function of a stable continuous data system and G the corresponding transfer function, then Parceval's theorem (1.2.23) when couched in the s- plane reads as:

$$\int_0^{\infty} g^2(t)\,dt = \frac{1}{2\pi j} \int_{-j\infty}^{j\infty} G(-s)G(s)\,ds \qquad (4.8.8)$$

The analogous result for a stable discrete data system is derived in Appendix A4.8 as:

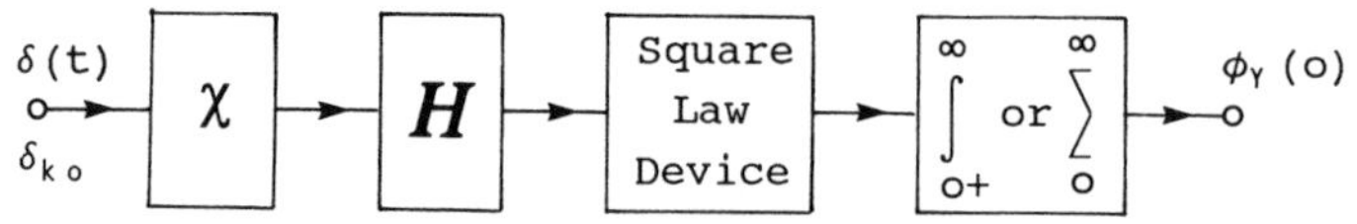

χ - shaping filter for input spectrum Φ_X

H - relevant transfer function

Fig 4.8.3 A Deterministic Measurement of an Ensemble Mean Square Value

$$\sum_{0}^{\infty} g_i^{\,2} = \frac{1}{2\pi j} \int_{\Gamma} G(z^{-1})G(z)z^{-1}dz \qquad (4.8.9)$$

By defining:

$$G(s) = H(s)\,\chi(s) \quad \text{and} \quad G(z) = H(z)\,\chi(z) \qquad (4.8.10)$$

a comparison with equation (4.8.7) shows that the required ensemble mean square value is the integral (sum) of the square of weighting function (sequence terms) for the cascaded transfer functions H and χ. Fig 4.8.3 illustrates this deterministic evaluation of the ensemble mean square value of a system variable. Sometimes a number of uncorrelated wide sense stationary inputs exists in a system, as exemplified by the multiplicative roundoff error sources in the realisation of a digital controller or filter. Under these circumstances, equation (4.7.18) enables the ensemble mean square value of the total response to be computed as above by simply adding up the separate contributions from each source acting separately.

It should be noted that although an actual signal and the synthesised signal have the same power spectrum function, their respective time behaviours can be quite different. For example, if the random telegraph signal illustrated in Fig 4.8.4 (on the next page) has on average two transversals per unit time, its

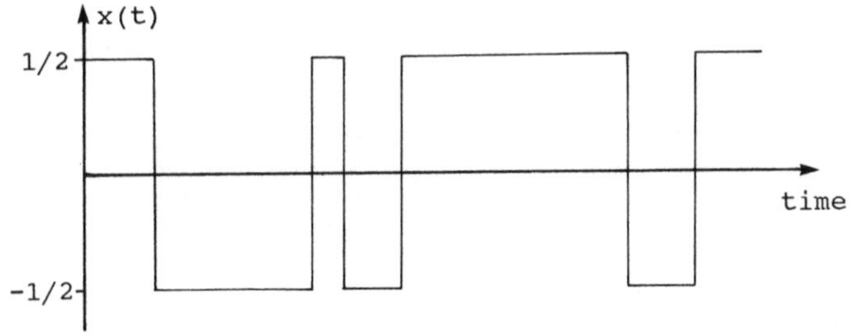

Fig 4.8.4 A Form of Random Telegraph Signal

autocorrelation function is (88):

$$\phi_x(\tau) = 0.25\exp(-4|\tau|) \tag{4.8.11}$$

The above analysis together with equations (4.3.7) and (4.3.8) show that the same power spectrum is possessed by the continuous output from a shot noise diode (88, 112) after it is filtered by a simple-lag network of the form c/(1+0.25s), where c is the appropriate real constant. It seems paradoxical that these inputs with such different temporal behaviour can produce the same mean square control error in a linearised control system model. However, it should be remembered in this respect that correlation functions and sequences describe average properties of ensembles, so that vast amounts of detail about their realisations are lost. A trivial example of this averaging effect is provided by a periodic square wave and sinusoid, both of which have zero mean.

With the completion now of Chapter 4, almost all the really fundamental relationships have been established. No more results for continuous data systems are necessary, and the strong parallelism between Laplace and z- Transformation analyses is considered to have been amply demonstrated. Detailed mathematical rigour is from now left generally to individual motivation. For instance, when valid, the order of integration and the order of summation may merely be changed without any formal justification.

Indeed because technological innovation is personally regarded as paramount, the Preface advises engineers to read the book just accepting such mathematical niceties. The next chapter addresses the first real engineering design problem of a DDC system; the choice of sampling frequency.

CHAPTER 5

The Choice of Sampling Frequency

...... a time for every purpose under heaven ......
a time to keep and a time to cast away. - Ecclesiastes

5.1 A PERSPECTIVE

Irrespective of whether a control system utilises a continuous or a digital data controller, it is generally required to meet an accuracy criterion founded on the behaviour of an error signal in continuous time. The intrinsic accuracy of a continuous data control system is determined only by the speed of response, environmental stability and relative topology of its components. However, there are additional specific features that influence the performance of DDC systems. As described earlier in Sections 1.3 and 2.5, the periodic sampling of continuous data creates a spectrum that contains high frequency sidebands, as well as a scalar multiple of that for the unsampled signal (the baseband). Due to their incomplete attenuation by a plant, these sidebands appear in the control error signal as a *ripple component* (28) which can vary appreciably between sampling instants. In addition to a loss of control accuracy, excessive ripple components can introduce an unacceptable loss of power or undue wear in actuators. The accuracy of a DDC system also suffers from *rounding error noise*, which is generated in its controller by the truncation of double length arithmetic products into the single length working format. Quantification of these degenerate effects under steady state conditions is sufficient for DDC system design purposes, because plant dynamics completely dominate the magnitude of transient control errors.

It is shown in Section 2.3 that realisations of digital controllers or filters rely on the multiplication of delayed data sequences by fixed constants. As established later in Chapter 6, the associated rounding error processes can be considered as uncorrelated with their double length 'parents' and between themselves. Consequently, the input signal and rounding error noise sources in a DDC system are all mutually uncorrelated, so that the total steady state control error in its linearised model can be characterised according to Section 4.7 by:

$$\phi_{total} = \phi_i + \phi_{fw} \tag{5.1.1}$$

where:

$$\phi_{fw} = \sum_p \phi_{fw}^p \tag{5.1.2}$$

and:

ϕ_i - steady state mean square control error for any particular input with an infinite wordlength controller

ϕ_{fw}^p - steady state mean square control error due to rounding error noise from the pth multiplier unit alone

Moreover, because a linear system responds independently to each spectral component of its sampled input, equation (5.1.2) can be further expanded as:

$$\phi_{total} = \phi_s + \phi_r + \phi_{fw} \tag{5.1.3}$$

where:

$$\phi_i = \phi_s + \phi_r \tag{5.1.4}$$

and:

ϕ_s - steady state mean square control error due to the baseband only of a sampled input signal

ϕ_r - steady state mean square control error due to all the sidebands of a sampled input signal (*the ripple component*)

Equation (5.1.3) formally asserts the important simplification that the steady state control error components ϕ_s, ϕ_r and ϕ_{fw} can be evaluated separately from a linear model. However, such a calculation of the total error (ϕ_{total}) should be accorded in practice an appropriate margin to allow for intermodulation products generated by the non-linearities of an actual system.

The baseband component of the steady state control error (ϕ_s) evidently originates from time lags corresponding to plant dynamics (inertia etc). It varies relatively slowly over a sampling period, because the bandwidth of a plant is necessarily much smaller than the sampling frequency so as to secure a small enough ripple component*. Consequently, the steady state (and transient) magnitude of a baseband control error component can be properly assessed sample-wise using z- Transformation techniques. On the other hand, the ripple component stems from spectral components at frequencies around ± n/T so that, as illustrated by Fig 2.5.1b, a fundamentally different calculation in continuous time is required. Because the ripple component (ϕ_r) clearly reduces with increasing sampling frequency and decreasing plant bandwidth, an analysis of their steady state interdependence enables the selection of a sampling frequency that is compatible with control accuracy specifications. Indeed, as discussed in Chapter 7, a start cannot even be made on assigning suitable pole-zero locations for a controller until the sampling frequency has been initially decided. The following brief review of previous algorithms used for selecting the sampling period provides an appropriate introduction to the author's subsequent analysis (51). Quantification of the steady state multiplicative rounding error

---

*Equivalently, in order to achieve adequate accuracy in reconstructing the continuous data - see Section 1.3

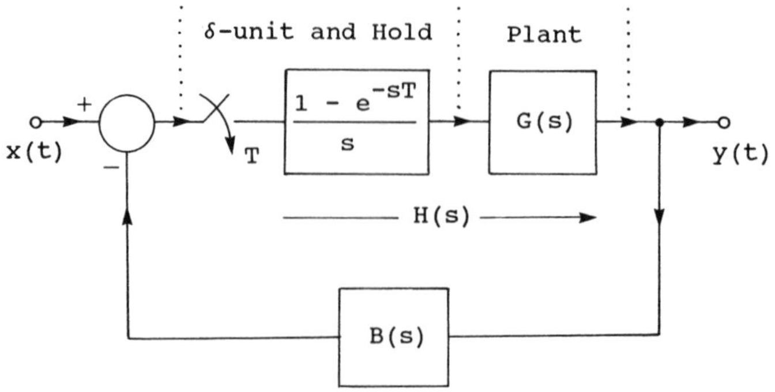

Fig 5.1.1 The Simple System of Sklansky and Ragazzini (116)

component is given later in Chapter 6.

Early publications baldly assert that the sampling frequency of a DDC system should be five- to ten-times greater than the bandwidth of its plant or the highest frequency component of its input (69, 70). These recommendations are clearly speculative and nebulous. The ripple error component for the DDC system in Fig 5.1.1 with an effectively all-pass controller is given by Sklansky and Ragazzini (116) as:

$$\phi_r = \frac{1}{2\pi}\int_{-\infty}^{\infty}\left[\frac{1}{T^2}\sum_{\substack{-\infty \\ \neq 0}}^{\infty}\left|H(\omega + n\omega_0)\right|^2 \Big/ \left|1 + HB^*(\omega)\right|^2\right]\Phi_x(\omega)\,d\omega \qquad (5.1.5)$$

where:

$HB^*(\omega)$ - Pulse Fourier Transformation of H(s)B(s) as in Section 2.4

and

$\omega_0 = 2\pi/T$; $\Phi_x(\omega)$ - spectrum of input signal in continuous time

Apart from two numerically awkward infinite limits, equation

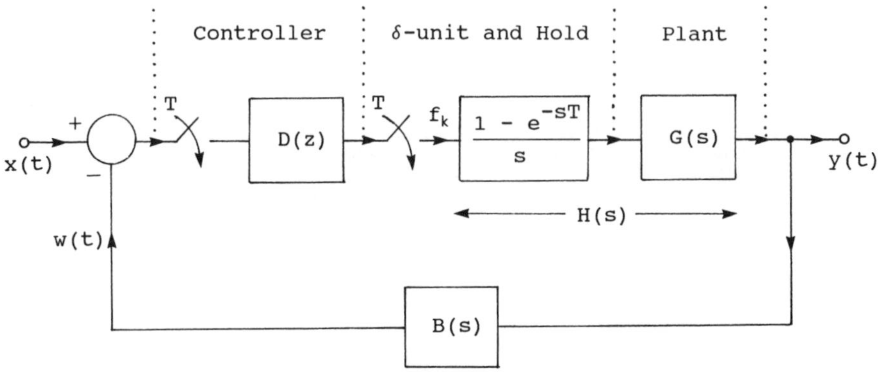

Fig 5.1.2 A more practical form of DDC System

(5.1.5) does not cater for the important practical situation in which there is a digital controller with a frequency dependent pulse transfer function D(z). To identify the analytical development (51) required in this case, an equivalent open loop system is now derived. The forcing sequence $\{f_k\}$ delivered to the zero-order held by the controller in Fig 5.1.2 has the z- transformation:

$$F(z) = D(z)\left[X(z) - W(z)\right] \tag{5.1.6}$$

According to equation (2.4.12), the z- transformation of the cascaded Laplace transformations H(s) and B(s) is given by:

$$HB(z) = \sum \text{Residues of} \left(\frac{H(s)B(s)}{1 - e^{sT}z^{-1}}\right) \text{ at poles of } H(s)B(s) \text{ only} \tag{5.1.7}$$

$$\neq H(z)B(z) \text{ in general}$$

so that from equation (2.2.6):

$$W(z) = HB(z)F(z) \tag{5.1.8}$$

It follows therefore from equation (5.1.6) that:

$$F(z) = D_E(z)X(z) \tag{5.1.9}$$

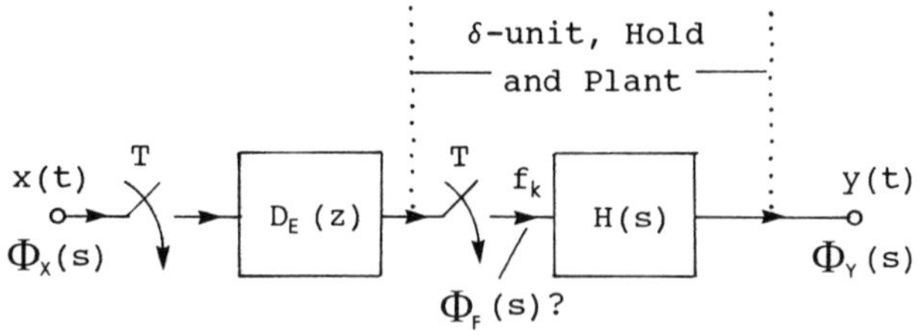

Fig 5.1.3 An Open Loop System equivalent to that in Fig 5.1.2

where:

$$D_E(z) = \frac{D(z)}{1 + D(z)HB(z)} \qquad (5.1.10)$$

Subject to equation (5.1.10), the systems in Figs 5.1.2 and 5.1.3 are equivalent in the sense that they produce the same output response (y) in continuous time given the input function x(t). The required steady state mean square ripple error corresponds to the sideband components in the spectral response function $\Phi_Y(s)$. If the output from the δ- unit driving the zero order hold has a spectrum $\Phi_F(s)$ when viewed in continuous time, then equation (4.6.9) yields:

$$\Phi_y(s) = H(-s)H(s)\,\Phi_F(s) \qquad (5.1.11)$$

However, it is the pulse spectrum of the discrete time sequence $\{f_k\}$ that is derived from equations (5.1.9) and (4.6.19) as:

$$\Phi_F^*(s) = D_E^*(-s)D_E^*(s)\,\Phi_X^*(s) \qquad (5.1.12)$$

Consequently a relationship is required between the output spectrum of a δ-unit when viewed as a number sequence in discrete time (Φ_F^*) and that (Φ_F) when viewed as a function in continuous time. Section 5.2 next establishes the surprisingly simple correspondence:

$$\Phi_F(s) = (1/T)\Phi_F^*(s) \tag{5.1.13}$$

obtains for practical purposes, so facilitating in Section 5.3 the derivation of a more generally applicable and tractable version of equation (5.1.5).

5.2 SAMPLING CONTINUOUS STOCHASTIC DATA

The sampling operation effected by analogue to digital converters is studied earlier in Section 1.3 as an amplitude modulation process in which samples (x_p) of some continuous data (x) are generated according to:

$$x_p(t) = p(t)x(t)$$

where each element of the pulse train p(t) has unit height, a narrow width (γ) and a mutual separation of T. Even if the continuous data are realisations of a wide sense stationary ensemble (X), pulse trains centred about the time origin as in Fig 1.3.3 produce an ensemble of sample pulses which viewed in continuous time is non-stationary. Trivially:

$$\overline{x_p^2(o)} = \overline{x^2}$$

while:

$$\overline{x_p^2(T/2)} = 0$$

To employ stochastic concepts in a design philosophy for DDC systems, ensembles of sample pulses must be somehow rendered wide sense stationary. The arbitrary switching-on instant of a control system evidently justifies the association of some statistically distributed delay time with the sampling pulse train. Allotting the delay time (ψ) a uniform probability over the pulse period (T):

$$P(\psi) = \begin{cases} \dfrac{1}{T} & \text{for} \quad o \le \psi \le T \\ = 0 & \text{otherwise} \end{cases} \qquad (5.2.1)$$

is now shown to render the ensemble of sample pulses wide sense stationary in continuous time.

From equations (1.3.8) and (1.3.10) the Fourier Series expansion of a practical train of sampling pulses $p(t + \psi)$ is derived as:

$$p(t + \psi) = \sum_{-\infty}^{\infty} P'_n(\psi) \exp(j2\pi nt/T) \qquad (5.2.2)$$

where:

$$P'_n(\psi) = \exp(-j2\pi n\psi/T)(\gamma/T) \left(\frac{\text{sinc}(n\gamma/T)}{1 + j2\pi n\tau_R/T} \right) \qquad (5.2.3)$$

with:

$$\text{sinc } x = \sin(\pi x)/\pi x$$

and:

τ_R - time constant of the sampling pulses with $\tau_R << \gamma << T$

Because the delay time (ψ) is assigned independently of the data ensemble, the autocorrelation function of the sampled process (X_p) is given by:

$$\overline{x_p(t)x_p(t + \tau)} = \overline{p(t + \psi)p(t + \tau + \psi)} \; . \; \overline{x(t)x(t + \tau)} \qquad (5.2.4)$$

where the 'bar' notation denotes the statistical expectation as defined by the integral relationship in equation (4.1.6). For the reasons described in Section 1.3, inclusion of a rise time constant (τ_R) for the pulses renders the above double series of functions involving $P'_m(\psi)P'_n(\psi)$ normally summable with respect to the switch-on delay variable (ψ). Consequently, the order of integration and summation in equation (4.6.4) can be interchanged to yield:

$$\overline{p(t + \psi)p(t + \tau + \psi)}$$

$$= \sum_{-\infty}^{\infty}\sum \overline{P_m'(\psi)P_n'(\psi)} \exp\left[j2\pi(m + n)t/T\right] \exp(j2\pi n\tau/T) \qquad (5.2.5)$$

where from equation (5.2.1):

$$\overline{P_m'(\psi)P_n'(\psi)} = (\gamma/T)^2 \left[\frac{\mathrm{sinc}(m\gamma/T)\,\mathrm{sinc}(n\gamma/T)}{(1 + j2\pi m\tau_R/T)(1 + j2\pi n\tau_R/T)}\right].$$

$$\frac{1}{T}\int_0^T \exp\left[-j2\pi(m + n)\psi/T\right] d\psi \qquad (5.2.6)$$

It is readily verified that:

$$\left.\begin{aligned}\frac{1}{T}\int_0^T \exp\left[-j2\pi(m + n)\psi/T\right] d\psi &= 1 \text{ for } m = -n\\ &= 0 \text{ otherwise}\end{aligned}\right\} \qquad (5.2.7)$$

and therefore equation (5.2.5) reduces to just the diagonal terms:

$$\overline{p(t + \psi)p(t + \tau + \psi)}$$

$$= (\gamma/T)^2 \sum_{-\infty}^{\infty} \left(\frac{\mathrm{sinc}(n\gamma/T)}{\left|1 + j2\pi n\tau_R/T\right|}\right)^2 \exp(j2\pi n\tau/T) \qquad (5.2.8)$$

Substituting the above into equation (5.2.4) gives:

$$\overline{x_p(t)x_p(t + \tau)} =$$

$$(\gamma/T)^2 \sum_{-\infty}^{\infty} \left(\frac{\mathrm{sinc}(n\gamma/T)}{\left|1 + j2\pi n\tau_R/T\right|}\right)^2 \phi_x(\tau)\exp(j2\pi n\tau/T) \qquad (5.2.9)$$

which shows that a wide sense stationary input to an analogue to digital converter produces a train of output pulses that are also wide sense stationary in continuous time. Moreover, because the series of functions in equation (5.2.9) are normally summable with respect to the temporal separation variable (τ), their sum can be integrated term by term to yield the spectrum of the pulse train as:

$$\Phi_{Xp}(s) = (\gamma/T)^2 \sum_{-\infty}^{\infty} \left(\frac{\text{sinc}(n\gamma/T)}{|1 + j2\pi n\tau_R/T|}\right)^2 \Phi_X(s - j2\pi n/T) \tag{5.2.10}$$

Because the time intervals required for data conversion are so much shorter than the sampling periods of contemporary DDC systems and filters, equation (5.2.10) is closely approximated over the frequency-ranges of practical interest by:

$$\Phi_{Xp}(s) = (\gamma/T)^2 \sum_{-\infty}^{\infty} \Phi_X(s - j2\pi n/T) \tag{5.2.11}$$

Sections 1.5 and 2.4 describe the dynamic characterisation of a digital to analogue converter and its staticisor register in terms of a δ-unit cascaded with the transfer function:

$$H_o(s) = K_{DA}\left(1 - e^{-sT}\right)/s \tag{5.2.12}$$

where

K_{DA} - converter gain constant (mA/l.s. bit or mV/l.s. bit)

The conceptual δ-unit produces impulses of area equal to the amplitudes of the output data samples from a digital controller, so that the corresponding form of equation (5.2.11) is therefore*:

---

*Note that unmodulated sampling pulses in equation (5.2.11) have an area of γ, while those for a δ-unit have unit area

$$\Phi_{x_p}(s) = (1/T)^2 \sum_{-\infty}^{\infty} \Phi_x(s - j2\pi n/T) \qquad (5.2.13)$$

Now the pulse spectrum of a wide sense stationary sequence, as derived by periodically sampling wide sense stationary continuous data and viewing the result in discrete time, is specified by equation (4.5.35) as:

$$\Phi_x^*(s) = (1/T) \sum_{-\infty}^{\infty} \Phi_x(s - j2\pi n/T) \qquad (5.2.14)$$

By comparing equations (5.2.13) and (5.2.14), one concludes that:

$$\Phi_{x_p}(s) = (1/T)\,\Phi_x^*(s) \qquad (5.2.15)$$

as asserted earlier in equation (5.1.13).

A uniformly distributed phase angle (ψ) attributed to a random switching-on instant is frequently used to convert a deterministic signal into a wide sense stationary ensemble. A simple example which finds application later in Section 5.3 is:

$$\{x(t)\} = \left\{\sin(\Omega t + \psi)\right\} \qquad (5.2.16)$$

The reader should verify that the autocorrelation function and power spectrum of x(t) are:

$$\phi_x(\tau) = (1/2)\cos(\Omega\tau) \qquad (5.2.17)$$

and:

$$\Phi_x(\omega) = (\pi/2)\left[\delta(\omega - \Omega) + \delta(\omega + \Omega)\right] \qquad (5.2.18)$$

respectively, from which the correct square value is derived as:

$$\overline{x^2} = \frac{1}{2\pi}\int_{-\infty}^{\infty} \Phi_x(\omega)\,d\omega = 1/2 \tag{5.2.19}$$

As an illustration of the principles described so far in Chapters 4 and 5, the continuous time spectrum of a periodically sampled sinusoid is now derived ab initio. The sampling pulse centred on the temporal origin is specified by:

$$\left.\begin{aligned} p(t) &= A \quad \text{for } |t| \le \gamma/2 \\ &= 0 \quad \text{otherwise} \end{aligned}\right\} \tag{5.2.20}$$

so that the sinusoidal samples are given by:

$$x_p(t) = \sum_{-\infty}^{\infty} p(t - nT)\sin(\Omega t + \psi) \tag{5.2.21}$$

with the practical constraint that

$$\gamma << T << 2\pi/\Omega \tag{5.2.22}$$

It is clearly not unreasonable to assume that the pulse train ensemble is ergodic, and that a member with a phase angle (ψ) of $\pi/2$ is totally representative. Accordingly, the required correlation function is derived from equations (4.1.11) and (4.1.13) as:

$$\phi_{xp}(\tau) = \lim_{N\to\infty} \frac{1}{(2N+1)T} \sum_{-N}^{N} \int_{(2n-1)T/2}^{(2n+1)T/2} x_p(t)\,x_p(t+\tau)\,dt \tag{5.2.23}$$

For any finite value of the displacement variable (τ), a positive or negative integer (k) exists* such that:

---

* By the Archimedean property of the real numbers and the well-ordered property of the integers (10, 29)

$$(2k - 1)T/2 \le \tau < (2k + 1)T/2 \qquad (5.2.24)$$

and exploiting the inequality (5.2.22) yields:

$$\phi_{x\,p}(\tau) = 2B_1(\tau - kT)\lim_{N\to\infty} \frac{1}{(2N + 1)} \sum_{-N}^{N} \cos(\Omega nT)\cos(\Omega\overline{n + kT}) \qquad (5.2.25)$$

where the single triangular pulse $B_1(\tau)$ is defined by:

$$B_1(z) = \left(\frac{A^2\gamma}{2T}\right)\left[1 - 2|\tau|/\gamma\right] \qquad \text{for} \quad |\tau| \le \gamma/2$$

$$= 0 \qquad \text{for} \quad \gamma/2 < |\tau| \le T/2 \qquad (5.2.26)$$

The trigonometrical identities:

$$\cos(\Omega\overline{n + kT}) = \cos(\Omega nT)\cos(\Omega kT) - \sin(\Omega nT)\sin(\Omega kT)$$
$$2\cos^2(\Omega nT) = 1 - \cos 2\Omega nT$$
$$2\cos(\Omega nT)\sin(\Omega nt) = \sin(2\Omega nT) \qquad (5.2.27)$$

are now applied to equation (5.2.25). Because the number of samples in a half period of $\cos(2\Omega\tau)$ and $\sin(2\Omega\tau)$ are less than $2\pi/\Omega T$, it follows that:

$$\frac{1}{2T}\left(1 - \pi/\Omega TN\right) \le \frac{1}{(2N + 1)T}\sum_{-N}^{N}\cos^2(\Omega nT) \le \frac{1}{2T}\left(1 + \pi/\Omega TN\right)$$

and

$$-\pi/2\Omega T^2 N \le \frac{1}{(2N + 1)T}\sum_{-N}^{N}\cos(\Omega nT)\sin(\Omega nT) \le \pi/2\Omega T^2 N$$

Therefore for each positive or negative integer k:

$$\phi_{xp}(\tau) = B_1(\tau - kT)\cos(\Omega kT) \qquad \text{for } |\tau - kT| \le \gamma/2$$

$$= 0 \qquad \text{for } \gamma/2 < |\tau - kT| \le T/2$$

(5.2.28)

By virtue of the inequality (5.2.22), the above equation (5.2.28) can be written as:

$$\phi_{xp}(\tau) = B_\infty(\tau)\cos(\Omega\tau) \tag{5.2.29}$$

The triangular pulse train $B_\infty(\tau)$ has period T and it can be expanded as the Fourier series:

$$B_\infty(\tau) = \sum_{-\infty}^{\infty} C_n \exp(jn\omega_o\tau)$$

where

$$\omega_o = 2\pi/T \quad ; \quad \Psi = n\omega_o\gamma/2 \tag{5.2.30}$$

$$C_n = (A^2\gamma^2/T^2).(\cos\Psi - 1)$$

If the sampling is implemented by a δ-unit, then:

$$A\gamma = 1 \tag{5.2.31}$$

and over the range of practically significant frequencies:

$$C_n = 1/2T^2 \tag{5.2.32}$$

Substituting equations (5.2.30) and (5.2.32) into equation (5.2.29) yields:

$$\phi_{xp}(\tau) = \sum_{-\infty}^{\infty} (1/4T^2).\left[\exp(j\overline{n\omega_o + \Omega\tau}) + \exp(j\overline{n\omega_o - \Omega\tau})\right]$$

and because equation (4.3.1) requires:

$$\phi_{xp}(\tau) = \frac{1}{2\pi}\int_{-\infty}^{\infty} \Phi_{xp}(\omega)\ \exp(j\omega\tau)d\omega$$

the required spectrum evaluates as:

$$\Phi_{xp}(\omega) = \frac{1}{T^2}\sum_{-\infty}^{\infty}(\pi/2)\left[\delta(\omega - n\omega_0 - \Omega) + \delta(\omega - n\omega_0 + \Omega)\right] \qquad (5.2.33)$$

According to the general equation (5.2.13), the required spectrum is also given by:

$$\Phi_{xp}(\omega) = \frac{1}{T^2}\sum_{-\infty}^{\infty}\Phi_{x}(\omega - n\omega_0)$$

and after substituting equation (5.2.18), the same result is obtained. Although this agreement is reassuring*, the principal conclusion here is the marked simplification afforded by the general formula.

5.3 OUTPUT RIPPLE POWER ANALYSIS

The previous results in Sections 5.1 and 5.2 are now used to quantify the ripple error component for the typical feedback system in Fig 5.1.2. Equation (5.1.11) gives the spectrum of its continuous data output as:

$$\Phi_y(\omega) = \left|H(\omega)\right|^2 \Phi_F(\omega) \qquad (5.3.1)$$

where $\Phi_F(\omega)$ denotes the spectrum of the plant forcing sequence viewed in continuous time. The pulse spectrum of this sequence is obtained from equations (5.1.10) and (5.1.12) as:

---

*Patently, a newly derived general result should always be assessed against a simple case.

$$\Phi_F^*(\omega) = \left| D^*(\omega) \Big/ \left[1 + HB^*(\omega)\right] \right|^2 \Phi_X^*(\omega) \tag{5.3.2}$$

By means of equation (5.2.15) in the form:

$$\Phi_F(\omega) = (1/T)\, \Phi_F^*(\omega) \tag{5.3.3}$$

the above results can be combined to yield:

$$\Phi_Y(\omega) = (1/T) \left| H(\omega) D^*(\omega) \Big/ \left[1 + HB^*(\omega)\right] \right|^2 \Phi_X^*(\omega) \tag{5.3.4}$$

The overall pulse transfer function of the system under consideration is derived from the analysis in Section 2.2 as:

$$K(z) = D(z)H(z) \Big/ \left[1 + D(z)HB(z)\right] \tag{5.3.5}$$

which enables the simplification of equation (5.3.4) to:

$$\Phi_Y(\omega) = (1/T) \left| K^*(\omega) H(\omega) / H^*(\omega) \right|^2 \Phi_X^*(\omega) \tag{5.3.6}$$

Using equation (4.5.35) to express the pulse spectrum Φ_X^* in terms of that for the unsampled signal results in:

$$\Phi_y(\omega) = \left| (1/T) K^*(\omega) H(\omega) / H^*(\omega) \right|^2 \sum_{-\infty}^{\infty} \Phi_X(\omega - n\omega_0) \tag{5.3.7}$$

where:

$$\omega_0 \triangleq 2\pi/T \tag{5.3.8}$$

According to equation (5.3.5):

$$K^*(\omega)/H^*(\omega) = D^*(\omega) \Big/ \left[1 + D^*(\omega)HB(\omega)\right] \tag{5.3.9}$$

where

$$\left.\begin{aligned} HB^*(\omega) &= \frac{1}{T}\sum_{-\infty}^{\infty} H(\omega - n\omega_0)B(\omega - n\omega_0) \\ &\neq H^*(\omega)B^*(\omega) \quad \text{in general} \end{aligned}\right\} \tag{5.3.10}$$

To achieve adequate closed loop stability, the Nyquist diagram must be made to collapse rapidly on to the origin of the D(z)HB(z)- plane for frequencies greater than the required bandwidth (ω_B) so that:

$$\left| D^*(\omega)HB^*(\omega) \right| << 1 \quad \text{for } \omega \gtrsim 2\omega_B \tag{5.3.11}$$

and therefore:

$$H(\omega)K^*(\omega)/H^*(\omega) \simeq H(\omega)D^*(\omega) \quad \text{for } \omega \gtrsim 2\omega_B \tag{5.3.12}$$

Within the required closed loop bandwidth, the control error is engineered to meet a performance specification by ensuring that:

$$\left| K^*(\omega) \right| \simeq 1 \quad \text{for } \omega \lesssim \omega_B \tag{5.3.13}$$

and therefore:

$$K^*(\omega)/H^*(\omega) \simeq 1/H^*(\omega) \quad \text{for } \omega \leq \omega_B \tag{5.3.14}$$

Qualitatively, the plant bandwidth must be much less than the sampling frequency to reduce sufficiently the output ripple, so that for present purposes equation (2.4.18) can be closely approximated by:

$$H^*(\omega) \simeq \frac{1}{T} H(\omega) \quad \text{for } \omega \lesssim 3\omega_B \tag{5.3.15}$$

It follows that:

$$H(\omega)K^*(\omega)/H^*(\omega) \simeq T \quad \text{for } \omega \lesssim \omega_B \tag{5.3.16}$$

and therefore the factor:

$$\left|(1/T)K^*(\omega)H(\omega)/H^*(\omega)\right|^2$$

in equation (5.3.7) is always bounded in practice; even if a plant contains integrators. The input spectrum (Φ_X) can also be taken as bounded, and due to inherent inertia or stray capacitance etc it converges asymptotically to zero at least as fast as $1/\omega^2$. Consequently, the infinite series of functions:

$$\left\{\left|(1/T)K^*(\omega)H(\omega)/H^*(\omega)\right|^2 \Phi_X(\omega - n\omega_o)\right\}$$

is normally summable, so that the terms can be summed in any order and their summation commutes with their integration. Such subsequent manipulations of equation (5.3.7) are therefore fully justified.

Equation (5.3.7) is now expanded as:

$$\Phi_y(\omega) = C(\omega)\left\{\Phi_X(\omega) + \sum_{1}^{\infty}\left[\Phi_X(\omega - n\omega_o) + \Phi_X(\omega + n\omega_o)\right]\right\} \tag{5.3.17}$$

with

$$C(\omega) = \left|(1/T)K^*(\omega)H(\omega)/H^*(\omega)\right|^2$$

Differences between the input spectrum and the baseband term above are seen to arise from intrinsic limitations in the dynamic performance (bandwidth) of plant items. They correspond in fact to the component ϕ_s of the total mean square control error in equation (5.1.3). The sideband terms centred around integer multiples of the angular sampling frequency (ω_o) represent the ripple component of the control error, whose mean square value is evidently:

$$\phi_r = (1/2\pi)\int_{-\infty}^{\infty}\sum_{1}^{\infty} C(\omega)\left[\Phi_X(\omega - n\omega_o) + \Phi_X(\omega + n\omega_o)\right]d\omega \tag{5.3.18}$$

Interchanging the order of integration and summation, and noting that the integrands are even functions of frequency yields:

$$\phi_r = (1/\pi) \sum_{1}^{\infty} \int_{-\infty}^{\infty} \left| (1/T) K^*(\omega) H(\omega) / H^*(\omega) \right|^2 \Phi_X(\omega - n\omega_0) d\omega \qquad (5.3.19)$$

Because pulse real frequency response functions are periodic with period ω_0, a further obvious change of variable gives:

$$\phi_r = (1/\pi) \sum_{1}^{\infty} \int_{-\infty}^{\infty} \left| (1/T) K^*(\omega) H(\omega + n\omega_0) / H^*(\omega) \right|^2 \Phi_X(\omega) d\omega \qquad (5.3.20)$$

In order to restrict irreversible distortion due to spectral aliasing, it is established in Section 1.3 that practical sampling frequencies (ω_0) must satisfy:

$$\omega_0 >> \text{Bandwidth of the Input Spectrum} \qquad (5.3.21)$$

Consequently, the above limits of integration are effectively very much less than the sampling frequency. Furthermore, in order to achieve the small ripple error component demanded in practice, equation (5.3.20) implies that:

$$\text{Bandwidth of } H(\omega) << \omega_0 \qquad (5.3.22)$$

Hence the pulse frequency response of a zero order hold and linearised plant model which is defined by equation (2.4.18) as:

$$H^*(\omega) = \frac{1}{T} \sum_{-\infty}^{\infty} H(\omega + n\omega_0) \qquad (5.3.23)$$

can be approximated in the context of equation (5.3.20) by:

$$H^*(\omega) \simeq \frac{1}{T} H_0(\omega) G(\omega) \qquad (5.3.24)$$

where from equation (2.4.22):

$$H_0(\omega) = \left[1 - \exp(-j\omega T)\right] / j\omega \qquad (5.3.25)$$

and

$G(\omega)$ - real frequency response of the linearised plant model

Likewise in the context of equation (5.3.20):

$$H(\omega + n\omega_0) \simeq H_0(\omega + n\omega_0) G_\infty / (jn\omega_0)^R \qquad (5.3.26)$$

where the asymptotic high frequency behaviour of the plant model is represented by:

$$G(\omega) \simeq G_\infty / (j\omega)^R \quad \text{for large enough } |\omega| \qquad (5.3.27)$$

and

$$|H_0(\omega + n\omega_0)| \simeq |\omega / n\omega_0| \, |H_0(\omega)| \qquad (5.3.28)$$

Substituting the above approximations into equation (5.3.20) produces the following expression for the output ripple power:

$$\phi_r = \pi^{-1} G_\infty^2 \, \omega_0{}^{-(2R+2)} S_R \int_{-\infty}^{\infty} |K^*(\omega)/G(\omega)|^2 \omega^2 \Phi_X(\omega) \, d\omega \qquad (5.3.29)$$

where:

$$S_R = \sum_{1}^{\infty} (1/n)^{2R+2} \qquad (5.3.30)$$

Limits of the above series as a function of the plant's rank (R) are tabulated in Table 5.3.1, from which the final result is derived as:

$$\phi_r = \pi^{-1} G_\infty^2 \, \omega_0{}^{-(2R+2)} \int_{-\infty}^{\infty} |K^*(\omega)/G(\omega)|^2 \omega^2 \Phi_X(\omega) \, d\omega$$

(5.3.31)

The practical significance of Table 5.3.1 is that the output ripple power is largely concentrated in the first sidebands of the input spectrum that are centred about the sampling frequency ($\pm\omega_0$).

R	1	2	3	4	5	6
S_R	1.0823	1.0173	1.0041	1.0010	1.0002	1.0000(6)

Table 5.3.1 Limits of the Summation in Equation (5.3.30)

More sophisticated sampled data extrapolation networks achieve greater high frequency attenuation than the zero order hold (or box car circuit) (28). In consequence*, they offer a potentially improved ripple performance to a DDC system. However, because they also introduce greater phase lags (28) which reduce the attainable closed loop bandwidth, they are seldom used in practice. Accordingly, output ripple power is best restricted by an appropriate choice of sampling frequency, and attention is now directed at the application of equation (5.3.31) in this respect.

5.4 AN INITIAL CHOICE OF SAMPLING FREQUENCY AND PRACTICAL CONSIDERATIONS

Adequate stability for the DDC system in Fig 5.1.2 is achieved by manipulating the roots of its characteristic equation:

$$1 + D(z)HB(z) = 0 \tag{5.4.1}$$

with the pulse transfer function of the compensator D(z). Nyquist Diagram techniques are recommended for this purpose for the reasons stated in Section 3.2. As shown explicitly by equation (5.3.10) and the Gain and Phase curves in Chapter 7, the real frequency responses associated with HB(z) and D(z) are functions

---

*the index of ω_0 in equation (5.3.31) would become 2R + 2 + 2M, where M denotes the order of the hold network

of the sampling frequency. Thus the compensation of a DDC system cannot begin until its sampling frequency is defined. On the other hand, because:

$$K^*(\omega) = D^*(\omega)H^*(\omega) \Big/ \left\{1 + D^*(\omega)HB^*(\omega)\right\} \tag{5.4.2}$$

the sampling frequency (ω_0) cannot be selected using equation (5.3.31) to achieve an acceptably small ripple power until the compensator is known. An iterative design procedure is therefore necessary to achieve a sampling frequency that is compatible with a particular control accuracy specification. A close initial estimate of the required sampling frequency is now derived on the basis of equation (5.3.31). Even though this recommended value might require iterative refinement, it is certainly a more soundly based (accurate) starting point than 'between five and ten times the bandwidth of the plant or input spectrum' (69, 70).

The proposed design algorithm involves the assumption that the input signal belongs to the ensemble of servo-frequency sinusoids:

$$\left\{x(t)\right\} = \left\{A \sin(\Omega t + \psi)\right\} \tag{5.4.3}$$

whose amplitude (A) is constant and whose phase angle (ψ) is uniformly distributed according to:

$$\left.\begin{aligned} P(\psi) &= 1/2\pi \quad \text{for } 0 \le \psi < 2\pi \\ &= 0 \quad \text{otherwise} \end{aligned}\right\} \tag{5.4.4}$$

Equation (5.2.18) specifies the power spectrum of this ensemble as:

$$\Phi_x(\omega) = (A^2\pi/2)\left[\delta(\omega - \Omega) + \delta(\omega + \Omega)\right] \tag{5.4.5}$$

Because the magnitudes of pulse and Laplace transfer functions are even functions of real frequency, equation (5.3.31) reduces for the above input spectrum to:

$$\phi_r = A^2 G_\infty^2 \omega_0^{-(2R+2)} \Omega^2 \left| K^* (\Omega)/G(\Omega) \right|^2 \qquad (5.4.6)$$

Over the closed loop bandwidth the gain function $|G(\omega)|$ usually decreases in practice*, so that contributions to the output ripple power increase with the frequency of a spectral component. On this basis, it is suggested now that an initial choice of sampling frequency should make the output ripple power consistent with the control accuracy specification for an input sinusoid having the maximum amplitude** at the frequency of the required closed loop bandwidth (ω_B). Quantified by means of equation (5.4.6), the proposed design algorithm reads as:

$$\text{rms ripple specification} \geq A_{max} G_\infty \omega_0^{-(R+1)} \omega_B \ \ 0.707/\left|G(\omega_B)\right| \qquad (5.4.7)$$

If the output ripple is not specified, the assumption of an rms value equal to about 10% of the quoted steady-state error provides the opportunity to engineer a DDC system with an accuracy matching that of its continuous data counterpart.

At first sight equation (5.4.7) appears to dictate a conservative design, but the cut-off rate of the particular input spectrum is now shown markedly to affect the situation. Consider a unity feedback system with a zero order hold and the linearised plant model:

$$G(s) = \frac{1}{s^2(1+s)} \qquad (5.4.8)$$

When this servomechanism was under development in 1957, no

---

*resonances due to load inertias and drive shaft resilience usually lie outside the required control bandwidth

**a personal preference is to calculate all errors in terms of the computer range [-1, 1], and then afterwards scale to the particular plant units (eg m, °C, Pa etc)

analytical relationship between output ripple power and sampling rate was then available. With accuracy and wide bandwidth in mind, its sampling period of 0.1 second was decided largely by the speed of the fastest then available electronic hardware. As shown later in Chapter 7, the digital controller:

$$D(z) = \frac{10(z - 0.975)(z - 0.9)}{z(z - 0.8 - j0.3)(z - 0.8 + j0.3)} \tag{5.4.9}$$

achieves a reasonably damped closed loop frequency response ($M_p = 1.3$) with a bandwidth of 1.6 rad/s. Corresponding results for the output ripple power with the normalised input spectra:

$$\Phi_1(\omega) = \frac{2.7}{\omega^2 + (2\pi/5)} \quad ; \quad \phi_1(0) = 1 \tag{5.4.10}$$

or

$$\Phi_2(\omega) = \frac{6.35 \times 10^{-7}}{\{\omega^2 + (2\pi/100)^2\}^2} \quad ; \quad \phi_2(0) = 1 \tag{5.4.11}$$

are derived from equation (5.3.31) by numerical integration, and these are compared in Table 5.4.1 with the pertinent evaluation of equation (5.4.7):

$$\phi_r = \left(\sqrt{2} \times 1 \times (62.8)^{-4} \times 1.6 \times 0.707/0.13\right)^2 \tag{5.4.12}$$

As expected from the discussion of equation (5.4.6) the output ripple power for the input spectrum Φ_1, is much larger than that for the spectrum Φ_2, whose smaller bandwidth and faster cut-off rate produces smaller higher frequency components. These calculations also demonstrate that the design algorithm (5.4.7) is now always unequivocally conservative. However, blended with experience, it is now shown to provide an initial estimate of the sampling frequency that could well avoid the necessity for iterative refinement.

The dramatic effect of spectral cut-off rate on the reconstruction error of sampled data has been previously investigated by Wiener optimum filter calculations (117). When

Spectrum	Spectral Breakfrequency (rad/s)	Ripple Power for a Spectral Input	Ripple Power for a Sinusoidal Input
Φ_1	1.26	2.7×10^{-12}	6.0×10^{-13}
Φ_2	0.063	4.6×10^{-17}	6.0×10^{-13}

Table 5.4.1 Ripple Powers for Stochastic and Sinusoidal Inputs

the asymptotic behaviour of the continuous data spectrum is as ω^{-2}, it has been shown that the ratio of sampling frequency (ω_0) to spectral bandwidth (ω_s) must be:

$$\omega_0/\omega_s \simeq 10^4 \qquad (5.4.13)$$

for an rms reconstruction error of 1%. On the other hand, if the asymptotic behaviour of the continuous data spectrum is as ω^{-4}, then the same 1% rms reconstruction error is achieved quite remarkably with the *bandwidth ratio*:

$$\omega_0/\omega_s \simeq 15 \qquad (5.4.14)$$

The high sampling rates implied by equation (5.4.13) appear to militate against a successful implementation of DDC. Indeed, it seems surprising on this basis that some patently successful practical systems can adequately attenuate their sampling sidebands. The paradox is resolved by noting that a first order model is generally judged accurate enough just for system compensation. Input signal sources generally possess, in addition to a possibly dominant pole, higher frequency modes which induce faster spectral cut-off rates than 20 dB per decade. Consequently, relatively quite modest sampling rates like in equation (5.4.14) are adequate in practice. Furthermore, the speed and integration of present day electronics can often offer higher than the minimum necessary sampling frequency at no extra cost. Because ripple power decreases extremely rapidly with sampling frequency as ω_0^{-2R-2}, there is a real opportunity of ensuring at the first attempt a satisfactory ripple performance by the imposition of an acceptable margin on the estimate derived from equation (5.4.7). For instance, if the sampling period of the above example is decreased by 10%, the ripple power evaluated from equation (5.3.31) or (5.4.7) reduces by more than 50%.

Wiener optimum filter theory (86, 111) appears at first sight to offer an alternative approach for selecting the sampling frequency of a DDC system. The problem can be construed as the practical realisation of a linear system model $H_{opt}(s)$ that would operate on a sampled signal to produce a best least-squares approximation to some function of the original continuous data. For any rational input spectrum, the theory provides the optimum transfer function, whose mean square reconstruction error is a function of the sampling frequency. If an actual DDC system could be compensated so as to achieve the relevant optimum transfer function $H_{opt}(s)$, the system design would evidently be non-iterative. Unfortunately, it is generally inappropriate to create the optimum overall operator for a given plant. First of all, the z- or s-plane cancellation of plant poles or zeros may infringe constraints (eg acceleration, temperature rates etc) that cannot be included in its linear model. Secondly, such pole or zero cancellations may conflict with system stability* or stiffness** requirements. Finally, according to equation (2.4.17) a pulse Laplace transformation is periodic with period $j2\pi/T$. Consequently, a discrete compensator which provides a

---

*Compensator poles and zeros are realisable to an accuracy determined by the available wordlength or component tolerances, so perfect cancellations are impossible. An attempted cancellation outside the unit circle would therefore compromise overall stability. Refer to pages 156-157 of reference 28

**Control systems for large radar antennas,gun mountings etc must counter load disturbances induced by wind, waves or terrain changes as well as faithfully track their input signal. This so-called load disturbance rejection requirement is promoted by controllers having a large zero-frequency gain or 'stiffness'

required pole or zero at say s = α also introduces additional poles or zeros at $\alpha \pm j2\pi n/T$. These additional poles or zeros are almost certainly not those of the optimum filter, which must necessarily be low pass relative to the sampling frequency. In conclusion, the choice of sampling frequency for a DDC system is in principle iterative, but there is a real opportunity to avoid iterations by imposing a modest design margin on the initial estimate derived from equation (5.4.7).

5.5 EXPERIMENTAL VALIDATION AND PULSED RC NETWORKS

An experimental validation of the preceding ripple power analysis was completed by SM Patel (118). Judged in terms of presently available signal sources and spectral analysers, the measurements now appear to be quite primitive. It evolved that the ripple power for the system described by equations (5.4.8) and (5.4.9) was undetectable. Accordingly, the investigation addressed a contrived unity feedback system with zero order hold and plant:

$$G(s) = \frac{1}{s(1 + 0.05s)} \qquad (5.5.1)$$

With a sampling period of 0.04 second, the simple compensator:

$$D(z) = 3.1 \qquad (5.5.2)$$

produces a reasonably damped closed loop response with a bandwidth of around 4 rad/s. The above Laplace transfer function was simulated on an analogue computer, and because digital controllers as integrated electronic circuits were not available then, the compensator was realised as a *pulsed RC network*. These networks which can synthesise an arbitrary pulse transfer function are described for completeness at the end of this section, though they now seem unlikely to find many applications. Steady-state measurements of the ripple power for 25 V amplitude sinusoidal inputs with frequencies in the range 0.2 to 0.9 Hz were made using a tunable bandpass filter and a true rms voltmeter. The experiments confirmed the implication in Table 5.3.1 that the ripple power is very largely concentrated in the first sidebands centred about the sampling frequency ($\pm \omega_o$). Table 5.5.1 compares the measurements against values calculated from equation (5.4.6), which reads for this purpose as:

$$\phi_r = \left[25 \times 20 \times (2\pi/0.04)^{-3} \times \Omega \left|K^*(\Omega)/G(\Omega)\right|\right]^2 \qquad (5.5.3)$$

f (Hz)	0.20	0.25	0.30	0.35	0.40	0.50	0.70	0.80	0.90
$\|G(f)\|$	0.795	0.635	0.528	0.452	0.392	0.314	0.222	0.193	0.170
$\|K^*(f)\|$	0.955	0.933	0.908	0.878	0.847	0.787	0.664	0.609	0.588
Calc. ϕ_r (mV^2)	0.038	0.089	0.18	0.30	0.49	1.03	2.08	4.20	5.73
Exptl. ϕ_r (mV^2)	0.044	0.081	0.18	0.25	0.37	0.66	1.30	1.75	2.15

Table 5.5.1 Calculated and Experimental Values of the Steady State Ripple Power (ϕ_r)

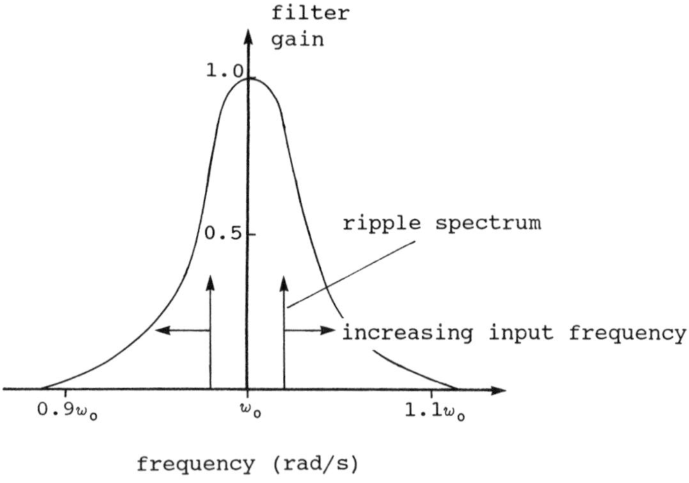

Fig 5.5.1 Gain Characteristic of the Spectrum Analyser

Although the agreement between theory and experiment is seen to deteriorate progressively with increasing input frequency, errors can be reconciled in terms of the gain characteristic of the bandpass filter shown in Fig 5.5.1. Readers may find their own derivation of the values in Table 5.5.1 a useful tutorial exercise.

The general structure of a pulsed RC network (28) in shown in Fig 5.5.2 in which the Laplace transfer functions P(s) and Q(s) have only real poles on the negative real axis. Consequently, they can always by synthesised as a passive lumped RC network (65, 66), but as digital controllers are often quite simple, they can be more conveniently implemented using operational amplifiers with parallel RC feedback. Either way, their transfer function can be expanded in the partial fraction form:

$$P(s) \text{ or } Q(s) = \sum_{i=1}^{N} \frac{c_i}{1 + s\tau_i} \tag{5.5.4}$$

where:

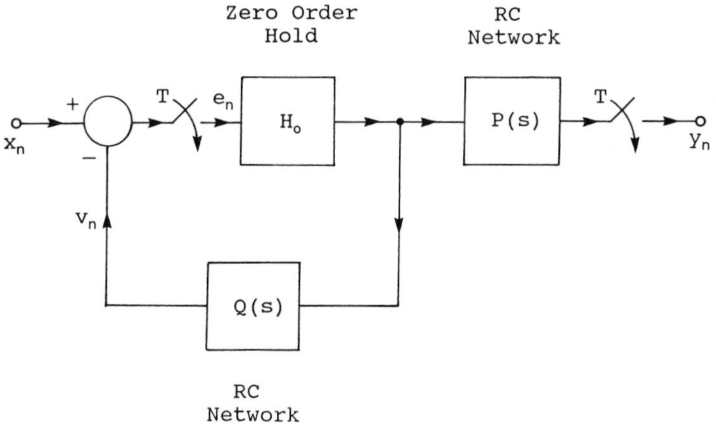

Fig 5.5.2 The General Structure of a Pulsed RC Network

c_i - arbitrary scalar gain τ_i - arbitrary RC time constant

In terms of the notation in Fig 5.5.2, the error sequence $\{e_n\}$ into the zero order hold is:

$$E(z) = X(z) - V(z) \tag{5.5.5}$$

with:

$$V(z) = H_oQ(z)\ E(z) \tag{5.5.6}$$

Straightforward manipulation of these equations yields:

$$E(z) = \frac{X(z)}{1 + H_oQ(z)} \tag{5.5.7}$$

and as:

$$Y(z) = H_oP(z)\ E(z) \tag{5.5.8}$$

the overall pulse transfer function of the system is derived as:

$$N(z) = \frac{(1 - z^{-1})Z[P(s)/s]}{1 + (1 - z^{-1})Z[Q(s)/s]} \tag{5.5.9}$$

where Z denotes the z- Transformation operation specified by equation (2.4.12). It is convenient to factor the above pulse transfer function as the quotient:

$$N(z) = N_s(z)/N_f(z) \tag{5.5.10}$$

where from equation (5.5.4):

$$\left.\begin{aligned} N_s(z) &= (1 - z^{-1})Z[P(s)/s] = (1 - z^{-1}) \sum_{i=o}^{N} \frac{c_i z}{z - \alpha_i} \\ N_f(z) &= 1 + (1 - z^{-1})Z[Q(s)/s] = 1 + (1 - z^{-1}) \sum_{i=o}^{N} \frac{c_i' z}{z - \alpha_i'} \end{aligned}\right\} \tag{5.5.11}$$

in which:

$$\alpha_o = \alpha'_o = 1 \qquad (5.5.12)$$

accounts for the poles at the origin of the s-plane. Observe that the poles of $N_s(z)$ and $N_f(z)$ are those of an RC network mapped into the z- plane, so that they must lie on the real line segment $0 < x < 1$. However, their zeros are determined by the arbitrary constants $\{c_i\}$ and $\{c'_i\}$, so that they too are completely arbitrary. It follows from equation (5.5.10) that the pulsed RC network can realise a pulse transfer function with arbitrary poles and zeros.

As an illustrative example consider the synthesis of the controller:

$$N(z) = \frac{z - 0.975}{z(z - 0.8)} \qquad (5.5.13)$$

by a pulsed RC network. According to equation (5.5.10), the pulse transfer function to be realised is first factored into the form:

$$N(z) = N_s(z)/N_f(z) \qquad (5.5.14)$$

with:

$$N_s(z) = \frac{1}{z - 0.8} \quad ; \quad N_f(z) = \frac{z}{z - 0.975} \qquad (5.5.15)$$

Note the pole of $N(z)$ at $z = 0$ is realised as a zero of $N_f(z)$ because it lies outside the real segment $0 < x < 1$ that contains the mapped poles of an RC network. Defining for ease of nomenclature:

$$a = 0.8 \quad \text{and} \quad b = 0.975 \qquad (5.5.16)$$

the required Laplace transfer functions $P(s)$ and $Q(s)$ are derived from equation (5.5.11) as:

$$Z\,[P(s)/s] = \frac{z}{(z-1)(z-a)} = \frac{1}{1-a}\left[\frac{z}{z-1} - \frac{z}{z-b}\right] \tag{5.5.17}$$

and:

$$Z\,[Q(s)/s] = \frac{z}{(z-1)}\left[\frac{z}{z-b} - 1\right] = \frac{b}{1-b}\left[\frac{z}{z-1} - \frac{z}{z-b}\right] \tag{5.5.18}$$

Published tables (28, 37) of z- transformations enable the above equations to be written as:

$$P(s)/s = \frac{1}{(1-a)}\left[\frac{1}{s} - \frac{1}{s+\alpha}\right] = \frac{1}{1-a}\left[\frac{\alpha}{s(s+\alpha)}\right] \tag{5.5.19}$$

$$Q(s)/s = \frac{b}{(1-b)}\left[\frac{1}{s} - \frac{1}{s+\beta}\right] = \frac{b}{1-b}\left[\frac{\beta}{s(s+\beta)}\right] \tag{5.5.20}$$

with

$$a = \exp(-\alpha T) \quad , \quad b = \exp(-\beta T) \tag{5.5.21}$$

Hence the required Laplace Transfer functions are:

$$P(s) = \frac{1}{1-a}\left[\frac{\alpha}{s+\alpha}\right] \tag{5.5.22}$$

and:

$$Q(s) = \frac{b}{1-b}\left[\frac{\beta}{s+\beta}\right] \tag{5.5.23}$$

each of which can be synthesised by an operational amplifier with parallel RC feedback.

To summarise Chapter 5: the output ripple power of a DDC system depends on its sampling frequency and digital compensator, whose dynamics in turn depend on the choice of sampling frequency. A direct calculation of the sampling frequency so as to satisfy precisely a control accuracy specification is therefore impossible, and an iterative design procedure is generally required. However, there is a real opportunity to avoid repetitive design by imposing a modest margin on the initial estimate of the sampling frequency derived from equation (5.4.7). Where on occasion this value does require iterative refinement, it is certainly a sounder (more accurate) starting point than some earlier qualitative recommendations (69, 70). Multiplicative rounding error noise is shown next in Chapter 6 to be a function of a controller's transfer function as well as the available wordlength (8 or 16 or 32 bits). In this respect, it is initially surprising to find later in Chapter 7 that rounding error noise can become unacceptably large if the sampling frequency is inordinately high. Similar to other engineering situations DDC system design involves satisfying constraints and compromises. This book attempts to provide the sound mathematical framework and physical insight by which the necessary experience for such judgements is acquired through practice.

CHAPTER 6

Finite Wordlength Effects

...... *pure reasoning can enable man to gain insight.* - *M Planck*

6.1 INTRODUCING THE STATISTICAL APPROACH

Sections 1.4 and 1.5 describe the elements of binary arithmetic in fixed point digital computers, and the conversion of continuous data into digital format. By virtue of the finite wordlength of these devices, errors are introduced into computer calculations. Data conversion errors are incurred when continuous data are read by A-D converters, and roundoff errors are generated when double length arithmetic products are automatically truncated to the single length working format. It should be recalled that the processes of arithmetic addition and subtraction do not cause further errors. Subsequent analyses demonstrate that data conversion and roundoff errors can be considered as originating from noise sources in an otherwise ideal calculation with a computer system of arbitrary long wordlength. A third, previously unmentioned, finite wordlength effect is the degeneration in real frequency response behaviour caused by errors in the coefficients of a pulse transfer function when quantised into the available format. Digital filters that are required to exhibit acute changes in their real frequency response characteristics are particularly vulnerable. However, compensators for DDC systems are generally* undemanding in this respect so that coefficient

---

* 'Notch filters' used to combat drive shaft resilience and load inertia are perhaps the exception. However, their simple structure permits the elementary ad hoc analysis of wordlength requirement in Chapter 7

quantisation is usually unimportant; especially as generous stability margins are deployed. Nevertheless, all the above finite wordlength effects are examined here in Chapter 6 in order to emphasise again the strong conceptual similarities between discrete and continuous data systems. In the present context, the quantum of one least significant bit(q) mimics the quantum of electronic charge in creating a 'noisy' output, and the quantisation of coefficients mimics the tolerance on R-L-C components in degenerating a real frequency response.

Because a double length variable is for practical purposes errorless compared to its single length truncation, all the above types of wordlength error can be considered conceptually as

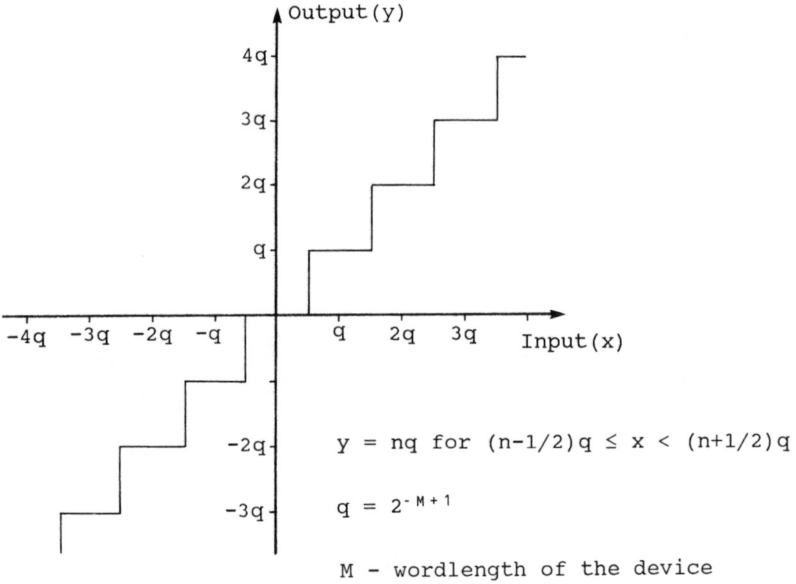

Fig 6.1.1 Uniform Rounded Quantiser

resulting from the processing of ideal (errorless) data by the stepped gain characteristic in Fig 6.1.1. With a priori knowledge of its input, the output of this *uniform rounded quantiser* (119) to a specific input is evidently without any statistical flavour. On the other hand, when viewed in terms of the output data, an output value of nq is produced by any input in the half open interval [(n-1/2)q ; (n+1/2)q[. It is this uncertainty that motivates the statistical analysis of amplitude quantisation in Sections 6.2 and 6.3. The statistical formulae in Chapter 4 are then deployed in Sections 6.4 to 6.5 to quantify the influence of these 'source terms' on the loss of control accuracy and the degeneration in the real frequency response of a digital filter. These analyses invoke the following approximations to the statistics of rounded quantisation errors (ϵ). Firstly, their ensemble mean is zero:

$$\overline{\epsilon} = 0 \tag{6.1.1}$$

and their auto correlation sequence behaves as:

$$\phi_E(k) = (q^2/12)\delta_{k0} \qquad \text{for all integer } \pm k \tag{6.1.2}$$

Thirdly, rounded quantisation errors are uncorrelated with their corresponding input (x):

$$\overline{x\epsilon} = 0 \tag{6.1.3}$$

and finally, the roundoff errors $\{\epsilon_1\}$ and $\{\epsilon_2\}$ from two different multiplications in the realisation of a pulse transfer function are uncorrelated:

$$\overline{\epsilon_1\epsilon_2} = 0 \tag{6.1.4}$$

It should be noted that an input whose probability density function consists of just a few δ-functions would not possess the uncertainty required to justify a statistical analysis of its amplitude quantisation errors. For example, in the case of the 'random' square wave with probability density function:

$$P(x) = 0.5 \left[\delta(x) + \delta(x - nq + q/2)\right] \tag{6.1.5}$$

the cross correlation:

$$\overline{x\epsilon} = (n - 0.5)q^2/4$$

increases with the signal amplitude $(n-\frac{1}{2})q$ in contradiction of equation (6.1.4). The validity of equations (6.1.1) to (6.1.3) for less clustered prequantised variables is investigated using *Characteristic functions* (87, 120), which for the N-dimensional variate:

$$x^T = (x_1, x_2, \ldots x_N) \tag{6.1.6}$$

exist* as:

$$F_X(\mathbf{u}) = \int_{-\infty}^{\infty}\cdots\int_{-\infty}^{\infty} \exp(-j\mathbf{u}^T\mathbf{x})\; P_X(\mathbf{x})\; dx_1 dx_2 \ldots dx_N \tag{6.1.7}$$

where the superscript ^{T} denotes the transpose of its associated vector and:

$$\mathbf{u}^T = (u_1, u_2 \ldots u_N) \tag{6.1.8}$$

Granted the existence of the two-dimensional Fourier integral for example:

$$\iint_{-\infty}^{\infty} \frac{\partial^{p+q}}{\partial u_1^p \partial u_2^q} \exp(-ju_1x_1 - ju_2x_2) . P_X(x_1, x_2) dx_1 dx_2$$

then by Leibnitz's Theorem (10) the above order of integration and differentiation can be interchanged to yield the moments of the probability density function as:

---

*Probability density functions are generally Fourier transformable - see page 91 of reference 87

$$\overline{x_1^p \; x_2^q} = (-j)^{-p-q} \left\{ \frac{\partial^{p+q}}{\partial u_1^p \partial u_2^q} F_X(u_1, u_2) \right\} \Bigg|_{u_1 = u_2 = 0} \tag{6.1.9}$$

Definitive texts (87, 120) employ the complex conjugate of equation (6.1.6) as the Characteristic function of a vector variate, but that given here is consistent with the previous definition of a Fourier integral in equation (1.2.21), and otherwise the change is of no consequence.

6.2 THE IDEALISED STATISTICS OF AMPLITUDE QUANTISATION ERRORS

The difference between the output and the input of the device shown in Fig 6.1.1 constitutes the error (ϵ) incurred in a rounded quantisation operation. Explicitly, for the given input (x):

$$\epsilon = nq - x \tag{6.2.1}$$

where the positive or negative integer n is specified by:

$$(n - 1/2)q \le x < (n + 1/2)q \tag{6.2.2}$$

An alternative viewpoint is that equation (6.2.1) defines for each positive or negative integer n the prequantised input (x) to produce a particular error (ϵ). Thus taking into account all values of the input that yield an error in interval $\epsilon' < \epsilon \le \epsilon' + \Delta\epsilon$ gives:

$$\text{Probability} \left\{ \epsilon' < \epsilon \le \epsilon' + \Delta\epsilon \right\}$$

$$= \sum_n \text{Probability} \left\{ - \epsilon' - \Delta\epsilon + nq \le x < nq - \epsilon' \right\} \tag{6.2.3}$$

where:

$$-2^{M-1} \le n \le 2^{M-1} - 1 \tag{6.2.4}$$

and

M - wordlength of the device (A-D converter or single length register)

Because appropriate variable scaling generally ensures that overflow is a statistically rare event, the above summation can be conveniently extended to doubly infinite limits. Expressed in terms of probability density functions equation (6.2.3) then becomes:

$$\left.\begin{aligned} P_E(\epsilon) &= \sum_{-\infty}^{\infty} P_X(nq - \epsilon) \qquad \text{for } -q/2 < \epsilon \leq q/2 \\ &= 0 \qquad \text{otherwise} \end{aligned}\right\} \qquad (6.2.5)$$

It is now assumed that the probability density function of the prequantised variable possesses a meromorphic two-sided Laplace transformation which converges asymptotically no slower than s^{-2}, and which has no poles in an infinite strip centred about the imaginary axis of the s- plane. Under these conditions it is shown in Appendix A6.2* that the Characteristic function of the rounded quantisation errors is:

$$F_E(u) = \sum_{-\infty}^{\infty} F_X(2\pi k/q) . S(u + 2\pi k/q) \qquad (6.2.6)$$

where:

$$S(w) = \frac{\sin(wq/2)}{wq/2} \qquad (6.2.7)$$

and by analogy with time domain data sampling:

$2\pi/q$ - *Amplitude Quantisation Frequency*

---

*Essentially, this formally justifies interchanging the order of integration and summation for equation (6.2.5) before applying the convolution theorem (1-2-25)

If the joint probability density function for two distinct input variates to the rounded quantiser is $P_X(x_1, x_2)$, the joint probability density function for the corresponding errors ϵ_1 and ϵ_2 can be derived by an exactly similar argument to equation (6.2.5) as:

$$\left. \begin{aligned} P_E(\epsilon_1, \epsilon_2) &= \sum_{-\infty}^{\infty}\sum P_X(mq - \epsilon_1, nq - \epsilon_2) \quad \text{for} \quad -q/2 < \epsilon_1, \epsilon_2 \leq q/2 \\ &= 0 \quad \text{otherwise} \end{aligned} \right\} \tag{6.2.8}$$

The corresponding second-order Characteristic function is:

$$F_E(u_1, u_2) = \int_{-q/2}^{q/2}\int \left\{ \sum_{-\infty}^{\infty}\sum P_X(mq - \epsilon_1, nq - \epsilon_2) \right\} \exp(-ju_1\epsilon_1 - ju_2\epsilon_2)\, d\epsilon_1\, d\epsilon_2 \tag{6.2.9}$$

Provided the function $P_X(x_1, x_2)$ has a meromorphic second-order two-sided Laplace transformation, which converges asymptotically no slower than $s_1^{-2} s_2^{-2}$ and which has no poles in infinite strips centred about the imaginary axes of the s_1, s_2- planes, then a straightforward extension of Appendix A6.2 establishes that:

$$F_E(u_1, u_2) = \sum_{-\infty}^{\infty}\sum F_X(2\pi k/q, 2\pi p/q) . S(u_1 + 2\pi k/q) . S(u_2 + 2\pi p/q) \tag{6.2.10}$$

As described in Section 4.1, the second-order joint probability density function $P_X(x_1, x_2)$ for a wide sense stationary process depends on the temporal separation of the variates:

$$x_1 = x(t_1) \quad \text{and} \quad x_2 = x(t_2)$$

and not on the absolute values of t_1 and t_2. It follows from equation (6.2.8) that a wide sense stationary prequantised process generates a wide sense stationary error process. Further attention is now directed at wide sense stationary prequantised ensembles.

The joint probability density function for an input (x) and its rounded quantisation errors (ϵ) can be expressed as:

$$P_{XE}(x,\epsilon) = P_E(\epsilon|x)P_X(x) \tag{6.2.11}$$

where the conditional probability density function $P_E(\epsilon|x)$ has the meaning:

$$P_E(\epsilon'|x')\Delta\epsilon \triangleq \text{Probability } \epsilon' < \epsilon \leq \epsilon' + \Delta\epsilon \text{ given that already } x' < x \leq x' + \Delta x \tag{6.2.12}$$

By virtue of the Archimedean property of the real numbers (10, 29) there exists an integer (n_x) such that:

$$(n_x - 1/2)q \leq x < (n_x + 1/2)q \tag{6.2.13}$$

and because a priori knowledge of an input exactly specifies the corresponding quantisation error, it follows that:

$$P_E(\epsilon|x) = \delta(\epsilon - n_x q + x) \tag{6.2.14}$$

Hence the joint Characteristic function for the variates x and ϵ is given by:

$$F_{XE}(u,v) = \int_{-\infty}^{\infty}\int_{-q/2}^{q/2} \delta(\epsilon - n_x q + x)\ P_X(x)\ \exp(-jux - jv\epsilon)d\epsilon\ dx$$

or

$$F_{XE}(u,v) = \int_{-\infty}^{\infty} P_X(x)\ \exp(-jux - jvn_x q + jvx)dx \tag{6.2.15}$$

Because the integer n_x depends on the prequantised input variable (x) as defined by equation (6.2.13), the above integral evaluates as:

$$F_{X E}(u,v) = \sum_{-\infty}^{\infty} \int_{-q/2}^{q/2} P'_X(nq - x) \exp(-jvx)dx \qquad (6.2.16)$$

where:

$$P'_X(x) = P_X(x) \exp(-jux) \qquad (6.2.17)$$

A comparison of equation (6.2.16) with equation (A6.2.4) followed by an identical argument to that in Appendix A6.2 yields:

$$F_{X E}(u,v) = \sum_{-\infty}^{\infty} F'_X(2\pi k/q).S(v + 2\pi k/q) \qquad (6.2.18)$$

where:

$$F'_X(v) = \int_{-\infty}^{\infty} P'_X(x) \exp(-jvx)dx \qquad (6.2.19)$$

Substituting equation (6.2.17) into (6.2.19) gives:

$$F'_X(v) = F_X(u + v)$$

so that:

$$F_{X E}(u,v) = \sum_{-\infty}^{\infty} F_X(u + 2\pi k/q).S(v + 2\pi k/T) \qquad (6.2.20)$$

The required cross correlation $\overline{x\epsilon}$ is derived according to equation (6.1.9) as:

$$\overline{x\epsilon} = \sum_{-\infty \atop =0}^{\infty} (-1)^{k+1}.(q/2\pi k).dF_X(2\pi k/q)/du \qquad (6.2.21)$$

By means of the above Characteristic functions and cross correlation, the required error statistics can be established in several different ways. For example, reference (124) follows a formulation of classical Sheppard's Corrections (87) that are used to compensate for errors in statistics estimated from the grouping (amplitude quantisation) of recorded data. Here, it is considered more appropriate to continue to exploit data communication principles, and the Characteristic functions are first of all assumed to satisfy the bandlimited conditions:

$$\left.\begin{aligned} F_X(u) &= 0 && \text{for } |u| \geq 2\pi/q \\ F_X(u_1,u_2) &= 0 && \text{for } |u_1| \geq 2\pi/q \text{ and } |u_2| \geq 2\pi/q \end{aligned}\right\} \quad (6.2.22)$$

Subject to the above constraints equations (6.2.6), (6.2.10) and (6.2.21) reduce to:

$$\left.\begin{aligned} F_E(u) &= F_X(o)\ S(u) \\ F_E(u_1,u_2) &= F_X(o,o)\ S(u_1)\ S(u_2) \end{aligned}\right\} \quad (6.2.23)$$

$$\overline{x\epsilon} = 0 \quad (6.2.24)$$

Because:

$$\left.\begin{aligned} F_X(o) &= \int_{-\infty}^{\infty} P_X(x)\,dx = 1 \\ F_X(o,o) &= \int_{-\infty}^{\infty}\int P_X(x_1,x_2)\,dx_1\,dx_2 = 1 \end{aligned}\right\} \quad (6.2.25)$$

and:

$$\int_{-\infty}^{\infty} \exp(-ju\epsilon)\,d\epsilon = q\ S(u) \quad (6.2.26)$$

it follows that the probability density functions corresponding to equations (6.2.23) are:

$$\left.\begin{aligned} P_X(\epsilon) &= 1/q \qquad \text{for} \quad -q/2 < \epsilon \le q/2 \\ &= 0 \qquad \text{otherwise} \end{aligned}\right\} \tag{6.2.27}$$

and:

$$\left.\begin{aligned} P_X(\epsilon_1, \epsilon_2) &= 1/q^2 \qquad \text{for} \quad -q/2 < \epsilon_1, \epsilon_2 \le q/2 \\ &= 0 \qquad \text{otherwise} \end{aligned}\right\} \tag{6.2.28}$$

The ensemble mean and autocorrelation function of these uniform distributions are derived directly as:

$$\overline{\epsilon} = \int_{-q/2}^{q/2} (\epsilon/q)\,d\epsilon = 0 \quad ; \quad \overline{\epsilon^2} = \int_{-q/2}^{q/2} (\epsilon^2/q)\,d\epsilon = q^2/12$$

$$\overline{\epsilon_1 \epsilon_2} = \int\int_{-q/2}^{q/2} (\epsilon_1 \epsilon_2 / q2)\,d\epsilon_1\,d\epsilon_2) = 0 \qquad \text{for } \epsilon_1 \neq \epsilon_2$$

which exactly match the values in equations (6.1.1) to (6.1.3). Alternatively, these results can be derived by differentiation of the appropriate Characteristic functions but care is required in evaluating the indeterminate forms. In spite of the fact that perfectly bandlimited Characteristic functions appear to be quite exceptional (125), the utility of the above analysis is left largely* as an open question in the original publications (121, 122, 123). Accordingly the adequacy of these idealised error statistics as practical approximations is addressed next in Section 6.3.

6.3 ACTUAL STATISTICS OF AMPLITUDE QUANTISATION ERRORS

Here first of all the bandwidths of the Characteristic functions $F_X(u)$ and $F_X(u_1, u_2)$ are quantified, so that a comparison against

---

*They examine Normally distributed prequantised inputs which possess particularly tractable Characteristic functions (87, 120)

typical amplitude quantisation frequencies can indicate whether or not the idealised error statistics form reasonable practical approximations. As with electrical networks several intuitively appealing definitions of bandwidth exist, but for present purposes it is most appropriate to deploy a mathematically similar quantity to the radius of gyration of a rigid body (126). Thus the bandwidth (B_1) of a one-dimensional Characteristic function is defined as:

$$B_1^2 = \int_{-\infty}^{\infty} u^2 \left|F(u)\right|^2 du \Big/ \int_{-\infty}^{\infty} \left|F(u)\right|^2 du \tag{6.3.1}$$

and that of a two-dimensional function by:

$$B_2^2 = \iint_{-\infty}^{\infty} (u^2 + u^2) \left|F(u_1, u_2)\right|^2 du_1 du_2 \Big/ \iint_{-\infty}^{\infty} \left|F(u_1, u_2)\right|^2 du_1 du_2 \tag{6.3.2}$$

For any one-dimensional real function g(x) with Fourier transformation G(u), Parceval's theorem (1.2.23) reads as:

$$\int_{-\infty}^{\infty} g^2(x) dx = (1/2\pi) \int_{-\infty}^{\infty} \left|G(u)\right|^2 du \tag{6.3.3}$$

and the corresponding result for a two-dimensional real function $g(x_1, x_2)$ is:

$$\iint_{-\infty}^{\infty} g^2(x_1, x_2) dx_1 dx_2 = (1/2\pi)^2 \iint_{-\infty}^{\infty} \left|G(u_1, u_2)\right|^2 du_1 du_2 \tag{6.3.4}$$

Noting the Fourier transform pairs:

$$\left.\begin{array}{l} dg/dx \longleftrightarrow juG(u) \\ \partial g/\partial x_1 \longleftrightarrow ju_1 G(u_1, u_2) \quad ; \quad \partial g/\partial x_2 \longleftrightarrow ju_2 G(u_1, u_2) \end{array}\right\} \tag{6.3.5}$$

and utilising Parceval's theorem yields the following estimates of bandwidth:

$$B_1^2 = \left[\int_{-\infty}^{\infty} (dP/dx)^2\, dx\right] \cdot \left[\int_{-\infty}^{\infty} P^2\, dx\right]^{-1} \tag{6.3.6}$$

and:

$$B_2^2 = \left\{\iint_{-\infty}^{\infty} \left[(\partial P/\partial x)^2 + (\partial P/\partial x_2)^2\right] dx_1\, dx_2\right\} \cdot \left\{\iint_{-\infty}^{\infty} P^2\, dx_1\, dx_2\right\}^{-1} \tag{6.3.7}$$

where:

P - the probability density function corresponding to F

This translation of the bandwidths into the more tangible probability domain now enables a derivation of their approximate values. For instance, if samples from a wide sense stationary ensemble* have a standard deviation or 'spread' of σ, then:

$$(dP/dx)^2 \simeq \left[P(o)/\sigma\right]^2$$

so that:

$$B_1^2 \simeq 2\sigma\left[P(o)/\sigma\right]^2 \cdot \left[2\sigma\ P^2(o)\right]^{-1}$$

giving:

$$B_1 \simeq 1/\sigma \tag{6.3.8}$$

To estimate the second order bandwidth quantity (B_2) requires the more recondite approach in Appendix A6.3 which yields:

---

*Stochastic processes in linearised DDC systems can be assumed to have zero mean, because their response to this zero frequency component can be determined separately by means of the principle of linear superposition

$$B_2 \simeq (1/\sigma) \left[1 + (1 - \rho^2)^{-1}\right]^{1/2} \qquad (6.3.9)$$

where the correlation coefficient (ρ) is defined in equation (4.2.4) as:

$$\rho = \overline{X_1 X_2} / \sigma^2 \qquad (6.3.10)$$

Too frequent overflows outside the range $[-1, 1-q]$ of a fixed point machine are avoided by the scaling:

$$5\sigma \simeq 1 \qquad (6.3.11)$$

so that equations (6.3.8) and (6.3.9) simplify to:

$$B_1 \simeq 5 \qquad (6.3.12)$$

and

$$B_2 \simeq 5 \left[1 + (1 - \rho^2)^{-1}\right]^{1/2} \qquad (6.3.13)$$

Thus provided the statistics of a prequantised variate satisfy:

$$5 \ll 2\pi/q \qquad (6.3.14)$$

and:

$$5 \left[1 + (1 - \rho^2)^{-1}\right]^{1/2} \ll 2\pi/q \qquad (6.3.15)$$

for μ-processors with the usual wordlengths of:

$$M = 8, \ 12, \ 16, \ 32 \text{ bits} \qquad (6.3.16)$$

then the Characteristic functions $F_X(u)$ and $F_X(u_1, u_2)$ could be considered close enough to bandlimited for equations (6.1.1), (6.1.2), and (6.1.3) to approximate the statistics of actual rounded quantisation or multiplicative rounding errors. Table 6.3.1 establishes that typical amplitude quantisation frequencies $(2\pi/q)$ are very much greater than 5, so in practice it can be concluded that:

$$\overline{\epsilon} \simeq 0 \quad ; \overline{\epsilon^2} \simeq q^2/12 \qquad \text{and} \qquad \overline{x\epsilon} \simeq \overline{x}.\overline{\epsilon} = 0 \qquad (6.3.17)$$

On the other hand if the prequantised samples (x_1, x_2) are strongly correlated, as can easily be envisaged with high accuracy (sampling rate) DDC systems or filters, the right-hand side of equation (6.3.13) could become extremely large. Sidebands in the Characteristic function domain would then overlap to cause a broadening of the error correlation sequence, though its value at the origin $(\overline{\epsilon^2})$ would in practice remain around $q^2/12$. These changes would be reflected into the real frequency domain as an increase in low frequency intensity and a decrease in bandwidth of the error pulse power spectrum, which in consequence could no longer be considered constant (white). Further understanding is gained from the following examples, in which the prequantised wide sense stationary variate (x) consists of a signal (s) that is additively contaminated by an uncorrelated wide band noise process (n) to the extent:

$$(\text{ms value of noise}) \Big/ (\text{ms value of signal}) = \eta^2 << 1 \qquad (6.3.18)$$

Wordlength (M)	8	12	16	32
Amplitude Quantisation Frequency	8.0E2	1.3E4	2.1E5	1.3E10

Table 6.3.1 Amplitude Quantisation Frequency for typical μ- Processor Wordlengths

In the first example, the signal and noise ensemble are taken to be Markovian (89, 93) so that the correlation coefficient of the prequantised variate (x) can be written as:

$$\rho_X(\tau) = \left[\exp(-2\pi T_N) + \eta^2 \exp(-2\pi T_N \omega_n/\omega_s)\right] . \left[1 + \eta^2\right]^{-1} \qquad (6.3.19)$$

where:

ω_s and ω_n - signal and noise bandwidth respectively

and the normalised time variable (T_N) is defined according to:

$$\omega_s \left|\tau\right| = 2\pi T_N \qquad (6.3.20)$$

Because the sampling frequency of a DDC system* is necessarily much greater than the signal bandwidth for acceptably small data reconstruction errors, it follows that the temporal separations of neighbouring data samples satisfy:

$$2\pi T_N \quad << 1 \qquad (6.3.21)$$

so that equation (6.3.19) closely approximates to:

$$\rho_X(T_N) = \left(1 - 2\pi T_N + \eta^2 \exp(-2\pi T_N \omega_n/\omega_s)\right) \cdot \left(1 + \eta^2\right)^{-1} \qquad (6.3.22)$$

Straightforward manipulation of the above yields:

$$1 - \rho_X^2(T_N) = \left(4\pi T_N + 2\eta^2 - 2\eta^2 \exp(-2\pi T_N \omega_n/\omega_s)\right) \cdot \left(1 + \eta^2\right)^{-1} \qquad (6.3.23)$$

and if the ratio of noise to signal bandwidths is say typically 100:1, the above becomes:

$$1 - \rho_X^2(T_N) = \left(4\pi T_N + 2\eta^2 - 2\eta^2 \exp(-200\pi T_N)\right) \cdot \left(1 + \eta^2\right)^{-1} \qquad (6.3.24)$$

Substituting the above into equation (6.3.13) enables the bandwidth (B_2) of the corresponding second order Characteristic function $F_X(u_1, u_2)$ to be plotted in Fig 6.3.1 as a function of

---

*Some digital filters (differentiators) have bandwidths that approach 90% of their sampling frequency (67)

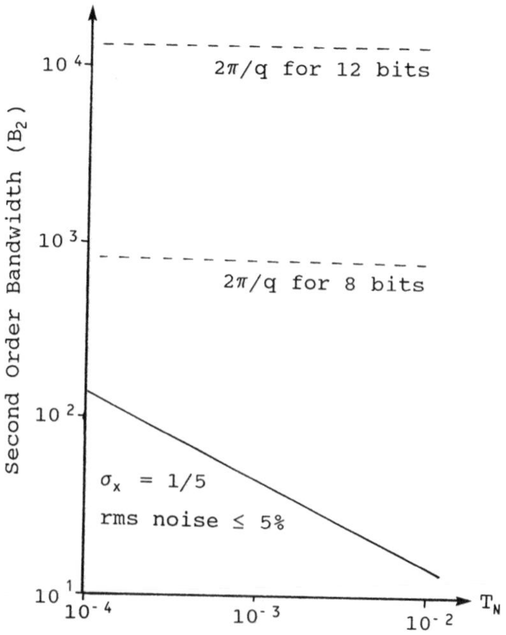

Fig 6.3.1 Second Order Bandwidth (B_2) against Normalised Time (T_N) for a Markovian signal contaminated with a similar noise process having 100-times the bandwidth.

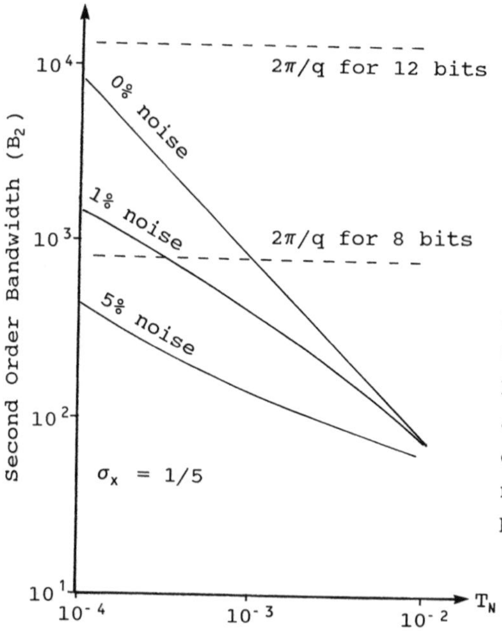

Fig 6.3.2 Second Order Bandwidth (B_2) against Normalised Time (T_N) for a Sinusoidal signal contaminated with Markovian noise having 100-times the bandwidth

normalised time (T_N) for rms noise amplitudes (η) of up to 5%. However, in this situation, the low level noise exercises no real influence. According to equation (4.5.29) the power spectrum of a wide sense stationary ensemble can be factored as:

$$\Phi_X(s) = \Psi(s) + \Psi(-s) \tag{6.3.25}$$

where $\Psi(s)$ has real or complex conjugate pairs of poles in the left-half s- plane only. Thus a partial fraction expansion of $\Psi(s)$ for a rational power spectrum involves terms of the form:

$$a/(s + b) \qquad \text{and} \qquad d/(s^2 + 2bs + c^2)$$

where the real numbers a, b, c and d are strictly positive. The former factor contributes a component exp $(-b|\tau|)$ to the even valued autocorrelation function, while the latter provides an oscillatory trigonometrical component (93) which is also damped by $\exp(-b|\tau|)$. From the discussion of equations (5.4.13) and (5.4.15), there are generally additional poles to the possibly dominant one(s) of an input spectrum to a DDC system. Their effect is to reduce the necessary* ratio of sampling frequency to signal bandwidth to less than around 100:1, so that normalised time values (T_N) in excess of 10^{-2} are almost certain to apply in practice to the above exponential factors. Fig 6.3.1 shows that typical amplitude quantisation frequencies are then at least some two decades larger than the second order bandwidth of the input data's Characteristic function. By invoking the analogous factorisation (4.5.4) of rational pulse power spectra, the same deduction can be made for the discrete data sequences in a μ-controller; particularly for phase lead compensators which extend their real frequency bandwidth. It can therefore be concluded that the Characteristic functions of all prequantised variates are essentially bandlimited with respect to the amplitude quantisation frequencies of typical μ- processors when the input spectrum is a rational function. Consequently, quantisation error

---

*to achieve adequately accurate data reconstruction

spectra can be reasonably approximated as constant (white) under these conditions.

Sinusoids with constant amplitude and frequency are frequent test inputs, but because their power spectra consist of δ-functions, a separate analysis to that above is required for the quantisation error statistics in DDC systems. If such a sinusoid:

$$s(t) = \sin(\omega_s t + \psi) \quad \text{with} \quad 0 \le \psi < 2\pi$$

is additively contaminated with a wide band Markovian noise process as in the first example, the correlation coefficient of the prequantised sum expressed in normalised time (T_N) reads as:

$$\rho_X(T_N) = \left(\cos 2\pi T_N + \eta^2 .\exp(-2\pi T_N \omega_n/\omega_s)\right).\left(1 + \eta^2\right)^{-1} \tag{6.3.26}$$

Sensibly accurate data reconstruction requires that neighbouring data samples have normalised temporal separations satisfying:

$$2\pi T_N << 1 \tag{6.3.27}$$

and similar manipulation as that leading to equation (6.3.23) gives:

$$1 - \rho_X^2(T_N) = \left(4\pi^2 T_N + 2\eta^2 - 2\eta^2 \exp(-200\pi T_N)\right).\left(1 + \eta^2\right)^{-1} \tag{6.3.28}$$

for a not untypical noise to signal bandwidth ratio of 100:1. Substituting the above into equation (6.3.13) enables the bandwidth (B_2) of the corresponding second order correlation function $F_X(u_1, u_2)$ to be plotted in Fig (6.3.2) as a function of normalised time for rms noise amplitudes (η) of up to 5%. Even with these low levels of added noise the second order bandwidth is seen to be substantially reduced in this case; thereby promoting the whiteness of its quantisation error spectrum. Nevertheless the second order bandwidths remain at least a decade larger than that for the Markovian signal in Fig 6.3.1. It can be concluded that if relatively noise-free sinusoidal testing of an 8 bit DDC system involves data samples separated in normalised time by less than 10^{-2}, then the whiteness of quantisation error spectra could not be wholly relied upon.

Developments in earlier Sections have proceeded fairly rigorously and they exemplify what might be described as 'mathematical analyses'. In the 'engineering analysis' of this Section, experienced judgement has been applied to formulate a broadly applicable approximation. A mathematical analysis is generally sufficient in itself, if its basic assumptions are correct, but an engineering analysis should always seek support from experiments or from consistency with rigorously analysed special cases. Experimental support for the proposed quantisation error statistics is mainly provided later by measurements of the degeneration in control accuracy or real frequency response characteristics. Here their consistency is investigated with respect to two strictly analysed special cases (127, 128).

When a Normally distributed variate (x) is subjected to sensibly fine quantisation, the corresponding error correlation function has been rigorously derived by Bennett (127) as:

$$\phi_E(\tau) = (q^2/2\pi) \sum_1^{\infty} (1/k^2) \exp\left[-4\pi k^2 \sigma_x^2 (1 - \rho_x)/q^2\right] \qquad (6.3.29)$$

Noting that:

$$\sum_1^{\infty} (1/k^2) = \pi^2/6$$

then equation (6.3.29) provides the anticipated result that:

$$\phi_E(0) = q^2/12 \qquad (6.3.30)$$

Granted the signal amplitude scaling in equation (6.3.11), which broadly prevents too frequent overflows, then the error correlation coefficient is derived from equation (6.3.29) and (6.3.30) as:

$$\rho_E(\tau) = (6/\pi^2) \sum_{1}^{\infty} (1/k^2) \exp\left[-0.04\pi k^2 (1 - \rho_X) 2^M\right] \qquad (6.3.31)$$

where M denotes the wordlength of the digital device. If the prequantised Normal variate is also Markovian (89, 93), then its correlation coefficient expressed in normalised time is:

$$\rho_X(T_N) = \exp(-2\pi T_N)$$

and subject to the sampling rate constraint in equation (6.3.21) the above reduces to:

$$\rho_X(T_N) \simeq 1 - 2\pi T_N \qquad (6.3.32)$$

Substituting this approximation into equation (6.3.31) yields:

$$\rho_E(T_N) \simeq (6/\pi^2) \sum_{1}^{\infty} (1/k^2) \exp\left[-0.79 k^2 T_N 2^M\right]$$

so that for digital devices with 8 or more bits:

$$\rho_E(T_N) \lesssim (6/\pi^2) \sum_{1}^{\infty} (1/k^2) \exp\left[-51773 k^2 T_N\right] \qquad (6.3.33)$$

Therefore, even if Normally distributed data samples are separated by as little as 10^{-4} in normalised time, their quantisation error spectrum is still white for all practical purposes. The previous engineering analysis is therefore fully corroborated by this rigorously analysed special case, and furthermore by direct computer simulations (129).

Next consider the situation for the maximum amplitude sinusoidal signal:

$$x(\Theta) = (1 - q/2)\sin(\Theta) \qquad (6.3.34)$$

Because its rounded quantisation error behaves as an odd function about π radius:

$$\int_{0}^{2\pi} \epsilon \, d\Theta = 0$$

and a similar ergodic argument to that deployed in deriving equation (5.2.33) from (5.2.23) proves that:

$$\overline{\epsilon} = 0 \qquad (6.3.35)$$

Likewise because the error is also an even function about $\pi/2$ radians, it follows that:

$$\overline{x\epsilon} = (2/\pi) \int_{0}^{\pi/2} \epsilon \sin(\Theta) d\Theta \qquad (6.3.36)$$

and

$$\overline{\epsilon^2} = (2/\pi) \int_{0}^{\pi/2} \epsilon^2 d\Theta \qquad (6.3.37)$$

For present purposes, these statistics are conveniently normalised with respect to the mean square values:

$$\sigma_E = q^2/12 \qquad \text{and} \qquad \sigma_X = \left(1 - q/2\right)^2/2$$

so that in accordance with equation (4.2.10) the corresponding values of the correlation coefficients are:

$$\rho_{XE}(0) = \overline{x\,\epsilon}/\sigma_X\sigma_E \qquad \text{and} \qquad \rho_E(0) = \overline{\epsilon^2}/\sigma_E$$

Equations (6.3.36) and (6.3.37) can be straightforwardly evaluated to yield:

$$\rho_{XE}(o) = \sqrt{12}\left[\sqrt{2}(2/\pi)\sum_{1}^{N}\cos(\Theta_n) - (\lambda/\sqrt{2})\right] \qquad (6.3.38)$$

and:

$$\rho_E(o) = 12\left[\lambda^2/2 - (4\lambda/\pi)\sum_{1}^{N}\cos(\Theta_n) + (2/\pi)\sum_{1}^{N}n^2(\Theta_{n+1} - \Theta_n)\right] \qquad (6.3.39)$$

where:

$$\left.\begin{array}{l} N = (1/q) - 1 \quad ; \quad \lambda = N + 1/2 \quad ; \quad \Theta_{N+1} = \pi/2 \\ \Theta_n = \sin^{-1}\left(\dfrac{2n - 1}{2\lambda}\right) \end{array}\right\} \qquad (6.3.40)$$

Computed values of the statistics $\rho_{XE}(o)$ and $\rho_E(o)$ for various wordlengths are specified in Table 6.3.2. Typical digital controllers and filters usually have wordlengths of 8 or more bits, so that an adequate approximation for the mean square quantisation error of a pure sinusoidal test input is seen to be

Wordlength (M)	3	4	5	8	12	16
$\rho_{XE}(o)$	-0.486E0	-0.333E0	-0.282E0	-0.812E-1	-0.203E-1	-0.507E-2
$\rho_E(o)$	1.195	1.134	1.093	1.033	1.008	1.002

Table 6.3.2 The Wordlength dependence of two Rounded Quantisation Error Statistics for a Pure Sinusoid of Maximum Amplitude

$q^2/12$. Under these same practical conditions:

$$\left|\rho_{XE}(0)\right| << 1$$

so that:

$$\overline{x(t)\ \epsilon(t)} = \overline{x(t)}.\overline{\epsilon(t)} = 0$$

These particular results for a pure sinusoid are clearly consistent with the general conclusions made earlier from a comparison of the first order bandwidth quantity B_1 with typical quantisation frequencies $(2\pi/q)$. Knowles and Edwards (128) have derived the autocorrelation function for these quantisation errors by expanding the second order amplitude probability density function for a pure sinusoid as a series of Chebyshev polynomials. Expressed in terms of normalised time, the error correlation coefficient reads as:

$$\rho_E(T_N) = 12\left[(\lambda^2/2 - \sqrt{2}\ \lambda C_1)\right] \cos 2\pi T_N + \sum_{\substack{1 \\ odd}}^{\infty} C_k^2 \cos 2\pi k T_N \qquad (6.3.41)$$

where

$$\left.\begin{aligned} C_k &= -(2\sqrt{2}/k\pi) \sum^{N} n \left[\sin(k\Theta_{n+\frac{1}{2}}) - \sin(k\Theta_{n-\frac{1}{2}})\right] \\ \Theta_{n\pm 1/2} &= \cos^{-1}\left(\frac{2n \pm 1}{2\lambda}\right) \end{aligned}\right\} \qquad (6.3.42)$$

and other variables are as defined in equation (6.3.40). Equation (6.3.41) has been validated by simulations representing quantisers of various wordlengths (128, 129). Its infinite series exhibits very slow convergence, which is aggravated by increasing wordlength, and over a thousand terms are required to compute the slowly decaying peak values in Fig 6.3.3. These results again substantiate the previous conclusion based on a comparison of the

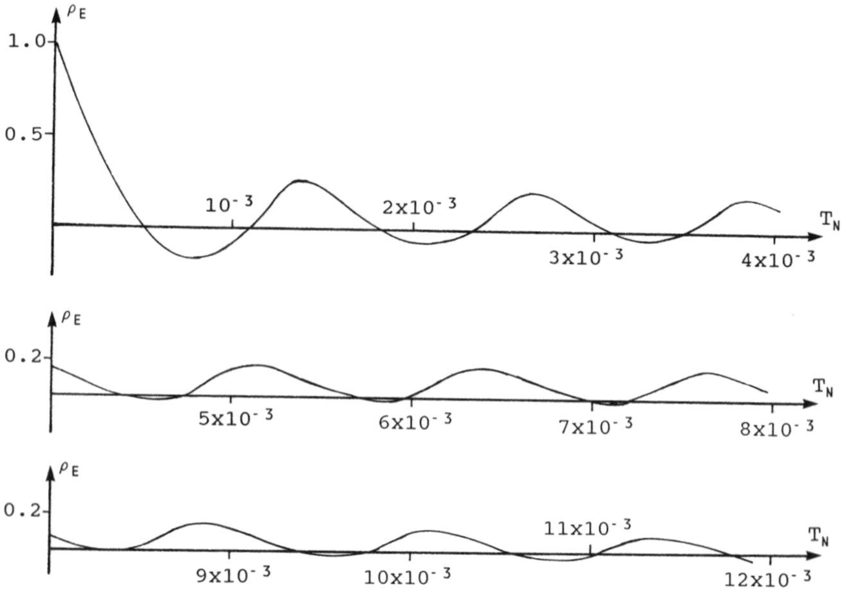

Fig 6.3.3 Error Correlation Coefficient for a Pure Sinusoid as a function of Normalised Time with an 8 bit Wordlength device

second order bandwidth quantity B_2 with the relevant amplitude quantisation frequency.

The realisation of a pulse transfer function involves subjecting series-parallel streams of data samples to the operations of: multiplication by a constant, quantisation of a double length product to single length format, addition or subtraction and then delaying for a complete sample period. Assertion (6.1.4) is that the roundoff errors from any two such different series or parallel multiplications are mutually uncorrelated. Although a mathematical analysis is apparently not currently available, the conjecture has been experimentally investigated (129) for the following types of sample sequences: Normal Markovian, constant amplitude and frequency sinusoids, and white noise with uniformly distributed amplitudes. Within the accuracy limits imposed by the sizes of the sample populations considered, the measurements were found to substantiate assertion (6.1.4) for typical sampling rates and widths of quantisation.

Because a DDC system must match the performance of a continuous data counterpart to be a viable engineering proposition, the loss of control accuracy due to its finite wordlength must be substantially less than the error set by plant dynamics. Accordingly, uncertainties in estimating the rms loss of steady state accuracy due to a finite wordlength controller could be as large as say ±20%. Subsequent simulations described later in Section 6.5 indicate that such predictive accuracy is easily achievable using analyses based on the idealised quantisation error statistics in equation (6.1.1) to (6.1.4); even with sinusoidal samples separated by less than 10^{-3} in normalised time. However, it has been shown here that actual quantisation error statistics are closely approximated in practice by these idealisations, so that adequate predictions are only to be expected.

6.4 AN UPPER-BOUND ON THE CONTROL ERROR DUE TO A FINITE WORDLENGTH

A useful upper-bound on the steady state control error component due to finite wordlength operations in a DDC system is derived below. This non-statistical analysis (130, 131) involves the triangular inequality (10, 29):

$$|C_1 + C_2| \leq |C_1| + |C_2| \quad \text{for all real numbers } C_1 \text{ and } C_2 \tag{6.4.1}$$

and the correspondence between pulse transfer functions and linear difference equations, which is described in Section 2.3.

Pessimism and analytical simplicity result from ignoring the statistical distribution of finite wordlength truncation errors,

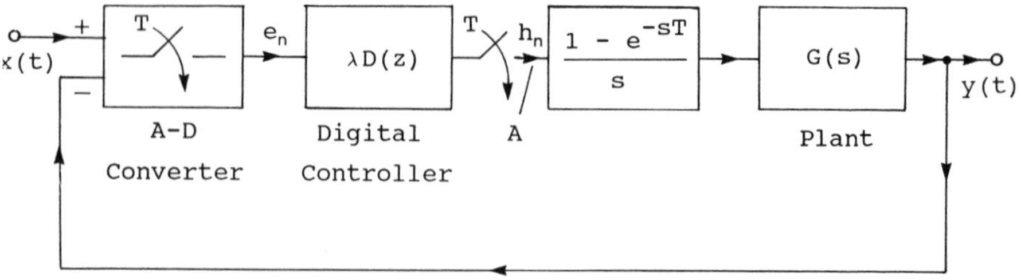

Fig 6.4.1 A Single Input Unity Feedback System

which on occasion can almost cancel. As well as enabling a quick 'back of envelope calculation' of approximate wordlength requirements, the analysis provides initial insight into optimising the pole-zero locations of digital controllers. To promote understanding, the experimental background leading to its development is first briefly described.

If the linearised unity feedback system in Fig 6.4.1 involves the plant transfer function:

$$G(s) = \frac{1}{s^2(1 + s)} \tag{6.4.2}$$

then theoretically it can be stabilised for a sampling period of 0.1 second by the digital controller (114):

$$D_1(z) = 4.8\,(z - 0.975)^2 / z(z - 0.8)^2 \tag{6.4.3}$$

However, simulations on a hybrid digital-analogue computer indicated complete instability (114). Finally, after breaking the feedback loop at the point A, the steady state output of the then isolated 9 bit controller was found* to vary disproportionately with the amplitude of various step function inputs. Indeed, the truncation of double length arithmetic products within the

---

*by the author's erstwhile colleague Dr V Latham at Ferranti Ltd, Wythenshawe, Manchester

controller produced zero output for quite significant inputs. Under these conditions the feedback loop could be reconnected at A without any corrective action being applied to the plant. Consequently, the loss of steady state control accuracy due to finite wordlength operations can be characterised by the largest constant input that produces zero steady state response from an isolated controller. As stated earlier in Section 5.1, it is sufficient to examine just steady state losses of control accuracy due to ripple and a finite wordlength because plant dynamics dominate the magnitude of transient control errors.

If the linearised DDC system in Fig 6.4.1 were to possess an arbitrary long wordlength, its control error sequence $\{e_n\}$ would be defined by:

$$e_n = x_n - y_n \tag{6.4.4}$$

However, the actual error sequence due to amplitude quantisation in A-D conversion is:

$$e'_n = x'_n - y'_n \tag{6.4.5}$$

where

$$x'_n = x_n + r_x(n) \quad ; \quad y'_n = y_n + r_y(n) \tag{6.4.6}$$

and

$$r_x,\ r_y \text{ - data quantisation errors} \tag{6.4.7}$$

Subtracting equations (6.4.4) and (6.4.5) yields:

$$e'_n = e_n + r_q(n) \tag{6.4.8}$$

where

$$r_q(n) = r_x(n) - r_y(n) \tag{6.4.9}$$

Assuming rounded quantisation throughout the system so as to reduce finite wordlengths errors, then:

$$\text{Max}\{r_q(n)\} \le q \tag{6.4.10}$$

The digital compensator is expected to realise a pulse transfer function of the general form:

$$\lambda D(z) = \lambda A(z) / B(z) \tag{6.4.11}$$

where

$$A(z) = \sum_{1}^{K} a_k z^{-k} \quad \text{and} \quad B(z) = \sum_{o}^{K} b_k z^{-k} \tag{6.4.12}$$

with

$$b_o = 1$$

According to Section 2.2, the coefficient a_o is usually zero in practical systems as a result of unavoidable computational delays. In terms of the nomenclature specified in Fig 6.4.1, the ideal realisation of the compensator by direct programming would correspond to the recursion formulae:

$$\left.\begin{aligned} f_n &= \lambda e'_n \\ h_n &= \sum_{1}^{K} a_k f_{n-k} - \sum_{1}^{K} b_k h_{n-k} \end{aligned}\right\} \tag{6.4.13}$$

However, due to multiplicative rounding errors, the sequences actually computed are:

$$f'_n = \lambda e'_n + r_\lambda(n) \qquad \text{with} \qquad |r_\lambda(n)| \le q/2 \tag{6.4.14}$$

$$h'_n = \sum_{1}^{K} a_k f'_{n-k} - \sum_{1}^{K} b_k h_{n-k} + r_D(n) \tag{6.4.15}$$

where r_λ and r_D are the error incurred in evaluating $\lambda e'_n$ and

$$\sum_{1}^{K} a_k f'_{n-k} - \sum_{1}^{K} b_k h'_{n-k} \tag{6.4.16}$$

If the above computation involves a total of μ multiplications excluding zero, unity and positive integral powers of 2, then at any time instant (nT):

$$\left| r_D(n) \right| \leq \mu q/2 \tag{6.4.17}$$

Substituting equations (6.4.8) and (6.4.14) into (6.4.15) gives:

$$h'_n = \sum_{1}^{K} a_k \left\{ \lambda e_{n-k} + \lambda r_q(n-k) + r_\lambda(n-k) \right\} - \sum_{1}^{K} b_k h'_{n-k} + r_D(n) \tag{6.4.18}$$

If a step function of amplitude ϵ is applied to the system input (x) with the control loop broken at A, the steady state response of the controller is given by:

$$h'_n = \sum_{1}^{K} a_k \left\{ \lambda \epsilon + \lambda r_q(n-k) + r_\lambda(n-k) \right\} - \sum_{1}^{K} b_k h'_{n-k} + r_D(n) \tag{6.4.19}$$

Because in cases of practical interest:

$$\epsilon >> q$$

equation (6.4.19) is closely approximated by:

$$h'_n = \sum_{1}^{K} a_k \lambda \epsilon - \sum_{1}^{K} b_k h'_{n-k} + r_D(n) \tag{6.4.20}$$

If the isolated controller is to apply no corrective control action to the plant, then necessarily:

$$\lambda \epsilon \sum_{1}^{K} a_k = - r_D(n) \qquad \text{for all } n \tag{6.4.21}$$

which by equations (6.4.12) and (6.4.17) implies:

$$\left| \lambda \epsilon A(1) \right| \leq \mu q/2$$

or

$$|\epsilon| \leq \mu \Big/ \left| 2^M \lambda A(1) \right| \tag{6.4.22}$$

Thus the condition of zero steady state output from the isolated controller necessitates a step amplitude less than $\mu/\left|2^M \lambda A(1)\right|$. Equivalently, a step input exceeding $\mu/\left|2^M \lambda A(1)\right|$ elicits corrective action from the controller. An upper bound ($\hat{\epsilon}$) on the steady state loss of control accuracy due to the finite wordlength (M) of a directly programmed digital controller is therefore taken as:

$$\hat{\epsilon} = \mu \Big/ \left| 2^M \lambda A(1) \right| \tag{6.4.23}$$

For the controller in equation (6.4.3), the above evaluates as:

$$\hat{\epsilon} = 1333/2^M \tag{6.4.24}$$

which is compared in Fig 6.4.2 on page 237 against results from a simulation of the feedback system with a 0.1 Hz sinusoidal input. Apart from confirming the inbuilt pessimism of the analysis, the diagram clearly explains the apparent instability of the system with a 9 bit controller. Wordlength requirements are next generally interpreted in terms of the zero locations of a digital controller.

Fig 3.3.2 illustrates a graphical construction for the real frequency response of a pulse transfer function based on its pole-

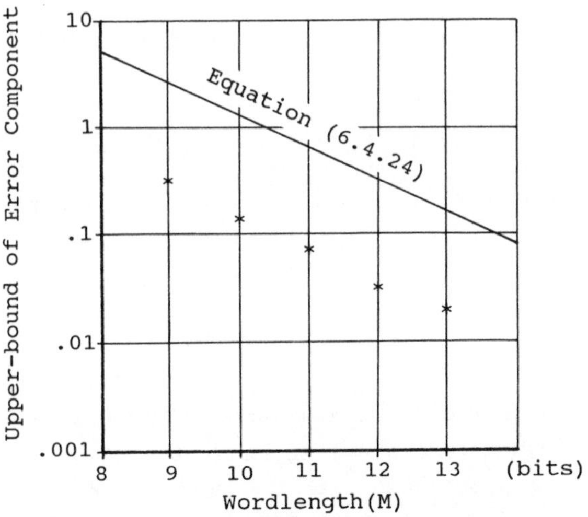

Fig 6.4.2 Predicted and Simulated Upper-bounds of the Finite Wordlength Control Error component as a function of Wordlength for the directly programmed $D_1(z)$ with a 0.1 Hz sinusoidal input

zero locations. More phase lead is seen to be created over the frequency range affecting closed loop stability as the real zeros of a digital compensator are moved closer to the point $1\underline{/0}$. In doing so, however, the magnitude of its numerator polynomial at this point:

$$A(1) = \prod_{1}^{K} (1 - \alpha_k) \qquad (6.4.25)$$

decreases, and according to equation (6.4.23) a consequential deterioration in control accuracy is incurred. Wordlength requirements are therefore eased if controller zeros can be positioned as far as practicable from the point $1\underline{/0}$. It is shown in Chapter 7 that such advantageous locations for compensator zeros can be engineered through the use of complex poles in a design (114). This is exemplified by the alternative controller:

$$D_2(z) = \frac{10(z - 0.975)(z - 0.9)}{z(z - 0.8 - j0.3)(z - 0.8 + j0.3)} \tag{6.4.26}$$

for the linearised model in equation (6.4.2). As a result of shifting the zero locations the upper-bound on the error component due to a finite wordlength is reduced to:

$$\hat{\epsilon} = 160/2^M \tag{6.4.27}$$

which is compared in Fig 6.4.3 against results from a simulation of the control system with a 0.1 Hz sinusoidal input. An additional benefit associated with the controller $D_2(z)$ is its enhanced zero frequency gain (stiffness), which provides a better rejection of load disturbances.

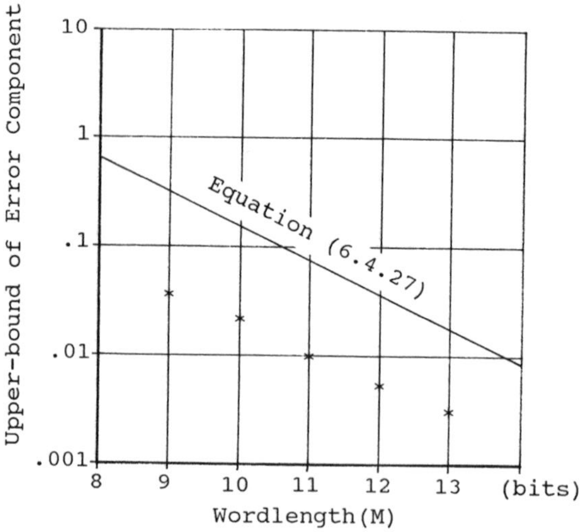

Fig 6.4.3 Predicted and Simulated Upper-bounds of the Finite Wordlength Control Error component as a function of Wordlength for the directly programmed $D_2(z)$ with a 0.1 Hz sinusoidal input

The above relationship between zero locations and the amplification of rounding error noise can be translated into more familiar time domain concepts. An approximation for the derivative of continuous data (u) in terms of its periodically sampled values (u_n) can be derived using Taylor series as:

$$du/dt \simeq (u_n - u_{n-1})/T \qquad \text{at } nT \qquad (6.4.28)$$

Differentiation therefore corresponds to processing the periodically sampled data by the pulse transfer function $(1 - z^{-1})/T$. Thus a compensator zero approaching the point $1\underline{/0}$ produces a progressively better approximation to derivative action*, and from equation (6.4.23), a greater amplification of multiplicative rounding errors. This result parallels the well-known behaviour of R-C networks in which more phase lead (differentiation over a wider bandwidth) exacerbates the amplification of electronic noise.

The programming structure used to realise a pulse transfer function affects the output noise due to rounding errors to a greater or lesser extent. In the case of digital filters designed to have acute real frequency response characteristics, the noise level is crucially dependent on the adopted programming structure, and parallel realisations are found to be particularly effective (36, 43, 124). On the other hand, the relatively simple pulse transfer functions of digital controllers often offer no real opportunity beyond direct programming. Derivations of similar upper-bounds for the loss of steady state control accuracy with parallel or other programming structures are therefore considered as academic. Indeed, the analytical detail involved (130, 131)

---

*Alternatively, the real frequency response of $(1 - \alpha z^{-1})$ at high sampling rates approximates with $z = \exp(j\omega T)$ to:

$$j\omega\alpha + (1 - \alpha)/T$$

which is asymptotic to that of a perfect differentiator ($j\omega$) as α approaches $1\underline{/0}$.

exceeds that for the more rigorous and accurate statistical analysis (41) next in Section 6.5, which copes readily with all types of programming structure at all levels of complexity. Nevertheless, equation (6.4.23) has served present purposes well by illustrating the advantages of complex poles in digital compensator design.

6.5 A STATISTICAL ANALYSIS OF FINITE WORDLENGTH CONTROL ERRORS

The steady state control errors incurred due to A-D conversion and multiplicative rounding in an actual DDC system are now shown to be characterised by a number of mutually uncorrelated noise sources at the input of an idealised system having an arbitrary long wordlength. Moreover, using the elementary quantisation error statistics in equations (6.1.1) to (6.1.4), a straightforward computation evolves that defines the rms steady state value of this finite wordlength control error component as a function of wordlength. Towards this goal, consider the linearised DDC system in Fig 6.4.1, whose true control error sequence $\{e_n\}$ in the absence of amplitude quantisation errors is given by:

$$e_n = x_n - y_n \tag{6.5.1}$$

However, owing to the finite wordlength of A-D conversions, the actual error sequence $\{e_n\}$ is:

$$e_n' = x_n' - y_n' \tag{6.5.2}$$

where

$$x_n' = x_n + r_x(n) \quad ; \quad y_n' = y_n + r_y(n) \tag{6.5.3}$$

and r_x and r_y denote the conversion errors. Substituting equations (6.5.3) and (6.5.1) into (6.5.2) yields:

$$e_n' = e_n + \left[r_x(n) - r_y(n)\right] \tag{6.5.4}$$

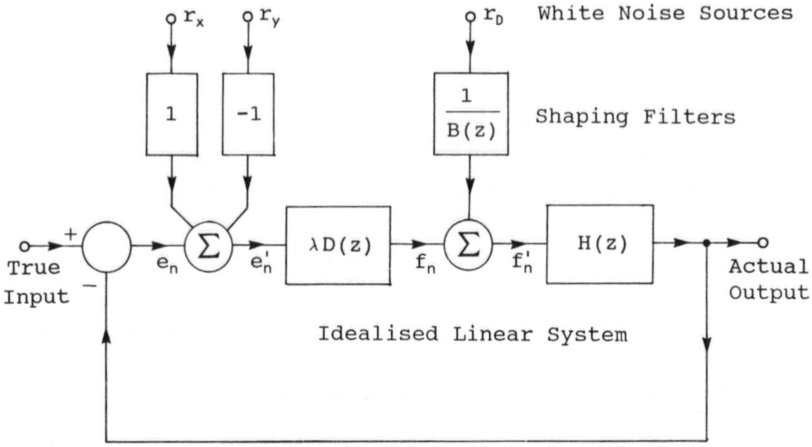

Fig 6.5.1 Actual DDC System as an Idealised System with White Noise Sources to represent Finite Wordlength Error Components

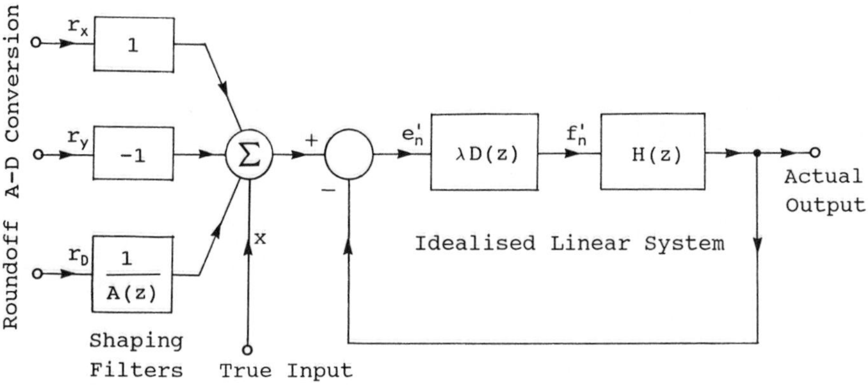

Fig 6.5.2 Actual DDC System with White Noise Sources referred to the Input of an Idealised Linear System Model

Thus the computed error sequence $\{e_n\}$ may be regarded as the true error sequence $\{e_n\}$ additively contaminated by the noise process $\{r_x(n) - r_y(n)\}$. Amplitude quantisation of input and output data by A-D conversion can therefore be characterised by noise generators added into the error channel of a DDC system. With a linear system model like that in Fig 6.5.1, elementary manipulation of its pulse transfer functions enables these noise sources to be represented as additive inputs to the idealised system as shown in Fig 6.5.2. As a consequence of the previously derived statistics for data quantisation or rounding errors, it follows that these sources are mutually uncorrelated and each generates a constant (white) spectrum with intensity $q^2/12$.

If the digital compensator:

$$\lambda D(z) = A(z)/B(z) \tag{6.5.5}$$

where:

$$A(z) = \sum_{1}^{K} a_k z^{-k} \quad \text{and} \quad B(z) = \sum_{o}^{K} b_k z^{-1} \quad \text{with} \quad b_o = 1 \tag{6.5.6}$$

is realised by direct programming, the corresponding linear recurrence equation that would be evaluated in the absence of multiplicative rounding errors is:

$$f_n = \sum_{k=1}^{K} a_k e'_{n-k} - \sum_{k=1}^{K} b_k f_{n-k} \tag{6.5.7}$$

However, owing to the finite wordlength of the controller, the actual calculation performed is:

$$f'_n = \sum_{k=1}^{K} a_k e'_{n-k} - \sum_{k=1}^{K} b_k f'_{n-k} + r_D(n) \tag{6.5.8}$$

where r_D denotes the rounding error in computing:

$$\sum_{k=1}^{K} a_k e'_{n-k} - \sum_{k=1}^{K} b_k f'_{n-k}$$

If the above calculation involves a total of μ multiplications excluding zero, unity and positive integral powers of 2, then:

$$r_D(n) = \sum_{1}^{\mu} \rho_i(n) \tag{6.5.9}$$

where ρ_i represents the rounding error on one such calculation. Defining the computational error quantity:

$$f'_n = f_n + E_D(n) \tag{6.5.10}$$

it follows from equations (6.5.7) and (6.5.8) that:

$$\sum_{0}^{K} b_k E_D(n-k) = r_D(n) \tag{6.5.11}$$

or in terms of z-transformations:

$$\mathcal{E}_D(z) = R_D(z)/\ B(z) \tag{6.5.12}$$

Equation (6.5.12) specifies the computational noise sequence (E_D) which must be added to the ideal compensator response (f) to account for its realisation with finite wordlength arithmetic. Because the statistics of the individual multiplicative rounding errors $\{\rho_i\}$ can be described by equations (6.1.1) to (6.1.4) in practice, the autocorrelation function of their sum (r_D) is evidently:

$$\phi_D(n) = (\mu q^2/12)\delta_{no} \tag{6.5.13}$$

whose corresponding pulse power spectrum is:

$$\Phi_D(z) = \mu q^2/12 \tag{6.5.14}$$

Thus the power spectrum of the computational noise that additively contaminates the ideal controller's response in Fig 6.5.1 is derived from equations (6.5.12) and (4.6.19) as:

$$\Phi_{EC}(z) = (\mu q^2/12)\big/B(z)B(z^{-1}) \tag{6.5.15}$$

Elementary manipulation enables this noise source to be referred to the input of the control system in Fig 6.5.2 as the additive component:

$$\Phi_{EI}(z) = (\mu q^2/12)\big/A(z)A(z^{-1}) \tag{6.5.16}$$

The steady state control error component due to finite wordlength operations in a directly programmed compensator has therefore the pulse power spectrum:

$$\Phi_{fw}(z) = (q^2/12)\left[2 + \mu\big/A(z)A(z^{-1})\right]K(z)K(z^{-1}) \tag{6.5.17}$$

where K(z) denotes the overall pulse transfer function of the DDC system which is defined by equation (2.2.13) as:

$$K(z) = \frac{\lambda D(z)H(z)}{1 + \lambda D(z)H(z)} \tag{6.5.18}$$

On account of the complexity of practical DDC systems, the shaping filter technique in Section 4.8 offers a far more tractable calculation of the mean square finite wordlength error component (ϕ_{fw}) than the Two-sided Inversion Integral (4.4.11). Because factorisation of the normalised input error spectrum according to:

$$(1/12)\left[2 + \mu\big/A(z)A(z^{-1})\right] = \chi(z)\,\chi(z^{-1}) \tag{6.5.19}$$

where

$\chi(z)$ - poles and zeros within the unit circle only

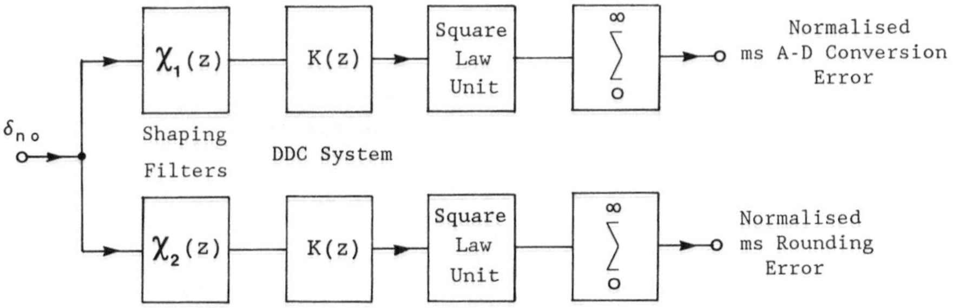

Fig 6.5.3 Scheme for calculating the normalised mean square component of control error due to finite wordlength arithmetic

is generally rather awkward, separate calculations for the additive contributions due to A-D conversions and multiplicative rounding errors are recommended using the shaping filters:

$$\chi_1(z) = 1/\sqrt{6} \quad \text{and} \quad \chi_2(z) = \sqrt{\mu/12}/A(z) \qquad (6.5.20)$$

The required computation, which is illustrated in Fig 6.5.3, can be created around one general purpose subroutine that realises an arbitrary pulse transfer function as its corresponding linear recurrence equation. In this way, a steady state rms component of control error due to finite wordlength operations is derived in the form*:

$$\sqrt{\phi_{fw}} = (2c)2^{-M} \qquad (6.5.21)$$

where c denotes the constant value evaluated as in Fig 6.5.3.

Fig 6.5.4 depicts the rms values computed by this technique for the DDC system with the plant in equation (6.4.2) whose

---

*If the A-D converter(s) and controller have different wordlengths, the computation scheme in Fig 6.5.3 copes with this situation

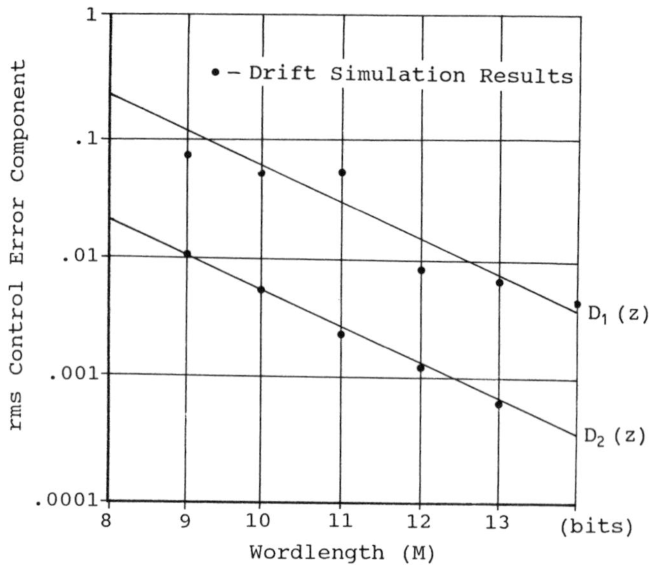

Fig 6.5.4 Predicted and Simulated rms values of the Finite Wordlength Control Error Component as a function of Wordlength for direct programming of $D_1(z)$ and $D_2(z)$

alternative controllers $D_1(z)$ and $D_2(z)$ in equations (6.4.3) and (6.4.26) are directly programmed. Excitation of a DDC system by a large enough single pulse input can create a sizeable* self-sustaining steady state output that is referred to as *drift* (41, 132). Finite sample-size estimates of the rms drift amplitude from mainframe computer simulations of these particular DDC systems are also plotted in Fig 6.5.4 for various wordlengths of A-D converter(s) and controller (41, 132). Within a predicted sample-size error bound of around 7%, the measurements confirm the

*With a 9 bit direct realisation of the controller $D_1(z)$, the drift amplitude observed on a hybrid simulator was more than forty least significant bits (41, 132)

above analysis as do additional simulations with sinusoidal inputs of 0.01 and 0.1 Hz (41, 132). However, it should be noted from the discussion preceding equation (6.1.5) that A-D conversion and multiplicative rounding errors become linearly independent of a particular input (ie loosely 'random'), only if its probable amplitudes involve many different quantum levels. Thus, the error statistics in equations (6.1.1) to (6.1.4) cannot be used to analyse limit cycle oscillations of a few quanta that occur in recursive digital filters (47, 48, 49, 133).

According to equation (6.5.17) or Fig 6.5.2, the finite wordlength control error component can be modelled by convolving a number of mutually uncorrelated quantisation error sequences with the weighting sequences of the low pass systems K(z) and K(z)/ A(z). Because the sampling frequency to bandwidth ratio of practical DDC systems is necessarily large to achieve adequate attenuation of sideband spectra, each weighted error from A-D conversion or multiplicative rounding makes only a relatively small contribution to the infinite* steady state convolution summation. Furthermore, each such contribution possesses the well defined set of statistical moments of a uniform probability density function. Therefore, apart from the full statistical independence of the quantisation errors (linear independence only established in Section 6.3), the conditions of the Central Limit Theorem (87, 120) are met. Consequently, it is reasonable to consider the possibility that finite wordlength control errors have an approximately Normal probability distribution. Asymmetry and 'peakiness' in the distribution of an arbitrary variate u can be assessed using the parameters:

$$\left.\begin{aligned} \text{Coefficient of Skew} \quad (S) &= \left(\overline{(u - \bar{u})^3}\right)\sigma^{-3} \\ \text{Coefficient of Excess} \quad (E) &= \left(\overline{(u - \bar{u})^4}\right)\sigma^{-4} \end{aligned}\right\} \qquad (6.5.22)$$

---

*In practice the steady state is achieved when this infinite sum is closely matched by an adequately large number of terms. The situation is analogous to 'a few time constants' for $1/(1 + \tau s)$

where:

$$\sigma^2 = \overline{(u - \overline{u})^2} \qquad (6.5.23)$$

For a Normal distribution:

$$S = 0 \qquad \text{and} \qquad E = 0 \qquad (6.5.24)$$

so that a comparison with corresponding values for an arbitrary distribution provides a sound* simply implemented test for Normality (87). Table 6.5.1 below contains estimates of these statistical parameters for the finite wordlength control error components in the previous simulations with drift and sinusoidal input excitations. These results are evidently consistent with the analytically plausible conjecture that steady state finite wordlength control errors are in general Normally distributed. As a result, a computed rms value ($\sqrt{\phi_{fw}}$) of their amplitude imposes precise confidence limits (87, 120) on the probability that they can exceed specific magnitudes (eg $3\sqrt{\phi_{fw}}$).

Mean	Coefficient of Skew (S)	Coefficient of Excess (E)
-0.0013	0.055	0.030

Table 6.5.1 Statistical Parameters of the Finite Wordlength Control Error Components derived from the Simulations

*A probability distribution is equivalent to its Characteristic function. Expanding the exponential factor in such a Fourier Integral as a power series shows that it is defined by a denumerable set of statistical moments. Because S and E involve the first four such moments, there is a sound basis for this comparison

The mean square value $\phi(o)$ corresponding to an arbitrary pulse power spectrum $\Phi(z)$ can be derived by means of the Two-sided Inversion Integral (4.4.11) as:

$$\phi(o) = \frac{1}{2\pi j} \int_{\Gamma} \Phi(z) z^{-1} dz \qquad (6.5.25)$$

where:

Γ - the unit circle in the z- plane

For present purposes, the above expression is conveniently translated into the real frequency domain by setting:

$$z = \exp(j\omega T) \qquad (6.5.26)$$

so that the steady state finite wordlength control error component (ϕ_{fw}) can be written as:

$$\phi_{fw} = \frac{T}{2\pi} \int_{-\pi/T}^{\pi/T} \Phi^{*}_{fw}(\omega) d\omega \qquad (6.5.27)$$

When the zeros of a digital compensator are moved closer to the point 1/0 to achieve a stronger derivative action, the low frequency intensity of the rounding error spectrum $\mu/A(z)A(z^{-1})$ in equation (6.5.17) is increased. Because the overall pulse transfer function K(z) of a DDC system is low pass, the control error component ϕ_{fw} is thereby increased. This aggravation of finite wordlength control errors with strengthening derivative action in a controller was previously deduced in Section 6.4 from a conceptually simpler upper-bound analysis. Attention there and in Chapter 7 is directed at the reduction in wordlength requirements that can be achieved by engineering controllers with complex poles.

Digital controllers have pulse transfer functions that are usually so simple as to be only directly programmable. However, sooner or later control engineers are likely to become involved with digital signal processing, where other structures enable more cost-effective designs as a result of smaller wordlength requirements (43). A brief description of applying the above analysis to other programming structures is therefore considered apposite; particularly as the required development verges on the trivial. Parallel realisations, which are generally close to the optimum (36, 43), require the expansion of an arbitrary pulse transfer function as:

$$D(z) = \sum_{k=1}^{K} D_k(z) = \sum_{k=1}^{K} A_k(z)/B_k(z) \qquad (6.5.28)$$

where $A_k(z)$ and $B_k(z)$ are appropriate real polynomials. For a cascade realisation, a pulse transfer function would be expanded in the form:

$$D(z) = \prod_{k=1}^{K} D_k(z) \qquad (6.5.29)$$

For both types of realisation, each component $D_k(z)$ is directly programmed and the individual multiplicative rounding errors can be modelled according to equation (6.5.16) as additive individual noise generators having the typical spectrum:

$$(q^2/12)\, \chi_k(z)\, \chi_k(z^{-1}) \qquad \text{with} \qquad \chi_k(z) = \sqrt{\mu_k}/A_k(z)$$

In the case of a parallel structure, these shaped white noise sources are already at the input of the complete idealised filter. With a cascade realisation, straightforward manipulation involving some of the pulse transfer function components $\{D_k(z)\}$ enables these sources to be similarly relocated. Thus in both cases finite wordlength errors can be modelled as in Fig 6.5.5a, and because the noise sources are mutually uncorrelated, the mean square value of the total output noise can be calculated in the manner of Fig 6.5.5b.

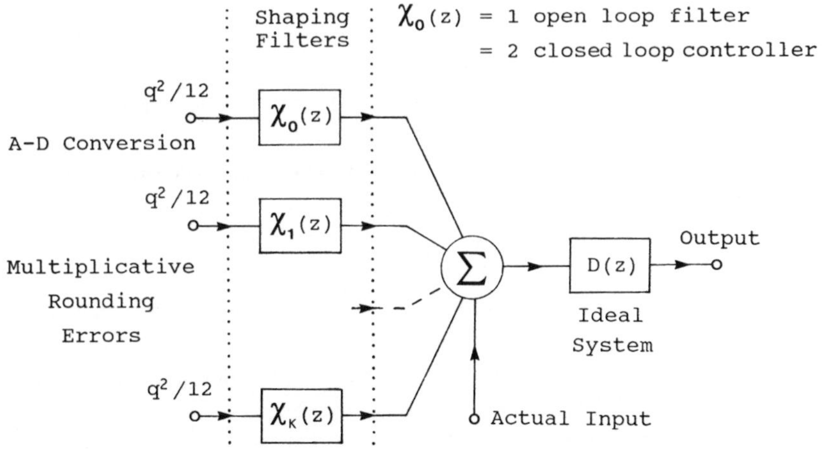

a) Modelling with White Noise Sources

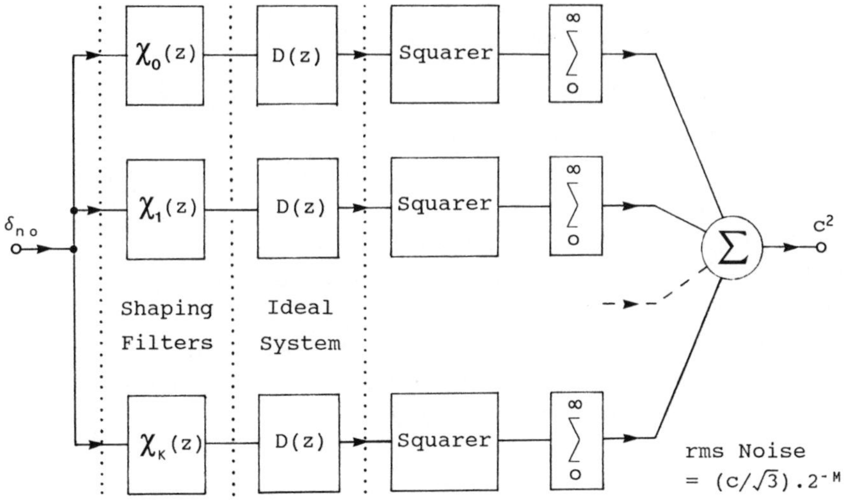

b) Computing the total output noise power as a function of Wordlength (M)

Fig 6.5.5 Quantifying the output noise due to Finite Wordlength Arithmetic in Parallel or Cascade Realisations

The distribution of scalar gain in a digital compensator also affects the finite wordlength control error component (ϕ_{fw}). If the proportion of scalar gain following the frequency dependent part D(z) is increased, errors from both the A-D conversion and arithmetic rounding are further accentuated. Whereas, if the scalar gain preceding the frequency dependent part is increased, then just the A-D conversion errors are aggravated. Although this latter distribution might therefore appear superior, the control error signal of a DDC system under these circumstances would be more likely to suffer *overflow* which could jeopardise its overall stability. A further discussion of overflow appears later in Section 7.7, but this somewhat arduous Chapter 6 is completed next by Section 6.6 which analyses the influence of coefficient rounding errors on the real frequency response of an arbitrary pulse transfer function.

6.6 COEFFICIENT QUANTISATION AND REAL FREQUENCY RESPONSE

When a pulse transfer function is realised as a linear recurrence equation on a digital computer, its coefficients suffer amplitude quantisation errors. In consequence, its real frequency response deviates to a greater or lesser extent from that which would have been idealistically obtained with an arbitrary long wordlength machine. Digital filters with acutely changing real frequency responses appear particularly vulnerable (43). Except for so-called 'notch networks' which are analysed specially in Section 7.4, coefficient quantisation makes little impact on the design of digital compensators because:

- their pulse transfer functions are relatively simple
- their real frequency response characteristics are relatively far less acute
- generous closed loop stability margins are allowed

Nevertheless, a general statistical analysis (43) is now briefly presented of the rms frequency response deviation induced by a finite wordlength because control engineers are at times involved with digital signal processing. Furthermore, the end-result demonstrates yet again the strong conceptual similarities between the analyses of linear discrete and continuous data systems.

The desired or ideal pulse transfer function to be realised by direct programming is:

$$D_I(z) = A_I(z)/B_I(z) = \left[\sum_{1}^{K} \bar{a}_k z^{-k}\right] \cdot \left[1 + \sum_{1}^{K} \bar{b}_k z^{-k}\right]^{-1} \quad (6.6.1)$$

but due to a finite wordlength, that actually realised is:

$$D(z) = \left[\sum_{1}^{K} a_k z^{-k}\right] \cdot \left[1 + \sum_{1}^{K} b_k z^{-k}\right]^{-1} \quad (6.6.2)$$

in which context the following amplitude quantisation errors are defined:

$$\alpha_k = a_k - \bar{a}_k \quad ; \quad \beta_k = b_k - \bar{b}_k \quad \text{and} \quad \beta_o \equiv 0 \quad (6.6.3)$$

It should be noted in the above context that coefficients like 0.0, 0.5 and 1.0 can be represented without error. Neglecting the second order interactions with multiplicative rounding errors, which are analysed separately as in Section 6.5, the direct realisation of equation (6.6.2) is according to:

$$y'_n = \sum_{k=1}^{K} a_k x_{n-k} - \sum_{k=1}^{K} b_k y'_{n-k} \quad (6.6.4)$$

whereas the ideal realisation would respond as:

$$y_n = \sum_{k=1}^{K} \bar{a}_k x_{n-k} - \sum_{k=1}^{K} \bar{b}_k y_{n-k} \quad (6.6.5)$$

Defining the computational error quantity:

$$E_n = y'_n - y_n \tag{6.6.6}$$

then from equations (6.6.3) to (6.6.6) it follows that:

$$E_n = \sum_{k=1}^{K} \left(\alpha_k x_{n-k} - \overline{b}_k E_{n-k} - \beta_k y_{n-k} - \beta_k E_{n-k} \right) \tag{6.6.7}$$

which by neglecting second order quantities reduces to:

$$E_n = \sum_{k=1}^{K} \left(\alpha_k x_{n-k} - \overline{b}_k E_{n-k} - \beta_k y_{n-k} \right) \tag{6.6.8}$$

Effecting the z- transformation of the above linear recurrence equation gives:

$$0 = \alpha(z)X(z) - B_I(z)\mathcal{E}(z) - \beta(z)Y(z) \tag{6.6.9}$$

where:

$$\alpha(z) = \sum_{k=1}^{K} \alpha_k z^{-k} \quad ; \quad \beta(z) = \sum_{k=o}^{K} \beta_k z^{-k} \quad \text{with } \beta_o = 0 \tag{6.6.10}$$

and

$$\mathcal{E}(z) = \sum_{n=o}^{\infty} E_n z^{-n} \tag{6.6.11}$$

Because:

$$Y(z) = D_I(z)X(z)$$

equation (6.6.9) can be written as:

$$\mathcal{E}(z) = \left\{ \frac{\alpha(z) - \beta(z) D_I(z)}{B_I(z)} \right\} X(z) \tag{6.6.12}$$

Hence the output of the actual finite wordlength realisation is derived from equations (6.6.6) and (6.6.12) as:

$$Y'(z) = \left\{ D_I(z) + \frac{\alpha(z) - \beta(z) D_I(z)}{B_I(z)} \right\} X(z) \tag{6.6.13}$$

so that:

$$D(z) = D_I(z) + \left[\alpha(z) - \beta(z) D_I(z)\right] / B_I(z) \tag{6.6.14}$$

Thus an actual pulse transfer function realised on a finite wordlength machine is equivalent to the ideal in parallel with a 'stray' that causes deviation from the required real frequency behaviour. In this sense, coefficient quantisation in digital filters creates the same effect as tolerances on discrete components (R,L,C) in electrical networks.

Though equation (6.6.14) has conceptual interest, its translation into a practically appropriate measure of the deviation between an actual and ideal frequency response is required for design purposes. Such is provided by the following mathematically convenient mean square convergence criterion*:

$$\sigma_\omega^2 = \overline{\frac{T}{2\pi} \int_{-\pi/T}^{\pi/T} \left| D^*(\omega) - D_I^*(\omega) \right|^2 d\omega} \tag{6.6.15}$$

where the 'bar' notation denotes the ensemble mean. It should be noted that the variance σ_ω^2 exists only if coefficient quantisation errors are small enough to preserve the stability of a

---

*Other deviation measures are clearly possible as illustrated by that for notch networks in Section 7.4 and that for stop-band filters in reference 138

realisation. In fact, no real limitation is posed by this constraint because the usual design requirement is for a frequency response deviation of just a few percent. That is to say, loss of stability is contended to occur well after the deviation between the actual and ideal responses has become totally unacceptable for practical purposes. Another point is that the deviation in a real frequency response characteristic is not really a statistical effect at all! Given any pulse transfer function and a particular wordlength, then the realisable coefficients and frequency response are both completely deterministic. However, viewed a posteriori like in the previous treatment of amplitude quantisation errors, a particular realisation D(z) could evolve from an ensemble of different ideal ('unquantised') parents. Equation (6.6.15) provides therefore a measure of the sensitivity of a realisation to small coefficient changes. Conveniently, it evolves that this sensitivity (σ_ω^2) can be determined for all wordlengths by a single computation, whereas the deterministic analysis would involve a separate calculation for each wordlength. Validation of the statistical analysis has been accomplished by simulations of complex pulse transfer functions in a variety of programming structures (43, 137).

For an arbitrary stable pulse transfer function G(z), it is shown in Appendix A4.8 that:

$$\frac{1}{2\pi j}\int_{\Gamma} G(z)G(z^{-1})z^{-1}dz = \sum_{0}^{\infty} g_k^2 \qquad (6.6.16)$$

and transforming the variable of integration according to:

$$z = \exp(j\omega T) \qquad (6.6.17)$$

yields:

$$\frac{T}{2\pi}\int_{-\pi/T}^{\pi/T} \left|G^*(\omega)\right|^2 d\omega = \sum_{0}^{\infty} g_k^2 \qquad (6.6.18)$$

Because both the functions D(z) and $D_I(z)$ can be assumed stable in the present context and because the joint probability density function of the variates $\{\alpha_k\}$ and $\{\beta_k\}$ is integrable, it follows that the integrand of the implicit multiple integral in equation (6.6.15) satisfies the conditions of the Lebesque-Fubini theorem (15, 108), so that the order of integration can be interchanged to give:

$$\sigma_\omega^2 = \frac{T}{2\pi} \int_{-\pi/T}^{\pi/T} \overline{\left| D^*(\omega) - D_I^*(\omega) \right|^2 d\omega}$$

which expressed in terms of z- transformations becomes:

$$\sigma_\omega^2 = \frac{1}{2\pi j} \int_\Gamma \overline{\left[D(z) - D_I(z)\right]\left[D(z^{-1}) - D_I(z^{-1})\right]} z^{-1} dz \tag{6.6.19}$$

and substituting equation (6.6.14) results in:

$$\sigma_\omega^2 = \frac{1}{2\pi j} \int_\Gamma \Big\{ \overline{\left[\alpha(z) - \beta(z) D_I(z)\right]\left[\alpha(z^{-1}) - \beta(z^{-1}) D_I(z^{-1})\right]} .$$

$$\left[B_I(z) B_I(z^{-1})\right]^{-1} \Big\} z^{-1} dz \tag{6.6.20}$$

Exploiting the statistical properties of amplitude quantisation errors in equations (6.1.1) to (6.1.4) reduces the above to:

$$\sigma_\omega^2 = (\mu_A q^2/12) \frac{1}{2\pi j} \int_\Gamma \left[B_I(z) B_I(z^{-1})\right]^{-1} z^{-1} dz$$

$$+ (\mu_B q^2/12) \frac{1}{2\pi j} \int_\Gamma \left[A_I(z) A_I(z^{-1})\right] . \left[B_I(z) B_I(z^{-1})\right]^{-2} z^{-1} dz \tag{6.6.21}$$

where μ_A and μ_B are the number of coefficients in the numerator and denominator polynomials $A_I(z)$ and $B_I(z)$ respectively that are not exactly representable in machine format. Reference to

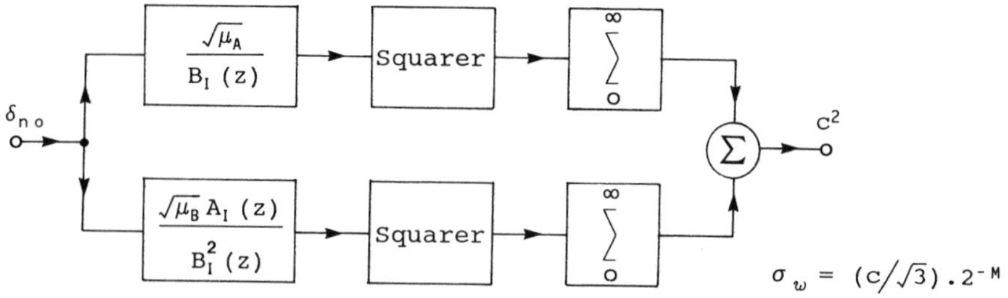

Fig 6.6.1 Computational Scheme for evaluating the rms Deviation in a Real Frequency Response due to a Finite Wordlength

equation (6.6.16) reveals that the above contour integrals equate with the sum of the squares of the weighting sequences for the pulse transfer functions $1/B_I(z)$ and $A_I(z)/B_I^2(z)$. Consequently, just one computation of the form shown in Fig 6.6.1 enables the rms deviation in the real frequency response to be evaluated as a function of computer wordlength (M). An adequate wordlength for this calculation on a mainframe, work station or personal computer should be established by repeated evaluations using say R*8 and the R*16 operations.

At the beginning of this Section, high order digital filters with acute real frequency response characteristics were baldly asserted as being particularly vulnerable to coefficient quantisation. Some justification of this remark can now be provided on the basis of equation (6.6.21). Because the numbers of non-exactly representable coefficients μ_A and μ_B usually increase with the order of a pulse transfer function, the greater sensitivity of high order filters is now self-evident. Furthermore, the graphical construction in Fig 3.3.2 indicates that acute real frequency characteristics stem from poles and zeros* close to the periphery of the unit circle (Γ). Under

---

*Both must be close 'to balance' the gain characteristic over the whole frequency band π/T, or to linearise the phase characteristic over the pass band

these conditions, the integrands in equation (6.6.21) both have relatively large magnitudes, so that the variance quantity (σ_ω^2) measuring the deviation of a real frequency response increases.

The above analysis is readily extended to other programming structures (43, 137), which broadly require the decomposition of a pulse transfer function into low order directly programmed elements. With parallel programming, a pulse transfer function is written in the form of a sum:

$$D_I(z) = \sum_{p=1}^{P} D_{Ip}(z) \qquad \text{with} \qquad D_{Ip} = A_{Ip}(z)/B_{Ip}(z) \qquad (6.6.22)$$

Coefficient quantisation in each of these components $\{D_{Ip}(z)\}$ is representable as a 'stray' parallel element according to equation (6.6.14). Because amplitude quantisation errors possess the elementary statistics in equations (6.1.1) to (6.1.4), it can be readily established (43) that the variance of the real frequency response deviation becomes in this case:

$$\sigma_\omega^2 = (q^2/12) \sum_{p=1}^{P} \frac{1}{2\pi j} \int_\Gamma \left[F_1(z) + F_2(z)\right] z^{-1} dz \qquad (6.6.23)$$

where

$$\left.\begin{aligned} F_1(z) &= \mu_{Ap}\left[B_{Ip}(z)B_{Ip}(z^{-1})\right]^{-1} \\ F_2(z) &= \mu_{Bp}\left[A_{Ip}(z)A_{Ip}(z^{-1})\right] \cdot \left[B_{Ip}(z)B_{Ip}(z^{-1})\right]^{-2} \end{aligned}\right\} \qquad (6.6.24)$$

and μ_{Ap} and μ_{Bp} denote the number of non-exactly representable coefficients in the numerator and denominator polynomials respectively of $D_{Ip}(z)$. Evaluation of equation (6.6.23) follows the same scheme as in Fig 6.6.1, except of course that there are now a total of P pulse transfer functions of the general form:

$$\sqrt{\mu_{Ap}}\left[B_{Ip}(z)\right]^{-1} + \sqrt{\mu_{Bp}}\left[A_{Ip}(z)\right]\left[B_I(z)\right]^{-2}$$

whose weighting sequences must be individually squared and summed before adding these sums together. With cascade programming, an arbitrary pulse transfer function is written in the form of a product:

$$D_I(z) = \prod_{p=1}^{P} D_{Ip}(z) \qquad (6.6.25)$$

Coefficient quantisation in each of these components $\{D_{Ip}(z)\}$ is again representable as a 'stray' parallel element according to equation (6.6.14). However, in this case the width of quantisation (q) does not linearly scale the variance of the real frequency response deviation which can be derived as (43):

$$\sigma_\omega^2 = \frac{1}{2\pi j} \int_\Gamma D_I(z) D_I(z^{-1}) \left\{ \left(\prod_{p=1}^{P} 1 + F_{Ap}(z) + F_{Bp}(z) \right) - 1 \right\} z^{-1} dz$$

(6.6.26)

where

$$F_{Ap}(z) = \mu_{Ap}(q^2/12) \left[A_{Ip}(z) A_{Ip}(z^{-1}) \right]^{-1}$$

$$F_{Bp}(z) = \mu_{Bp}(q^2/12) \left[B_{Ip}(z) B_{Ip}(z^{-1}) \right]^{-1} \qquad (6.6.27)$$

The zeros of the above integrand depend obviously on the width of quantisation so that separate calculations in the form of Fig 6.6.1 are required for each particular wordlength. Nevertheless the necessary factorisation of the numerator involves just relatively simple terms.

Details of comprehensive simulations, which validate the above statistical analysis for high order digital filters in various programming structures, are described in references 43 and 137. Chapter 7 next applies the considerable preceding body of analysis to the actual design of linearised DDC system models.

CHAPTER 7

Nyquist Based Design Principles

To drive the limited in pursuit of the limitless is fatal. - Chuang Tzu

7.1 PRELIMINARIES

A DDC system is engineered to achieve a specified transient and steady state performance (accuracy) within the constraints imposed by mechanical construction and operational environment in much the same way as a continuous data system. However, set performance objectives can only be achieved by appropriate choices of sampling frequency and wordlength in conjunction with the conventional use of compensating networks, minor feedback loops* (73, 74), divided reset (72, 75, 76) or feedforward (77, 78). Experience of a specific process or an operational constraint often decides a particular system configuration, so that practical DDC systems assume a variety of different forms. Nevertheless an adequate appreciation of the presently advocated design principles can be gained from studying the unity feedback arrangement in Fig 7.1.1.

Frequency domain techniques for compensating DDC systems were first exploited in 1951 by Linvill (148), who identified data sampling with pulse amplitude modulation and a plant's role in its reconstruction. Series compensation by a continuous network N(s) was considered for the system in Fig 7.1.1, whose overall Laplace transfer function was derived as:

---

*sometimes an intrinsic feature eg friction or temperature reactivity feedback in a nuclear reactor etc

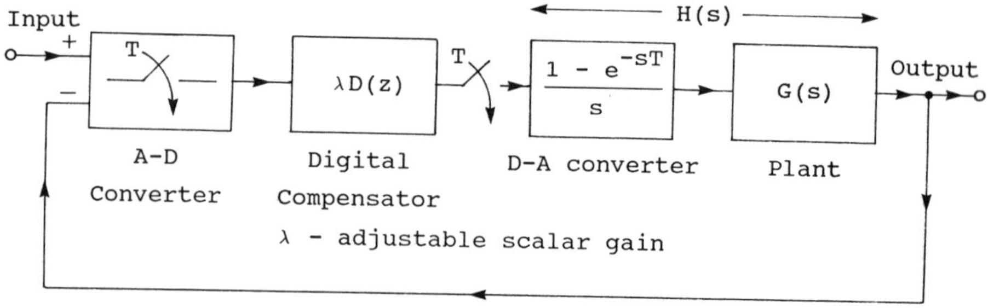

Fig 7.1.1 Linearised Unity Feedback DDC System

$$K(s) = \frac{N(s)H(s)}{1 + NH^*(s)} \tag{7.1.1}$$

Nyquist's theorem was then applied in the usual manner to relate closed loop stability to encirclements of the critical point (-1,0) by the locus $NH^*(\omega)$. For any particular sampling frequency (ω_0), a graphical construction was proposed for approximating:

$$NH^*(\omega) = \frac{1}{T}\sum_{-\infty}^{\infty} N(\omega - k\omega_0)H(\omega - k\omega_0) \tag{7.1.2}$$

in terms of the lowest order (k) sidebands that dominate the pertinent behaviour of this locus. In a later paper, Linvill and Salzer (149) characterised the pulse real frequency response of a linear recurrence relation with real poles and zeros by a Nyquist style polar plot, and they suggested its application to the design of digital compensators. A frequency domain approach to digital compensation is also recommended here for the reasons outlined in Section 3.2, but the alternative curves presented in Section 7.3 are considered to provide greater insight as well as an easier computation. Furthermore, they form just part of the comprehensive design procedure shown in Fig 7.1.2, which combines

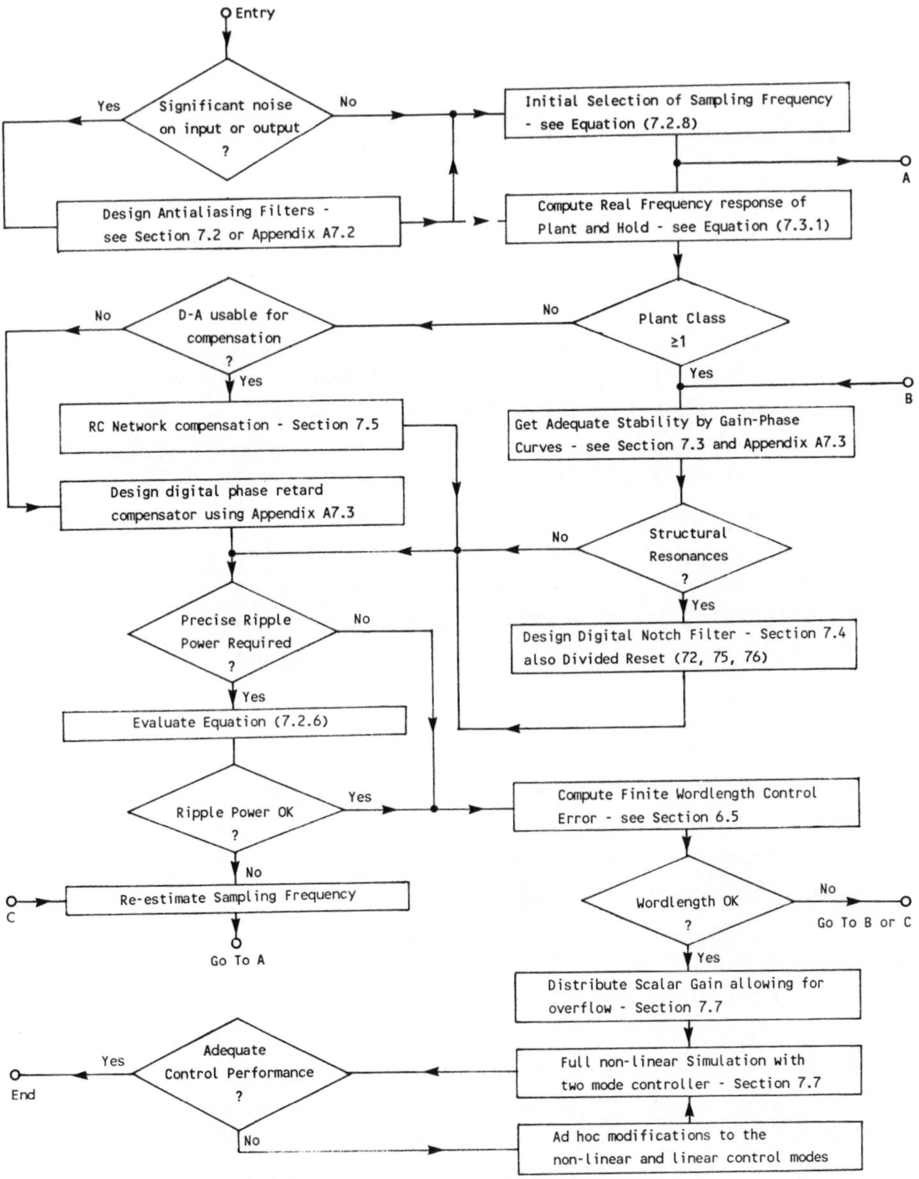

Fig 7.1.2 The Author's Design Procedure for a Unity Feedback DDC System

earlier mathematical analyses with engineering pragmatism. Detailed descriptions of its various aspects constitute the different Sections of this Chapter 7.

7.2 PREFILTERING AND THE CHOICE OF SAMPLING FREQUENCY

Fig 1.3.4 illustrates that the periodic sampling of continuous data generates a wanted spectral baseband and an infinite series of perverse sidebands centred about integral multiples of the sampling frequency (ω_0). Because the spectra of continuous data possess in practice finite asymptotic cut-off rates, these sidebands inevitable 'spill' into the baseband where they corrupt any attempted reconstruction of the unsampled information. A relatively wideband, yet low intensity, noise process superimposed on the continuous input or output data of a DDC system can therefore seriously aggravate such irreversible distortion unless proper action is taken to prevent significant amounts of noise sideband power being demodulated into the baseband frequencies. Prior to sampling the continuous data, it is therefore prudent to ascertain the possible benefits in this respect of low pass filtering.

In some special cases (105, 106, 107), it is either economic or strategically necessary to effect an optimum spectral separation of the signal and noise components by means of a low pass Wiener or Kalman filter. There exists a considerable body of helpful literature to assist a designer faced with such stringent performance requirements. However, in many cases sufficient improvement can be achieved by using traditional low pass filter designs. Realisations of Butterworth, Tchebychev or Bessel filters by operational amplifier circuits are well-documented (138, 139), but although they accommodate reasonable RC component tolerances, care must be exercised with regard to the possibility of high common mode noise levels (140). Personal experience indicates that the efficient spectral separation of a control signal from wideband noise by a linear filter depends mostly on its asymptotic cut-off rate and bandwidth. Minor passband variations associated with the particular traditional type (eg

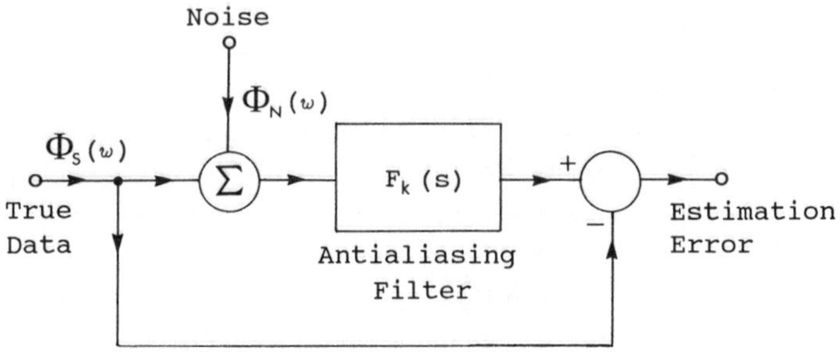

Fig 7.2.1 Calculation of the Mean Square Estimation Error for an arbitrary design of Lowpass Filter $F_k(s)$

maximally flat, equi-ripple etc) appear largely insignificant. Furthermore, step function response calculations reveal a delay time which increases with filter cut-off rate as a result of correspondingly greater phase lags in the real frequency response. It can therefore be concluded that any traditional low pass filter type is broadly suitable as an antialiasing filter for a DDC system, but its bandwidth and (low) order require further consideration.

Provided the signal and noise spectra are reasonably well-defined, suitable parameters for the recommended 1st, 2nd or 3rd order forms of a traditional low pass filter can be decided by computing the mean square error on its estimate of the signal as a function of its bandwidth. The proposed calculation is illustrated in Fig 7.2.1 in which $F_k(s)$ denotes the Laplace transfer function of the arbitrarily selected design of low pass filter. If the signal and noise ensembles are uncorrelated, the required mean square estimation error (ϕ_E) is derived from equation (4.7.14) as:

$$\phi_E = \frac{1}{2\pi} \int_{-\infty}^{\infty} \left\{ \left|1 - F_k(\omega)\right|^2 \Phi_s(\omega) + \left|F_k(\omega)\right|^2 \Phi_N(\omega) \right\} d\omega \qquad (7.2.1)$$

Though an evaluation of equation (7.2.1) by numerical integration poses no real difficulties, references (74) and (141) contain general algebraic formulae which enable much faster computations without any concern for 'step size' convergence. Plotting the mean square estimation error as a function of filter bandwidth for these low orders of a selected filter type usually enables a soundly engineered choice of filter order and bandwidth as indicated by Fig 7.2.2. Sometimes an identification of the signal and noise spectra is patently unnecessary, and then the simplest RC network having a break frequency compatible with an intuitive estimate of the signal bandwidth or a plant rate constraint is recommended. Appropriate instrumentation in a DDC system must ensure that noise processes superimposed on its true continuous input or output data have relatively much lower spectral intensities. Moreover, granted the introduction of antialiasing filters, continuous control data and their additive noise have broadly similar bandwidths. Consequently such noise components are not further involved then in the design procedure. However, in some quite special circumstances, prefiltering of the continuous input and output data is not physically possible and this different design scenario is briefly described for completeness in Appendix A7.2.

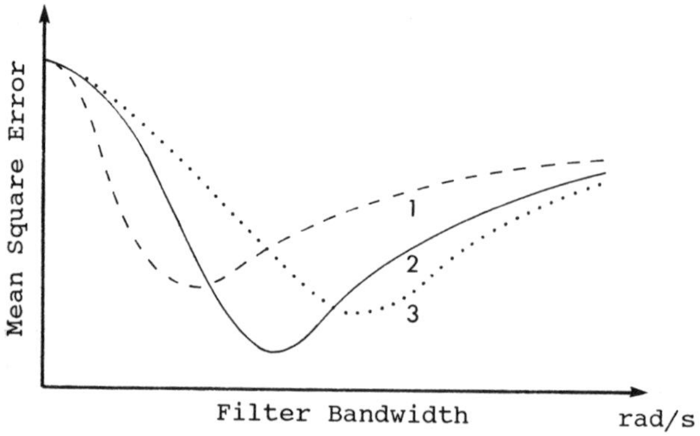

Fig 7.2.2 Selection of Bandwidth and Order for an Antialiasing Filter in terms of Mean Square Estimation Error

Adequate stability of the DDC system in Fig 7.1.1 is to be engineered by manipulation of its closed loop pulse transfer function:

$$K(z) = \frac{\lambda D(z)H(z)}{1 + \lambda D(z)H(z)} \tag{7.2.2}$$

with the digital controller $\lambda D(z)$. For the reasons outlined in Section 3.2, this compensation procedure should be implemented in the real frequency domain using Nyquist or Inverse Nyquist diagrams. The real frequency response of a linearised plant model cascaded with a zero order hold is derived from equation (1.5.4) as:

$$H(\omega) = K_c T \left| \sin(0.5\omega T)/(0.5\omega T) \right| \exp\left(-j\omega T(p + 0.5)\right) G(\omega) \tag{7.2.3}$$

where:

K_c - combined DC gain of the A-D and D-A converters

and the fixed computational delay of p (≥1) sample periods is included in the above expression for convenience in subsequent iterative calculations. Equations (2.4.18) and (3.3.12) yield the corresponding pulse real frequency response as:

$$H^*(\omega) = K_c \left| \sin(0.5\omega T)/(0.5\omega T) \right| \exp\left(-j\omega T(p + 0.5)\right) \sum_{-\infty}^{\infty} G(\omega - m\omega_0) \tag{7.2.4}$$

and that for a linear recursive digital controller is given by equation (3.3.7) as:

$$\lambda D^*(\omega) = \lambda \left(\sum_{0}^{K} a_k z^{-k}\right)\left(\sum_{0}^{K} b_k z^{-k}\right)^{-1} \quad \text{with } z = \exp(j\omega T) \tag{7.2.5}$$

Thus the shape of the all important Nyquist locus $\lambda D^*(\omega)H^*(\omega)$ in the D(z)H(z)- plane depends on the sampling frequency.

Consequently, the design of a suitable compensator as embodied in the constants $\{a_k\}$ and $\{b_k\}$ cannot begin until the sampling frequency is known. Although the ripple error component (ϕ_r) in equation (5.3.31):

$$\phi_r = \pi^{-1} G_\infty^2 \omega_0^{-(2R+2)} \int_{-\infty}^{\infty} \left| K^*(\omega) / G(\omega) \right|^2 \Phi_X(\omega) \omega^2 d\omega \qquad (7.2.6)$$

can at least be inferred from a steady state accuracy specification, it is an implicit function of both the digital compensator and the sampling frequency*. Hence the direct selection of a precisely consistent sampling frequency followed by the design of a suitable compensator is not an option, and an iterative procedure like that in Fig 7.1.2 is strictly involved. As described in Section 5.4, a sound initial estimate of an adequate sampling frequency can usually be based in practice on the input:

$$x(t) = A_{max} \sin(\omega_B t + \psi) \qquad (7.2.7)$$

where:

A_{max} - maximum input amplitude

ω_B - required closed loop bandwidth

ψ - uniformly distributed phase angle $[0, 2\pi]$

In this case, the steady state ripple error power is specified by equation (5.4.7) as:

$$\phi_r = \left\{ A_{max} G_\infty \omega_0^{-(R+1)} \omega_B \; 0.707 / \left| G(\omega_B) \right| \right\}^2 \qquad (7.2.8)$$

with:

---

* see equations (7.2.2) and (7.2.5)

$$G(\omega) \simeq G_\infty \omega^{-R} \qquad \text{for large } \omega \qquad (7.2.9)$$

The above ripple error power patently decreases very rapidly with sampling frequency. Because the speed and integration of present-day electronics can often offer higher than the minimum required at no extra cost, real opportunities exist therefore for avoiding the tedium of an iterative design by imposing a modest margin on an initial estimate from equation (7.2.8).

Section 7.3 next continues with the design scheme outlined in Fig 7.1.2 by developing the principles of a Nyquist diagram based compensation technique.

7.3 GAIN-PHASE CURVES FOR FREQUENCY DOMAIN COMPENSATION

Section 3.5 relates the stability of the linearised DDC system in Fig 7.1.1 to encirclements of the critical point (-1,0) by the image of the z- plane contour *C* in Fig 3.5.1 under the mapping $\lambda D(z)H(z)$. It is shown there that this z- plane contour reduces in practice to the unit circle; apart from a small semicircular detour around poles at $1\underline{/0}$. Albeit that the mapped locus (or Nyquist diagram) consists therefore of 'mainly' the open loop system's pulse real frequency response $\lambda D^*(\omega)H^*(\omega)$, stability determinations in terms of encirclements are demonstrably subtly dependent on the asymptotically infinite behaviour of $\lambda D(z)H(z)$ around this small detour. Although some control engineers might favour investigations exclusively in the frequency domain, the author prefers interplay with a Root locus to establish that a proposed form of compensator indeed endows closed loop stability. In essence, the gross stability of a DDC system with some conceived form of compensation is broadly confirmed by a manually sketched* Root locus, while the detailed tuning of its pole-zero locations for adequate stability is effected by an accurate Nyquist or Inverse Nyquist diagram.

If a sampling frequency (ω_0) is selected for compatibility with

---

*quite modest experience enables its speedy implementation

contemporary control accuracy requirements on the basis of equation (7.2.8), then it is also generally high enough for equation (7.2.4) to be approximated by:

$$H^*(\omega) = K_c \left| \sin(0.5\omega T)/(0.5\omega T) \right| \exp\left(-j\omega T(p + 0.5)\right) G(\omega) \tag{7.3.1}$$

over the frequency range affecting closed loop stability. The generous stability margins in Fig 3.5.5, as well as those incorporated to accommodate plant variations or uncertainties, further vindicate the approximation. For ease of nomenclature in the following discussion, the number of essentially* pure integrations in a linearised plant transfer function G(s) is termed the *Class* of the system. It is recommended now that the number of digital phase-advance stages in a compensator matches the Class number of a system. This design empiricism ensures that sufficient phase lead is created to bend a Nyquist locus far enough away from the critical point to achieve an adequate phase margin, whilst preserving sufficient higher frequency attenuation from the plant to endow an adequate gain margin. Thus the digital compensator for a Class-1 system has for example the general form:

$$\lambda D(z) = \lambda(z - a)/(z - b) \tag{7.3.2}$$

where:

$$o \le b < a < 1 \tag{7.3.3}$$

and

λ - an adjustable scalar gain constant

Digital realisations of the frequently deployed P+I and P+D

---

*$1/(1 + s\tau)$ behaves 'essentially' as a pure integrator, if in the vicinity of the critical point (-1,0) of a Nyquist plot it can be closely approximated (±10%) by $1/s\tau$

controllers:

$$K_P + K_I/(1 - z^{-1}) \quad ; \quad K_P + K_D(1 - z^{-1})$$

are also clearly representable in the form of equation (7.3.2). Graphs showing the gain and phase of $\exp(j2\pi\omega/\omega_0) - C$ to base of the normalised frequency ω/ω_0 are given in Appendix A7.3 for some real and complex values of the parameter C. In the following description* of their use in the design of digital compensators, the parameter C corresponds to a pole or zero location (114, 142).

As with RC networks, Figs A7.3.1 and A7.3.2 show that increasing amounts of phase lead require increasingly severe rising gain-frequency characteristics. A maximum phase advance of about 55° is presently suggested for each pole-zero pair or *stage* in a digital compensator. Otherwise, its rising gain-frequency characteristic becomes so severe that an open loop locus bulges too close to the critical point. Resonances then induced in the overall pulse real frequency response:

$$K^*(\omega) = \frac{\lambda D^*(\omega)H^*(\omega)}{1 + \lambda D^*(\omega)H^*(\omega)} \tag{7.3.4}$$

would become intense, and according to the analysis in Section 3.4, the corresponding transient response behaviour would be unacceptable (under-damped). Even though the pursuit of a maximum bandwidth control system design is not always appropriate, as when for example the continuous input or output data contains significant unfilterable wideband noise, it is a sufficiently useful starting point to be considered here. For this purpose, values of $H^*(\omega)$ from equation (7.3.1) are first used to determine the angular frequency ω_B for which:

---

*Tou (143) devised a similar basic frequency domain compensation procedure for just real poles and zeros. This fuller exploitation involving complex compensator poles was developed independently in the present author's doctoral thesis of 1959 - see reference 114

$$H^*(\omega_B) + \left\{55° \times \text{Class Number}\right\} = -130° \qquad (7.3.5)$$

Equation (7.3.5) indicates that the proposed compensation procedure is orientated towards establishing an adequate phase margin by an apposite choice of scalar gain constant (λ). By means of Fig A7.3.2, which graphs Arg $\left[\exp(j2\pi\omega/\omega_0) - C\right]$ to base of normalised frequency, real controller zeros are selected furthest from the point $1\angle 0$ to give each no more than 70° phase advance at ω_B/ω_0. Observe that once a phase lead of around 70° is produced, neighbouring curves for suitable* pole locations are all roughly parallel, so that the phase lead per stage cannot increase fast enough to keep the locus $\lambda D^*(\omega)H^*(\omega)$ far enough away from the critical point. In the case of a Class-1 system, a compensator's single pole is chosen from the same Fig A7.3.2 to provide between 15 to 20° phase lag at ω_B/ω_0, and subject to achieving an acceptable gain margin, the scalar gain constant (λ) is finally adjusted according to:

$$\left|\lambda D^*(\omega_B)H^*(\omega_B)\right| \simeq 1 \qquad (7.3.6)$$

A comparison of Fig A7.3.2 with the other phase-frequency curves in Appendix A7.3 reveals that complex pole pairs create less phase lag than their real counterparts over the frequency range affecting closed loop stability. Indeed, this feature is patently clear from the graphical construction of a pulse real frequency response illustrated in Fig 3.3.2. Consequently, compensators for Class-2 systems can have complex poles positioned relatively closer to the $1\angle 0$ point so as enhance their DC gain, and thereby their system's load disturbance rejection performance. Alternatively, compensators with complex poles can have their real zeros moved somewhat further back from the $1\angle 0$ point, so that the steady state finite wordlength control error bound ($\hat{\epsilon}$) is reduced according to equation (6.4.23) as:

---

* 'suitable' pole locations are as close as possible to these zeros, in order to maximise DC controller gain and to prevent a Nyquist locus bulging too close to the critical point

$$\hat{\epsilon} = \mu / \left| 2^M \lambda A(1) \right| \tag{7.3.7}$$

where:

M - wordlength of the controller ; $D(z) = A(z)/B(z)$

Equation (7.3.7) also partly motivates the above suggestion that the zeros of a controller should be "furthest from the $1\underline{/0}$ point to give each no more than 70° phase lead at ω_B/ω_o".

Finally, Fig A7.3.2 shows that if the sampling frequency (ω_B) is increased, the necessary phase advance at the bandwidth frequency (ω_B) can be provided only by moving the zeros of a digital compensator closer to the $1\underline{/0}$ point. It follows therefore from equation (7.3.7) that a sampling frequency can be set too high in practice because of the consequential increase in wordlength requirement. Physically, if the sampling frequency of a digital phase advance type compensator were to be progressively increased at a fixed wordlength, amplitude quantisation errors would eventually dominate its intrinsic differencing of successive data samples*, so the realised control action would deviate markedly from that required. At first sight it seems odd that sampling frequency and wordlength of a DDC system are dependent, but control accuracy is the essence of the matter. After selecting an adequately high sampling frequency for a low enough ripple error component, the deployment of a short wordlength incurring large arithmetic round-off errors would be patently inconsistent.

The tractability and effectiveness of the above design principles is now demonstrated by considering the system in Fig 7.1.1 with the plant:

$$G(s) = \frac{1}{s^2(1+s)} \tag{7.3.8}$$

and a sampling period of 0.1 second. The pulse real frequency

---

*as they progressively approach the same value

f (Hz)	0.03	0.05	0.07	0.10	0.20	0.30	0.50
ω (rad/s)	0.19	0.31	0.44	0.63	1.26	1.88	3.14
$H^*(\omega)$	$28\angle{-192°}$	$9.7\angle{-200°}$	$4.7\angle{-218°}$	$2.2\angle{-218°}$	$0.4\angle{-242°}$	$0.13\angle{-252°}$	$0.03\angle{-279°}$
$D^*(\omega)$	$0.019\angle{43°}$	$0.031\angle{62°}$	$0.042\angle{76°}$	$0.065\angle{91°}$	$0.16\angle{112°}$	$0.34\angle{116°}$	$1.00\angle{94°}$
$10D^*(\omega)H^*(\omega)$	$5.4/{-149°}$	$3.0\angle{-138°}$	$2.0\angle{-132°}$	$1.38\angle{-127°}$	$0.64\angle{-130°}$	$0.45\angle{-136°}$	$0.30\angle{-185°}$

Table 7.3.1 Pulse Real Frequency Responses for the Design Example

response values in Table 7.3.1 for this plant, zero order hold, and a computational delay of one sampling period are calculated from equation (7.3.1). These results enable the frequency ω_B defined in equation (7.3.5) to be deduced as approximately 1.26 rad/s so that:

$$\omega_B/\omega_o = 0.02 \tag{7.3.9}$$

Two stages of phase advance are required for this Class-2 example, and two identical zeros at z = 0.95 are seen from Fig A7.3.2 to create each a phase lead of 70° at 1.26 rad/s. According to Fig A7.3.12, two complex poles at $z = 0.8 \pm j0.3$ incur a total phase lag of 24° at 1.26 rad/s so that the digital compensator:

$$\lambda D(z) = \frac{\lambda(z - 0.95)^2}{(z - 0.8 - j0.3)(z - 0.8 + j0.3)} \tag{7.3.10}$$

is readily derived. By setting the scalar gain constant (λ) to 10, a satisfactory gain margin of 10 dB is achieved to complete the design.

When two cascaded RC phase advance networks are required in a continuous data control system, more favourable Nyquist diagrams are produced by using different attenuation ratios (ie zeros). The same appears true as well for digital controllers. In the present case, compensation by:

$$\lambda D(z) = \frac{10(z - 0.975)(z - 0.9)}{(z - 0.8 - j0.3)(z - 0.8 + j0.3)} \tag{7.3.11}$$

endows the same gain margin and load disturbance rejection performance as the design in equation (7.3.10), but the extra 6° of phase lead below 0.16 rad/s results in a superior transient control accuracy. The improved Nyquist diagram shown in Fig 7.3.1 has a closed loop peak magnification of 1.3 at a frequency of 0.44 rad/s. Corresponding ripple and finite wordlength error components are calculated in Section 5.4 and 6.5 respectively. Familiarity with the preceding principles can be promoted by establishing that the alternative compensator

$$\lambda D(z) = \frac{4.8(z - 0.975)^2}{(z - 0.8)^2} \tag{7.3.12}$$

provides adequate stability, but has a relatively inferior load disturbance rejection performance and wordlength requirement. This calculation therefore also serves to exemplify the benefits of using complex poles in a digital compensator.

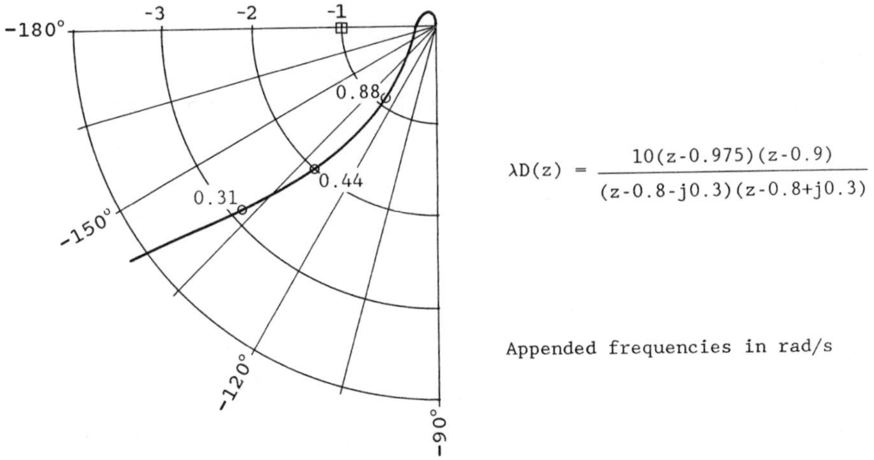

Fig 7.3.1 Nyquist Diagram for the Design Example

7.4 STRUCTURAL RESONANCES

A servomotor is often coupled to its load by intrinsically resilient drive shafts and a gear box*. Because normal dissipative forces are engineered to be as small as practicable, the combination possesses very strong selective resonances. These can occur at frequencies less that three times the required control system bandwidth in some high performance equipment, where they complicate the compensation procedure by causing a Nyquist diagram to loop out again from the origin to approach the critical point. Their influence is evidently aggravated by high values of scalar gain for enhancing load disturbance rejection, and the rising gain-frequency characteristic of phase advance type controllers. When certain causal effects of operation can be largely anticipated, as in firing a gun or missile, feedforward (77, 78) sometimes enables the required control accuracy to be achieved with a small enough scalar gain for structural resonances to be unimportant. In other circumstances, continuous data controllers achieve adequate stability by the divided reset technique** (75, 76) or by additional processing of the control error signal with a notch or anti-resonance filter (144). Excellent descriptions of the divided reset technique are given in the cited references, but here attention is directed at digitally implemented notch filters which block the excitation of oscillatory structural modes. A notch or anti-resonance at ω_N rad/s is provided by the simple digital filter:

$$N(z) = K_N \frac{(z - z_N)(z - \bar{z}_N)}{(z - r_N z_N)(z - r_N \bar{z}_N)} \qquad (7.4.1)$$

where

---

*to match load and motor inertias close enough to achieve maximum acceleration of the load, subject to the constraint of maximum motor speed (72, 144)

**also useful for countering backlash (hysteresis) in mechanical gearing

$$\left.\begin{array}{l} z_N = \exp(j\omega_N T) \quad ; \quad \overline{z}_N = \exp(-j\omega_N T) \\ 0 \leq r_N < 1 \end{array}\right\} \qquad (7.4.2)$$

Unity gain for the filter at zero frequency is clearly imposed by setting:

$$\overline{K_N} = \left|1 - r_N z_N\right| . \left|1 - z_N\right|^{-1} \qquad (7.4.3)$$

A structural resonance can be effectively nulled by matching its bandwidth and centre frequency with the corresponding parameters of the above filter, because then a Nyquist locus cannot loop out again from the origin.

The gain-frequency response of the proposed digital filter is derived from equation (3.3.7) as:

$$\left|N^*(\omega)\right| = \left|N(z)\right| \quad \text{with} \quad z = \exp(j\omega T) \qquad (7.4.4)$$

Defining the frequency deviation variable:

$$\Omega = \omega - \omega_N \qquad (7.4.5)$$

and expanding the exponential functions as Maclaurin series yields the first order approximation:

$$\left|N^*(\Omega)\right| = \frac{K_{N0}}{\sqrt{1 + \left(\dfrac{1 - r_N}{T\Omega}\right)^2}} \qquad (7.4.6)$$

where:

$$K_{N0} = \overline{K_N}\left|1 - \exp(-j2\omega_N T)\right| . \left|1 - r_N \exp(-j2\omega_N T)\right|^{-1} \qquad (7.4.7)$$

Equation (7.4.6) has complete similarity with the gain-frequency characteristic of a cascaded RLC filter in the vicinity of its anti-resonance. Accordingly, the bandwidth (B) of this digital counterpart is likewise defined as the frequency increment about its centre frequency within which the attenuation exceeds 3 dB, so:

$$\bar{B} = \pm(1 - r_N)/T \tag{7.4.8}$$

Calculated or measured plant frequency responses can therefore be used in conjunction with equations (7.4.2) and (7.4.8) to specify the poles and zeros of an appropriate notch filter, which is then included as an algebraic factor of the controller's pulse transfer function.

Though zeros of the above notch filter are precisely defined by the natural frequency of a structural resonance, some latitude apparently exists on its pole locations that determine bandwidth ($\bar{B}$). Maclaurin series expansions give the phase lag (ψ_p) produced by the poles alone as approximately:

$$\psi_p = 2\omega T . F(r_N) \qquad \text{for} \qquad |\omega T| << 1 \tag{7.4.9}$$

where:

$$F(r_N) = \frac{1 - r_N}{(1 - r_N)^2 + (r_N \omega_N T)^2} \tag{7.4.10}$$

Typical behaviour of the function $F(r_N)$ is illustrated in Fig 7.4.1 from which it is concluded that additional

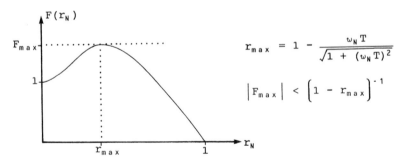

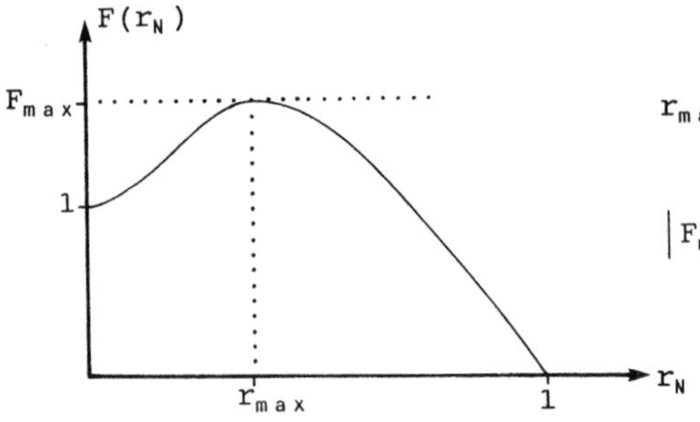

$$r_{max} = 1 - \frac{\omega_N T}{\sqrt{1 + (\omega_N T)^2}}$$

$$|F_{max}| < \left(1 - r_{max}\right)^{-1}$$

Fig 7.4.1 Typical Behaviour of the Function $F(r_N)$ in Equation (7.4.10)

phase lags (inaccuracies) over the control system bandwidth should be minimised by placing the poles as close as practicable to the unit circle. Translated into the frequency domain by means of equation (7.4.8), this objective equates to a notch bandwidth that is just sufficient to prevent a Nyquist locus bulging out towards the critical point. However, some margin may be required in practice to accommodate for example environmentally induced changes in the viscous damping, and therefore in the bandwidth, of a structural resonance.

Though coefficient rounding is usually unimportant as regards the realisation of digital phase advance and phase retard compensators, it is generally prudent for the reasons presented in Section 6.6 to examine its effect on a notch filter. With an arbitrary long wordlength, the pulse transfer function in equation (7.4.1) is written as:

$$\overline{N}(z) = \frac{z^2 - \overline{a}z + 1}{z^2 - \overline{rb}z + \overline{r^2}} \qquad (7.4.11)$$

where the ideal coefficients are derived from equations (7.4.1) and (7.4.2) as:

$$\left.\begin{aligned} &\overline{a} = 2\cos\omega_N T \quad ; \quad \overline{rb} = 2r_N \cos\omega_N T \\ &\overline{r^2} = r_N^2 \end{aligned}\right\} \qquad (7.4.12)$$

Representing the above coefficients as $\{a, rb, r^2\}$ in a finite wordlength format is seen therefore to modify both the filter's centre frequency and bandwidth. To quantify these changes define:

$$\left.\begin{aligned} &a = 2\cos\omega_1 T \quad ; \quad b = 2\cos\omega_2 T \\ &\Omega_1 = \omega - \omega_1 \quad ; \quad \theta = \omega_1 - \omega_2 \end{aligned}\right\} \qquad (7.4.13)$$

In this nomenclature the gain-frequency characteristic of an actual realisation N(z) is given by:

$$\left|N^*(\Omega_1)\right| = \overline{K_N} \left|\frac{\exp(j\Omega_1) - 1}{\exp(j\Omega_1 + j\theta) - r}\right| \left|\frac{1 - \exp(-j\omega T - j\omega_1 T)}{1 - r\exp(-j\omega T - j\omega_2 T)}\right| \tag{7.4.14}$$

Thus the actual centre frequency is now at ω_1, and this shift from ω_N can be evaluated directly from equations (7.4.12) and (7.4.13) as:

$$\Delta\omega_N = \frac{1}{T}\left\{\cos^{-1}(a/2) - \cos^{-1}(\overline{a}/2)\right\} \tag{7.4.15}$$

Setting:

$$\Delta a = a - \overline{a} \tag{7.4.16}$$

then because the ideal centre frequency (ω_N) is non-zero in the present context, a Maclaurin series expansion of the inverse cosine function about $\overline{a}/2$ provides the first order approximation:

$$\left|\Delta\omega_N\right| = \frac{\left|\Delta a\right|}{T\sqrt{4 - \overline{a}^2}} \tag{7.4.17}$$

Around the centre frequency and with fine enough quantisation:

$$T\Omega_1 << 1 \qquad \text{and} \qquad T\theta << 1 \tag{7.4.18}$$

so that similar Maclaurin expansions can be applied to the exponential functions in equation (7.4.14) to yield:

$$\left|N^*(\Omega_1)\right| \simeq K_N \left\{(1 + \theta/\Omega_1)^2 + (1 - r)^2/(T\Omega_1)^2\right\}^{-1/2} \tag{7.4.19}$$

where:

$$K_N = \overline{K_N} \left|\frac{1 - \exp(-j2\omega_1 T)}{1 - \exp(-j2\omega_2 T)}\right| \tag{7.4.20}$$

Thus the 3 dB bandwidth (B) of an actual realisation is specified by:

$$(1 + \theta/B)^2 + (1 - r)^2/(TB)^2 = 2 \qquad (7.4.21)$$

or

$$B \simeq 2\theta B + (1 - r)^2/T^2 \qquad (7.4.22)$$

After substituting:

$$\Delta B = B - \overline{B} \qquad \text{and} \qquad \Delta r = r - \overline{r} \qquad (7.4.23)$$

into equation (7.4.22), binomial expansions* then give the first order approximation:

$$|\Delta B| \le |\theta| + |\Delta r|/T \qquad (7.4.24)$$

Because the denominator coefficient $\overline{r^2}$ of N(z) is realised within the error bounds ± q/2, then:

$$|\Delta r| \le q/4\overline{r} \qquad (7.4.25)$$

and from equation (7.4.8):

$$|\Delta r| \le \frac{q}{4(1 - T\overline{B})} \qquad (7.4.26)$$

Furthermore, by the triangular inequality (6.4.1)

$$|\theta| \le |\Delta\omega_N| + |\omega_2 - \omega_N| \qquad (7.4.27)$$

where from equations (7.4.12) and (7.4.13):

$$\omega_2 - \omega_N = \cos^{-1}(rb/2r) - \cos^{-1}(\overline{a}/2) \qquad (7.4.28)$$

Because the denominator coefficient $\overline{rb}$ of $\overline{N}(z)$ is realised within error bounds of ± q/2, then a Maclaurin series expansion of the inverse cosine function about $\overline{a}/2$ provides the first order approximation:

---

*This ab initio treatment is considered more transparent here than the differentiation procedure

$$\left|\omega_2 - \omega_N\right| \leq \frac{2}{T\sqrt{4 - \bar{a}^2}} \left[0.5\bar{a} \left|\Delta r\right| + 0.25q/\bar{r}\right] \qquad (7.4.29)$$

Finally, by substituting equations (7.4.26), (7.4.27) and (7.4.29) into equation (7.4.24), the change in notch bandwidth (ΔB) induced by a finite wordlength representation of the filter coefficients is derived as:

$$\left|\Delta B\right| \leq \left|\Delta \omega_N\right| + 0.25q \left(1 + \frac{2 + \bar{a}}{\sqrt{4 - \bar{a}^2}}\right)\left(T - T^2 \bar{B}\right)^{-1} \qquad (7.4.30)$$

where as usual:

$$q = 2^{-M+1}$$

and M denotes the computer wordlength.

To illustrate the above analysis, consider the realisation of a digital notch filter having a centre frequency of 12 rad/s, a bandwidth of ± 1 rad/s and a sampling period of 0.1 second. Equation (7.4.12) defines the ideal numerator coefficient a as:

$$\bar{a} = 0.724716$$

but the coefficient actually realised in a fixed point 8 bit format is:

$$a = 0.718750 \qquad \text{or} \qquad 0.1011100 \text{ (binary)}$$

The corresponding deviations in centre frequency and bandwidth evaluate from equations (7.4.17) and (7.4.30) as

$$\left|\Delta \omega_n\right| = 0.032 \text{ rad/s}$$

and

$$\left|\Delta B\right| \leq 0.085 \text{ rad/s}$$

respectively. Thus in practical terms an 8-bit microprocessor is about sufficient for realising this particular filter specification.

7.5 COMPENSATION OF CLASS-0 OR REGULATOR SYSTEMS

The control accuracy or stability margin of a Class-0 system can usually be improved by a phase retard compensator, whose Laplace transfer function takes the general form:

$$R(s) = K_R(1 + s\alpha\tau)/(1 + s\tau) \qquad (7.5.1)$$

In practice the attenuation constant (α) is less than unity, but greater than about 1/12. To clarify subsequent developments, the corresponding real frequency response $R(\omega)$ is illustrated as a Bode diagram in Fig 7.5.1. Although A-D converters are sometimes time-division multiplexed between several control loops, each loop normally has exclusive use of a D-A converter which is now assigned an additional role. Fig 7.5.2 depicts a series resistor and capacitor connected in parallel with the feedback resisitor R_b of a D-A converter, which is the recommended hardware for implementing the phase retard compensation of Class-0 DDC systems. Unless marked variations in plant dynamics necessitate significant changes in controller parameters, this RC network (with possibly

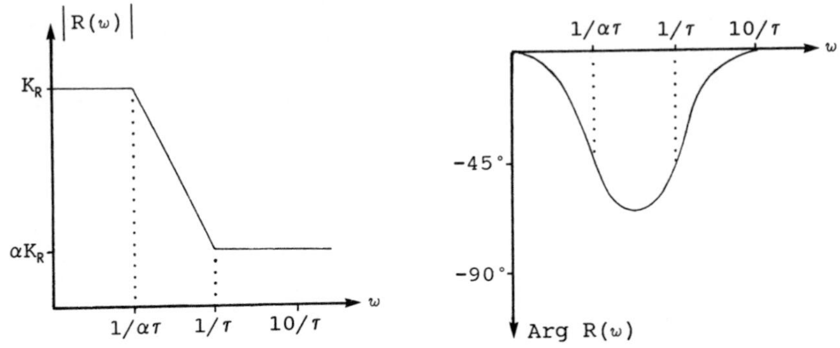

Fig 7.5.1 Bode Diagram of an RC Phase Retard Network

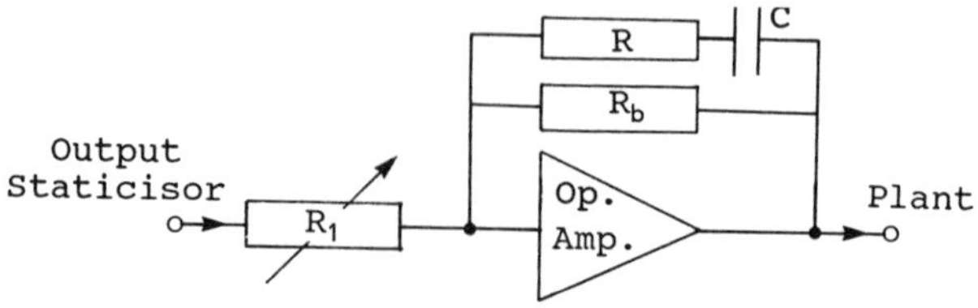

$K_R = -R_b/R_1 \quad ; \quad \alpha = R/(R + R_b) \quad ; \quad \tau = (R + R_b)C$

Fig 7.5.2 Phase Retard Network using D-A Converter

an adaptive gain constant in the digital controller) is considered a better option than a digital counterpart designed using the curves in Appendix A7.3. Experience shows that the RC network introduces an appreciable phase lag over a relatively narrower frequency band, and this can be appreciated as important from the following description of phase retard compensation.

After computing the pulse real frequency response function $H^*(\omega)$ for the plant and zero order hold from equation (7.3.1), the next step is to determine the angular frequency (ω_B) for which:

$$\underline{/H^*(\omega)} = -130° \tag{7.5.2}$$

If the gain and phase margins of the uncompensated closed loop system are satisfactory, but the problem is to improve its steady state control error for a step input to ϵ_o, then define:

$$10/\alpha\tau = \omega_B \tag{7.5.3}$$

and set:

$$\alpha = \left[\epsilon_o/(1 - \epsilon_o)\right]H^*(o) \quad ; \quad K_R = 1/\alpha \tag{7.5.4}$$

but subject to the empirical constraint of:

$$1/12 \lesssim \alpha \lesssim 1 \tag{7.5.5}$$

With these parameter values, the gain and phase of $R(\omega)$ for frequencies around and above ω_B are seen from Fig 7.5.1 to be about unity and zero respectively. Its gain and phase at zero frequency are $1/\alpha$ and zero, so that the steady state control error for step input is now:

$$\left[1 + (1/\alpha)H^*(o)\right]^{-1} = \left[1 + (1 - \epsilon_o)/\epsilon_o\right]^{-1} = \epsilon_o \qquad (7.5.6)$$

as required. Alternatively, if the steady state step response is acceptable but improvements in the stability margins of the closed loop system are sought, then the parameters of the network R(s) should be selected according to:

$$\left.\begin{array}{ll} K_R = 1 \quad ; & 10/\alpha\tau = \omega_B \\ \alpha = 1/\left|H^*(\omega_B)\right| & \text{for a } 50^\circ \text{ Phase margin} \end{array}\right\} \qquad (7.5.7)$$

With these parameter values, the gain and phase of $R(\omega_B)$ are α and zero respectively, so the gain margin is improved by the factor $20 \log_{10}(1/\alpha)$ dB and the phase margin remains unaltered at 50°. Both the above design procedures for phase retard compensators contrive to sacrifice phase in an unimportant portion of a Nyquist locus for gain in a separate region that significantly influences control accuracy or stability margins. Decreasing the attenuation constant (α) is seen from Fig 7.5.1 to widen progressively the frequency range over which the network R(s) introduces an appreciable phase lag. With a sufficiently small attenuation factor, this additional phase lag intrudes into low enough frequencies to compromise the control accuracy. Accordingly, the minimum value of the attenuation constant is empirically recommended as about 1/12.

If the continuous input or output data of a DDC system are significantly contaminated by wideband noise, and if prefiltering as described in Section 7.1 is impracticable, then an irrevocable loss of control accuracy can be prevented only by employing a high enough sampling frequency to adequately separate the noise sideband spectra. Under these circumstances, the operational amplifier in a D-A converter can be used in another form of signal conditioning. By placing a suitable capacitor in parallel with its feedback resistor (R_b), the plant actuator demand signal is smoothed thereby alleviating wear and tear associated with excessive operations. In addition, the steady state ripple error component is diminished by a greater attenuation of spectral sidebands with the increase of plant rank.

7.6 TIME DOMAIN SYNTHESIS OF DDC SYSTEMS

Time domain synthesis aims to compensate a linearised DDC system model so that certain time domain responses comply exactly with an a priori specification. An early publication on the topic by Smith et al at the 1951Cranfield Conference proposed that a DDC system should reproduce a kth degree polynomial extrapolation of its sampled input data. Denoting a control error sequence by $\{e_n\}$, the advocated design strategy was the selection of an overall pulse transfer function to minimise the steady state weighted error function:

$$S_N = \sum_{n=0}^{\infty} c^n e_{N-n} \qquad (7.6.1)$$

with:

$$|c| < 1 \qquad (7.6.2)$$

Values of the *staleness factor* (c) less than unity were recommended in order to restrict the influence of initial transient behaviour, which clearly cannot match the specified k^{th} degree polynomial form. On this basis the required overall pulse transfer function was derived as:

$$K(z) = \left(\frac{z - 1}{z - c} \right)^{k+1} \qquad (7.6.3)$$

so that the intuitively founded inequality (7.6.2) is actually necessary for system stability. Ragazzini and Franklin (28) later approached the same problem somewhat differently. They assumed that equation (7.6.3) defined a desirable *prototypical pulse transfer function*, whose pole position (c) should be adjusted to minimise response time or the sum of the squares of a control error sequence. However, digital compensation of a linearised model to impose a 'desirable' form of overall pulse transfer function or a strictly finite response time is now shown to be generally inappropriate.

Compensation to achieve a particular overall pulse transfer function like that in equation (7.6.3) could evidently require the exact cancellation of plant zeros outside the unit circle by controller poles (52, 53). It may also require the exact cancellation of plant poles outside the unit circle by controller zeros (28). Such cancellations would be inherently imperfect in practice due to the finite wordlength of a controller, and the system would be made unstable. Bertram (28, 147) devised an overall pulse transfer function synthesis that neatly avoids untenable pole or zero cancellations, and which theoretically achieves the fastest response to a conventional test input (step, ramp etc). However, linear compensation techniques that impose strictly finite response times are almost certainly incompatible with rate constraints in real non-linear plant, and in consequence they are non-viable. Another of their inherent deficiencies is best illustrated by example using the so-called *dead beat* or *ripple free** design procedure (28, 37). These idealised DDC systems track some specified continuous input polynomial exactly with no intersample ripple after a strictly finite number of sample periods. Suppose the unity feedback system in Fig 7.1.1 contains the plant:

$$G(s) = \frac{10}{s(1 + s)} \tag{7.6.4}$$

with a sampling period of 1 second, then the overall pulse transfer function for a dead beat response to a ramp input is

*Such systems are not really ripple free at all; except for the specific input polynomial around which a design is constructed

derived in reference (28) as:

$$K(z) = 0.73z^{-1} + 1.35z^{-2} - 0.90z^{-3} - 0.18z^{-4} \qquad (7.6.5)$$

Straightforward manipulation of the relevant transfer functions then yields the required digital compensator as:

$$\lambda D(z) = \frac{0.73 - 1.007z^{-1} + 0.445z^{-2} - 0.635z^{-3}}{1 + 0.27z^{-1} - 1.094z^{-2} - 0.176z^{-3}} \qquad (7.6.6)$$

The step, ramp and overall pulse real frequency responses of the above system are shown in Fig 7.6.1. Although a reasonably damped ramp response is achieved, the resonance in the overall real frequency is more strongly excited by the wider energy spectrum of a step input* to produce for many purposes an unacceptably large transient overshoot. Finite settling time systems are generally 'highly tuned' in this way; their response to the specific design input is extraordinary, while to others it is apt to be poor (28). A comparison of the above dead beat design with that effected using the recommended frequency domain procedure in Section 7.3 demonstrates the superiority of this previously recommended technique. For a closed loop bandwidth of 1 rad/s, it can be deduced in the first place from Fig A7.3.2 that stiffer compensation is on offer if the sampling period is decreased to 1/4 second**. The reader can then readily verify from such curves that:

$$\lambda D(z) = \frac{0.4(z - 0.9)}{(z - 0.4)} \qquad (7.6.7)$$

is a suitable compensator giving a gain margin of 10 dB. The corresponding step, ramp, and overall pulse real frequency responses shown in Fig 7.6.2 are all 'even-tempered' which presages well for actual operation with other than polynomial inputs.

---

*refer to the discussion on page 93

**the compensator's zero can then be placed further back from $1\angle 0$, because its pole then produces proportionately less phase lag

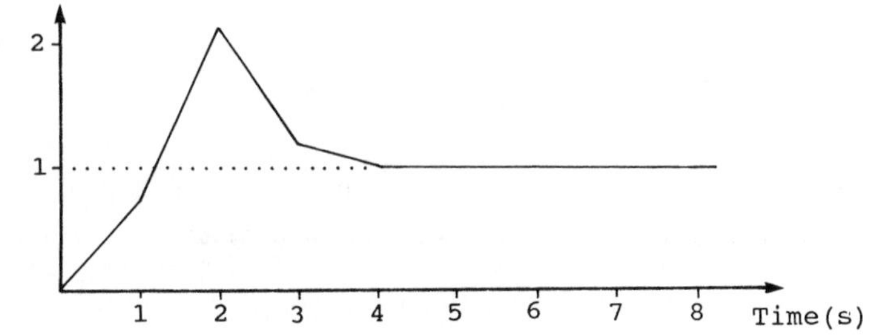

a) Step Response

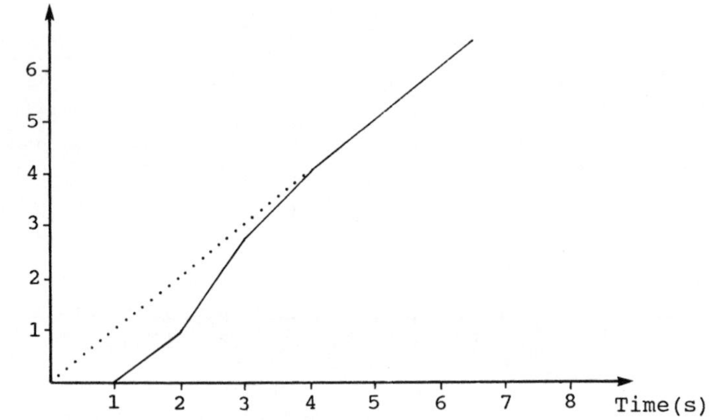

b) Ramp Response

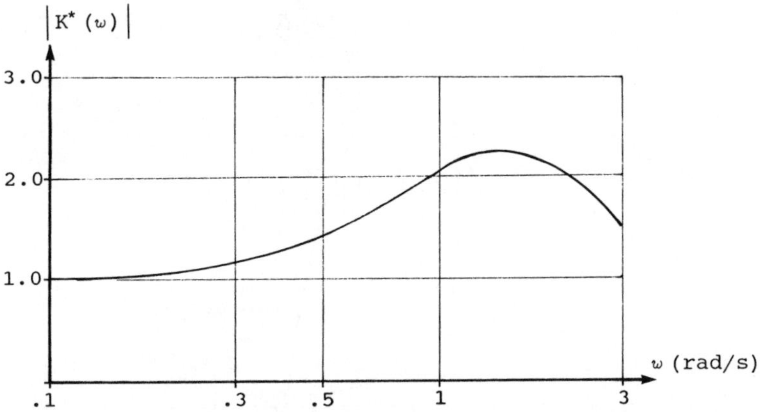

c) Closed Loop Pulse Real Frequency Response

Fig 7.6.1 Responses of the Dead Beat Design

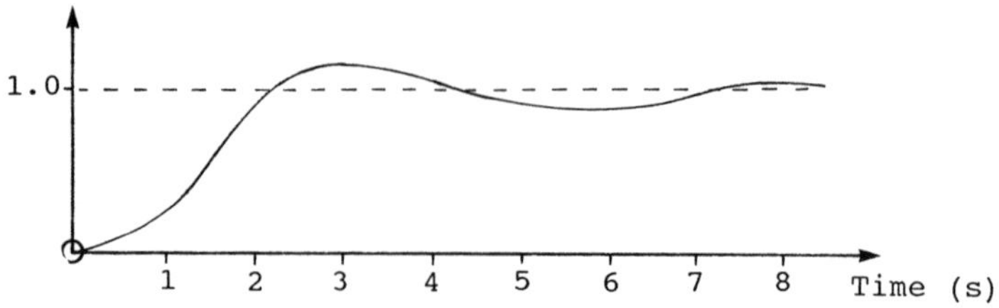

a) Step Response

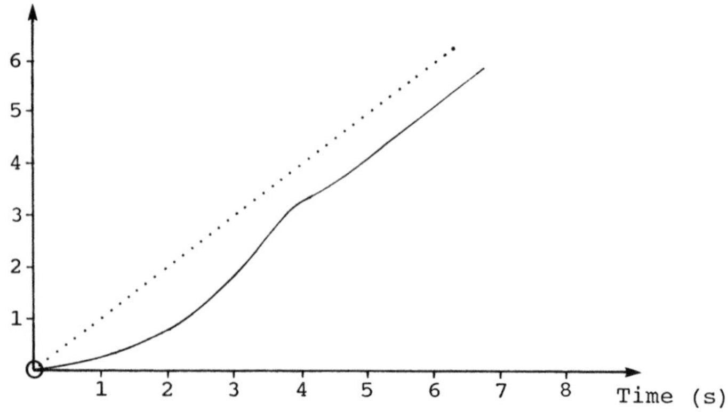

b) Ramp Response

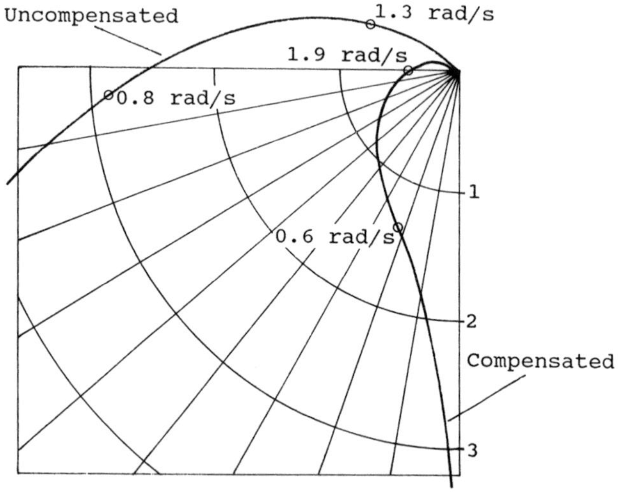

c) Nyquist Diagram

Fig 7.6.2 Responses of the Frequency Domain Computerised Design

Smelka (147) has proposed a less academic approach to time domain synthesis in which a digital compensator takes the general form:

$$\lambda D(z) = (d_0 + d_1 z^{-1} + d_2 z^{-2}) D_s(z) \tag{7.6.8}$$

with

$$D_s(z) = (1 - z^{-1})^k \tag{7.6.9}$$

The coefficients $\{d_0, d_1, d_2\}$ are uniquely defined by the requirement of zero steady state control error, and the specified values of transient overshoot (M_T) and rise time ($\tilde{n}T$) for a kth degree polynomial input. While rise times compatible with plant constraints can be inferred from simulation or experience, such independent choices are not generally consonant with the inherent dynamics of a linearised plant model. For example, granted a dominant or dominant pair of poles and a realistic sampling frequency, Section 3.4 establishes a one to one correspondence between M_T and $\omega_p \tilde{n}T$ where:

ω_p - the resonant frequency at which the peak magnification (M_p) occurs

Smelka's synthesis therefore imposes on the linearised feedback system a resonant frequency, which could easily be at variance with its open loop pole-zero pattern. On the other hand, the Nyquist Diagram technique in Sections 7.3 and 7.5 intrinsically includes the pole-zero pattern of the linearised plant model, and the correspondence in Section 3.4 provides a firm bridge into the time domain for checking compatibility with plant rate constraints. A compensator designed in the frequency domain is also not required to implement the numerical differentiation associated with the factor $D_s(z)$ in equation (7.6.9). This factor reduces the dynamic control accuracy achievable by accentuating superimposed noise or by necessitating a smaller controller gain to preserve gain margin. If a plant contains the requisite number of integrations, this factor is in any case superfluous. Finally, unlike the Nyquist Diagram method, the complexity of Smelka's and other time domain syntheses appears to increase disproportionately with that of the plant. For all the above reasons, time domain synthesis procedures for DDC systems are not recommended.

Although the relative simplicity of digital controllers designed in the frequency domain might raise doubts regarding untapped performance capabilities, they have enjoyed unequivocal success for over fifty years in satisfying the transient performance and robustness requirements of practical single- and multi-input control systems (150, 151, 152). Nyquist or Inverse Nyquist Diagram techniques, together with adequately detailed non-linear plant simulations for essential physical insight and validation, are therefore the presently advocated design route for DDC systems.

7.7 OVERFLOW

Amplitude saturation of the error signal in a feedback control system is often destabilising because its controller cannot then supply the derivative action necessary to compensate for time lags or delays in the plant. With DDC systems, the 'idiosyncrasies' of fixed point binary arithmetic further aggravate this stability problem. For example, consider the following simplified situation in which binary numbers are represented in the two's complement notation:

System Input = 0.11 (0.75 decimal)

System Output = 1.01 (-0.75 decimal)

Although the control error in a unity feedback situation is 1.5 (decimal), binary arithmetic on the basis of equation (1.4.6) actual yields 110 (binary) or -0.5 (decimal), so that the plant would be driven in the wrong direction. Overflow (saturation) in a DDC system must clearly be prevented as the situation threatens by switching from the linear mode to a non-linear mode of control; like a constant output velocity. Such coarse-fine servomechanisms

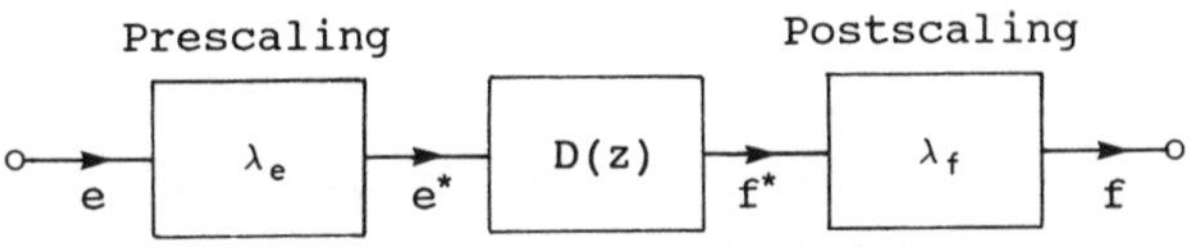

Fig 7.7.1 Typical Distribution of Scalar Gain in a Linear Controller

(134) find frequent industrial application, but their realisation in digital form clearly requires the onset of overflow to be forecast analytically. As described in Section 6.5, the specific distribution of scalar gain in a linear controller affects both the finite wordlength control error component and the onset of overflow. Consequently, this distribution features in the following analysis of the overflow problem that is centred around Fig 7.7.1.

A reasonable strategy is to allow operation of the linear control mode only when a system's input and output correspond to a control error in some range [-E, E]. Though clearly:

$$0 < E < 1 - q \tag{7.7.1}$$

more recondite constraints must generally be sought in practice. First and foremost, the range must be small enough to prevent overflow. However, it should also be large enough to enable the linear controller to achieve adequate stability after the imposition of perhaps a constant output velocity in the non-linear control mode. The question of the widest error range without overflow occurring in a linear control mode is greatly simplified by the special feature of fixed point two's complement arithmetic described in Section 1.4. Specifically, the addition of any finite sum of in-range* numbers is unaffected by intermediate overflows provided their correct total is also in-range. As an illustration of this remarkable and important property, it is

---

*within the interval [-1, 1-q]

instructive to evaluate in three bit format the sum:

$$8(1/2) - 4(3/4) - 1/2 \qquad (= 1/2)$$

by different strategies. Using left-shift operations to implement the scalar multiplications, the pertinent machine held numbers are:

$$8(1/2) \equiv 0.00 \quad ; \quad 4(3/4) \equiv 0.00 \quad ; \quad 1/2 \equiv 0.10$$

Effecting the required subtractions in binary by equation (1.4.6) then yields:

$$0.00 + (1.11 + 0.01) + (1.01 + 0.01) = 1.10 \qquad (= -1/2)$$

which is patently erroneous. On the other hand, its evaluation as a finite series using multiple additions proceeds as:

1/2 + 1/2 + 1/2 + 1/2 + 1/2 + 1/2 + 1/2 + 1/2	
	binary subtotal so far = 0.00
- 3/4 - 3/4 - 3/4 - 3/4	binary subtotal so far = 1.00
- 1/2	final binary total = 0.10

which is the correct result. Consequently, all multiplications by coefficients or scalar gains outside the machine range [-1, 1-q] should be programmed as repeated additions and not as left-shift operations. Adopting this programming convention enables the onset of significant overflows to be prevented by simply ensuring that the 'external' variables e, e^*, f^* and f always lie in the machine range [-1, 1-q].

If the extent of the linear control regime is assigned algebraically for the moment as [-E, E], then clearly:

$$|e_n| \leq E \qquad (7.7.2)$$

and

$$\left| e_n^* \right| \leq \lambda_e E \tag{7.7.3}$$

Expressing the output of the frequency dependent part of the linear controller D(z) as the convolution summation (2.4.2):

$$f_n^* = \sum_{0}^{n} d_k e_{n-k}^* \tag{7.7.4}$$

then application of the triangular inequality (6.4.1) gives:

$$\left| f_n^* \right| \leq \sum_{0}^{n} \left| d_k \right| \left| e_{n-k}^* \right| \leq \lambda_e E \sum_{0}^{\infty} \left| d_k \right| \tag{7.7.5}$$

where terms of the above infinite summation are computed from equation (2.2.2) as*:

$$d_n = \sum_{k=1}^{K} a_k \delta_{n-ko} - \sum_{k=1}^{K} b_k d_{n-k} \tag{7.7.6}$$

Finally:

$$\left| f_n \right| = \lambda_f \left| f_n^* \right| \leq \lambda_f \lambda_e E \sum_{0}^{\infty} \left| d_k \right| \tag{7.7.7}$$

so, provided the repeated addition programming convention is adopted, then overflow is prevented provided that:

$$\text{Max} \left\{ E,\ \lambda_e E,\ \lambda_e E \sum_{0}^{\infty} \left| d_k \right|,\ \lambda_f \lambda_e E \sum_{0}^{\infty} \left| d_k \right| \right\} \leq 1 - q \tag{7.7.8}$$

---

*In the nomenclature of Fig 7.7.1, the coefficients $\{a_k\}$ and $\{b_k\}$ include any distribution of scalar gain into the frequency dependent part of a compensator D(z)

Extension of the above analysis to other programming structures is quite straightforward (35). For example, each pulse transfer function component of a parallel realisation contributes a term involving the infinite absolute sum of its weighting sequence to the left-hand-side of the above inequality. Similarly with cascade programming, the products of several such infinite sums are involved in the consideration.

Section 6.5 establishes that a finite wordlength control error component decreases with increasing values of prescaling gain (λ_e). According to the above inequality (7.7.8), increasing the prescaling gain necessitates a proportionate decrease in the width (E) of a linear control regime. On the other hand, a wider linear regime provides more opportunity for the linear controller to effect adequate stability after the imposition of perhaps a constant output velocity in the non-linear control mode. Indeed, a linear regime which is too narrow can cause unstability by allowing for example too much kinetic energy to be developed by a load inertia. Consequently, a finalised control scheme must achieve a compromise in this respect using an iterative procedure that generally involves engineering judgement and non-linear simulation.

For the majority of practical cases, this text could end here with the completion of a comprehensive set of adaptable design principles. However, the published literature describes controllers whose input and output samplers operate at different, yet integral multiple, rates. At one time such schemes were thought to endow specific advantages (28, 154). Design techniques along similar lines to those presented in Chapters 5, 6 and 7 are next briefly presented to examine these claims more fully.

CHAPTER 8

Multirate and Subrate Systems

I am one of those made for exceptions. - O Wilde

8.1 STABILITY AND COMPENSATION OF MULTIRATE DDC SYSTEMS

A *multirate* DDC system is regarded here as one having a controller whose output sampler operates an integral number of times faster than its input unit. On the basis that a multirate controller appears to drive the plant with a closer approximation to continuous data, Kranc originally concluded that the ripple component of control error would be lower than that of the corresponding single rate system (154). However, this argument overlooks the fact that a multirate controller first produces sideband spectra shifted to around multiples of its input sampling frequency ω_o, and these certainly cannot be eliminated by the output sampler operating at $n\omega_o$*. Hence from the outset Kranc's conjecture is not necessarily correct. The preceding theory of single rate DDC systems is now formally extended so that a soundly based assessment of the multirate strategy can be made.

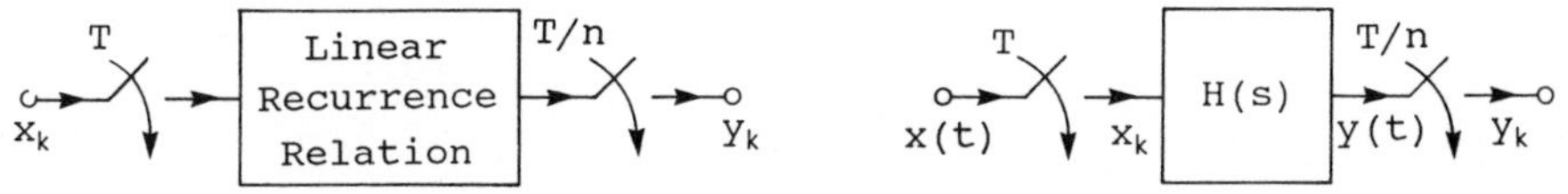

Fig 8.1.1 Open Loop Multirate Systems

---

*the input generated sideband centred on ω_o is for instance reproduced after output sampling at ω_o, $\pm n\omega_o$, $\pm 2n\omega_o$ etc

Analysis of multirate systems like in Fig 8.1.1 is rendered straightforward by *deterministically embedding* the slow rate input sequence $\{x_k\}$ into the fast rate sequence $\{\underline{x}_k\}$ defined by:

$$\left.\begin{array}{lll} \underline{x}_k & = x_p & \text{for } k = np \\ & = 0 & \text{otherwise} \end{array}\right\} \qquad (8.1.1)$$

With this notation for example, the input sequence:

$$x_p = 0 \quad 1 \quad 3 \quad 2 \quad 1 \quad \text{etc}$$

of a triple rate system becomes:

$$\underline{x}_k = 0\ 0\ 0\ 1\ 0\ 0\ 3\ 0\ 0\ 2\ 0\ 0\ 1 \quad \text{etc}$$

In this way, a multirate system is conceptually converted into a single rate system at the higher sampling rate, but to identify the multirate situation it is customary to write z-transformations as functions of the complex variable z_n. Accordingly, an output sequence in Fig 8.1.1 is transformed as:

$$Y(z_n) = \sum_{k=o}^{\infty} y_k z_n^{-k} \qquad (8.1.2)$$

and applying the input-output relationships in Sections 2.2 and 2.4 yields:

$$Y(z_n) = D(z_n)\underline{X}(z_n) \qquad \text{or} \qquad Y(z_n) = H(z_n)\underline{X}(z_n) \qquad (8.1.3)$$

where:

$$D(z_n) = \left[\sum_{o}^{P} a_p z_n^{-p}\right] \cdot \left[1 + \sum_{1}^{P} b_p z_n^{-p}\right]^{-1} \qquad (8.1.4)$$

$$H(z_n) = \sum_{o}^{\infty} h(kT/n) z_n^{-k} \qquad (8.1.5)$$

and

$$\underline{X}(z_n) = \sum_{0}^{\infty} \underline{x}_k z_n^{-k} \tag{8.1.6}$$

By virtue of equation (8.1.1) the above transformation of the embedded input sequence can be written as:

$$\underline{X}(z_n) = \sum_{0}^{\infty} x_p z_n^{-np} = X(z)\Big|_{z = z_n^{\,n}} \tag{8.1.7}$$

so that equations (8.1.3) become:

$$Y(z_n) = D(z_n)X(z_n^{\,n}) \quad \text{or} \quad Y(z_n) = H(z_n)X(z_n^{\,n}) \tag{8.1.8}$$

in which the pulse transfer function $H(z_n)$ is derived from equation (2.4.12) as:

$$H(z_n) = \sum \text{Residues of} \left\{ \frac{H(s)}{1 - z_n^{-1}\exp(sT/n)} \right\} \text{ at poles of } H(s) \text{ only} \tag{8.1.9}$$

More generally, the embedding artifice enables almost all the preceding mathematics of single rate systems to be applied directly to the multirate case. In particular an application of the Inversion Integral (2.1.9) to equation (8.1.8) proves that:

> A multirate system is stable if and only if its poles all lie strictly within the unit circle of the z_n-plane. (8.1.10)

Translation of this result into a Nyquist Diagram technique for engineering the adequate stability of a multirate feedback system involves, however, further mathematical developments and an approximation.

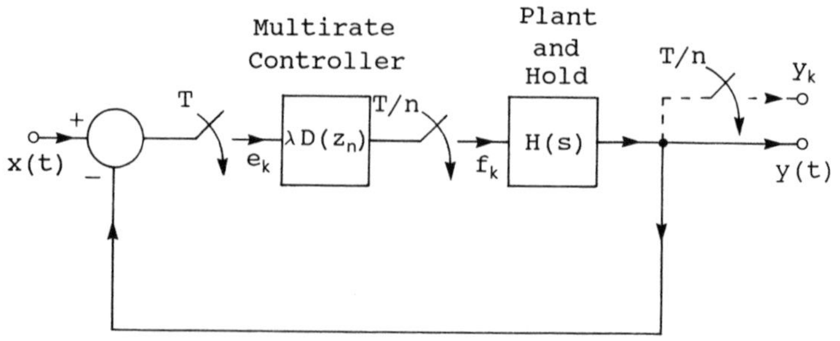

Fig 8.1.2 A Unity Feedback Multirate DDC System

Compensation of the multirate system in Fig 8.1.2 relates to its overall pulse transfer function $K(z_n)$, whose derivation necessitates a formal procedure for summing the subsequence of terms in an arbitrary z_n-transformation $C(z_n)$ that are indexed by integral multiples of n. That is, given:

$$C(z_n) = \sum_{0}^{\infty} c(kT/n)\, z_n^{-k} \tag{8.1.11}$$

the problem is to find:

$$\mathscr{S}\left[C(z_n)\right] = \sum_{0}^{\infty} c(pT)\, z_n^{-np} \tag{8.1.12}$$

from which is minted the z-transformation:

$$Z\left[C(z_n)\right] = \sum_{0}^{\infty} c(pT)\, z^{-p} \tag{8.1.13}$$

by simply setting:

$$z = z_n^{\,n} \tag{8.1.14}$$

The required result is derived in Appendix A8.1 as:

$$Z\left[C(z_n)\right] = \frac{1}{n}\sum_{m=0}^{n-1} C(p_m) \qquad (8.1.15)$$

with:

$$p_m = z^{1/n}.\exp(j2\pi m/n) \qquad \text{for } 0 \le m \le n-1 \qquad (8.1.16)$$

Now the slow rate control error sequence in Fig 8.1.2 has the z- transformation:

$$E(z) = X(z) - Y(z) \qquad (8.1.17)$$

and according to equation (8.1.8) fictitious fast rate sampling of the continuous output data yields the sequence:

$$Y(z_n) = \lambda D(z_n)H(z_n)E(z_n{}^n) \qquad (8.1.18)$$

Applying equation (A8.1.16) to derive the corresponding z- transformation gives:

$$Y(z) = \lambda Z\left[D(z_n)H(z_n)\right]E(z) \qquad (8.1.19)$$

which after substitution into equation (8.1.17) produces:

$$E(z) = \frac{X(z)}{1 + \lambda Z\left[D(z_n)H(z_n)\right]} \qquad (8.1.20)$$

Equations (8.1.12) and (8.1.13) imply for any arbitrary multirate pulse transfer function $C(z_n)$ that:

$$\mathcal{S}\left[C(z_n)\right] = Z\left[C(z_n)\right]\Big|_{z = z_n{}^n} \qquad (8.1.21)$$

and so substitution of equation (8.1.20) into (8.1.18) yields the required overall multirate pulse transfer function as:

$$K(z_n) = \frac{Y(z_n)}{X(z_n{}^n)} = \frac{\lambda D(z_n)H(z_n)}{1 + \lambda \mathscr{S}\left[D(z_n)H(z_n)\right]} \tag{8.1.22}$$

Statement (8.1.10) asserts that poles of a multirate pulse transfer function outside or on the unit circle:

$$\left| z_n \right| = 1 \tag{8.1.23}$$

represent unstable modes of oscillation. The presence or absence of such poles can clearly be established by applying the encirclement theorem (3.5.3) with a z_n- plane contour exactly like that shown in Fig 3.5.1 for a single rate system. On this basis it can be concluded that:

> Necessary and sufficient conditions for the multirate DDC system in Fig 8.1.2 to be stable are that its Nyquist diagram encircles the (-1, 0) point of the $\mathscr{S}\left[D(z_n)H(z_n)\right]$-plane a total of P-times anticlockwise, where P denotes the number of poles of this function outside the unit circle.
>
> (8.1.24)

Once again the most significant portion of a Nyquist diagram is the mapping of the unit circle:

$$\lambda \mathscr{S}\left[D(z_n)H(z_n)\right]_{z_n = \exp(j\omega T/n)} = \lambda Z\left[D(z_n)H(z_n)\right]\Bigg|_{z = \exp(j\omega T)} \tag{8.1.25}$$

which by virtue of equations (8.1.15) and (8.1.16) can be written as:

$$\mathscr{S}\left[D(z_n)H(z_n)\right]_{z_n = \exp(j\omega T/n)} = \left(\frac{1}{n}\right)\sum_{m=o}^{n-1} D^*(\omega + 2\pi m/T)H^*(\omega + 2\pi m/T) \tag{8.1.26}$$

with:

$$D^*(\omega) = D(z_n)\Big|_{z_n = \exp(j\omega T/n)} \qquad (8.1.27)$$

Though $H^*(\omega)$ is similarly related to $H(z_n)$, it is more helpful here to invoke equation (2.4.17) in the form:

$$H^*(\omega) = \left(\frac{n}{T}\right)\sum_{-\infty}^{\infty} H(\omega - 2\pi pn/T) \qquad (8.1.28)$$

where:

$$H(\omega) = H(s)\Big|_{s = j\omega} \qquad (8.1.29)$$

In practice a linearised plant model must severely attenuate ripple sidebands at multiples of $2\pi/T$ rad/s*, so that over the frequency band determining closed loop stability the summation in equation (8.1.26) is totally dominated by its first term. Therefore for stability calculations, equation (8.1.26) can be reduced to:

$$\mathscr{A}\left[D(z_n)H(z_n)\right] \simeq \left(\frac{1}{n}\right)D^*(\omega)H^*(\omega) \qquad (8.1.30)$$

If doubts exist about the adequacy of the above approximation in a particular case, (despite the generous stability margins normally allowed), a completely accurate plot could be computed from equation (8.1.26).

The tenability of frequency domain compensation techniques ultimately resides in the existence of a firm analytical bridge into the time domain. It is now shown that a practical multirate DDC system can be closely approximated by an easily formulated single rate system, so that in this respect the fundamentally sound relationships in Section 3.4 apply. If the continuous input data of the system in Fig 8.1.2 is set to:

---

*refer to the first paragraph of this Section and the more detailed exposition in Section 8.2

$$x(t) = \exp(j\omega t) \tag{8.1.31}$$

its fictitiously sampled output is specified by equation (8.1.22) as:

$$Y(z_n) = K(z_n)X(z_n{}^n) \tag{8.1.32}$$

with:

$$X(z_n{}^n) = \frac{z_n{}^n}{z_n{}^n - \exp(j\omega T)} \tag{8.1.33}$$

Granted closed loop stability, the steady state behaviour $\{\hat{y}_k\}$ of the output sequence is determined by just the poles of $X(z_n{}^n)$. Accordingly, application of the Inversion Integral (2.1.9) yields:

$$\hat{y}(kT/n) = \frac{1}{n}\left\{\sum_{m=o}^{n-1} K^*(\omega + 2\pi m/T)\exp(j2\pi km/n)\right\}\exp(j\omega kT/n) \tag{8.1.34}$$

where from equations (8.1.22) and (8.1.30):

$$K^*(\omega) \simeq \frac{\lambda D^*(\omega)H^*(\omega)}{1 + (\lambda/n)D^*(\omega)H^*(\omega)} \tag{8.1.35}$$

over the frequency band affecting closed loop stability*. The practical considerations that enable equation (8.1.30) to closely approximate equation (8.1.26) also allow the summation in equation (8.1.34) to be reduced to its first term, and therefore:

$$\hat{y}(kT/n) \simeq \left\{\frac{(\lambda/n)D^*(\omega)H^*(\omega)}{1 + (\lambda/n)D^*(\omega)H^*(\omega)}\right\}\exp(j\omega kT/n) \tag{8.1.36}$$

Digital compensators for multirate DDC systems like Fig 8.1.2 can therefore be designed on the basis of the fast single rate system

---

*equivalently, for frequencies within the bandwidth of a practical DDC system

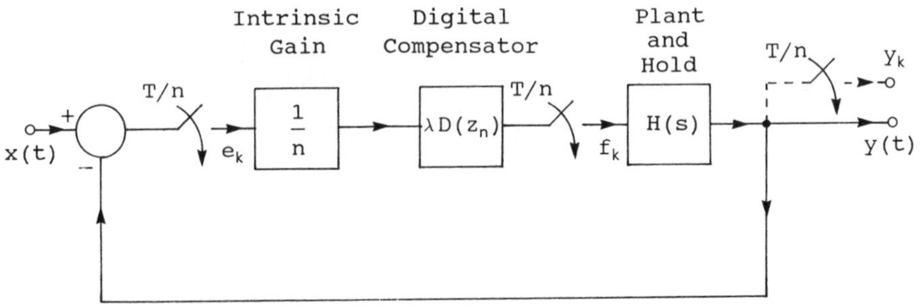

Fig 8.1.3 Single rate Approximation of the Multirate DDC System

in Fig 8.1.3 which merely involves the additional scalar gain factor 1/n. Consequently, the fundamentally sound relationships between time and frequency domain behaviour found in Section 3.4 also apply approximately to multirate DDC systems. Their Nyquist diagrams can therefore be shaped for adequate stability by the procedures described in Chapter 7. It is particularly relevant to note that the approximation in Fig 8.1.3 can be viewed as a preservation of the DC loop gain after speeding up the slower input sampler. A final observation on equation (8.1.34) is that a time invariant least frequency envelope does not characterise the steady state sinusoidal response of a multirate system.

As an example of the above frequency domain compensation technique, consider a multirate unity feedback system with linearised plant:

$$G(s) = \frac{1}{s^2(1 + s)} \qquad (8.1.37)$$

, zero order hold and sampling parameters defined by

$$T = 0.1 \text{ second} \quad ; \quad n = 3 \qquad (8.1.38)$$

Using the approximate model in Fig 8.1.3 and the design algorithms in Section 7.3 produces the multirate compensator (132):

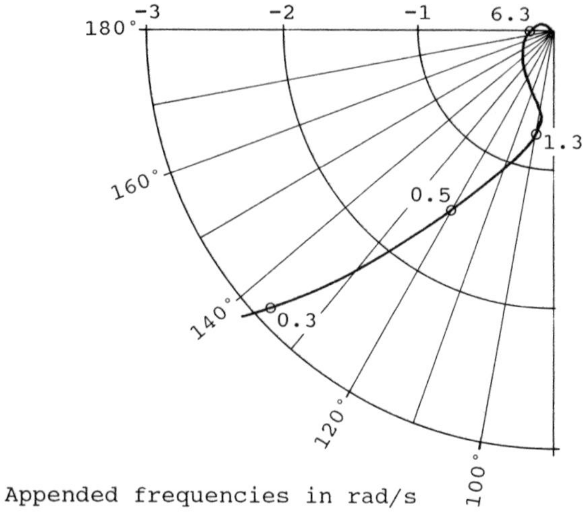

Fig 8.1.4 Approximate Nyquist Diagram of the Multirate System

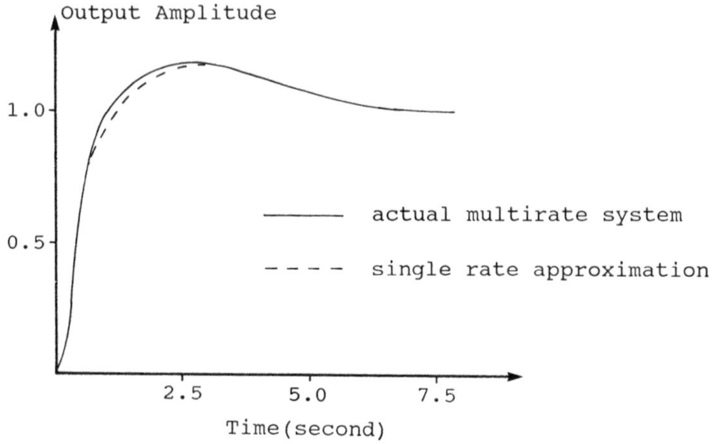

Fig 8.1.5 Step Responses of the Actual Multirate System and its Single rate approximation

$$\lambda D(z_3) = \frac{16.5(z_3 - 0.978)(z_3 - 0.955)}{z_3(z_3 - 0.78 - j0.15)(z_3 - 0.78 + j0.15)} \qquad (8.1.39)$$

into which the computational delay (1/30 second) has been properly relocated after its design. The corresponding Nyquist diagram in Fig 8.1.4, which of course allows for a loop gain reduction by 1/3, suffers perhaps from too little phase advance between 0.3 to 0.5 rad/s and too much after 0.63 rad/s. Nevertheless, the simulated step function responses of the actual multirate system and its approximation in Fig 8.1.5 are adequately damped and in excellent agreement; even for this broad bandwidth test input. It is suggested that an informative tutorial exercise would be for the reader to re-shape the Nyquist diagram in Fig 8.1.4 in accord with the above criticisms of phase advance distribution.

To complete the proposed design procedure for multirate systems, their steady state ripple and finite wordlength control error components are quantified next in Sections 8.2 and 8.3 respectively.

8.2 RIPPLE IN MULTIRATE SYSTEMS

A spectral input-output relationship like equation (4.6.19) for single rate systems is clearly a necessity for quantifying the ripple power in a multirate system (155). In this context, it is again analytically convenient to embed an input sequence $\{x_k\}$ into its faster rate equivalent $\{\underline{x}_k\}$. However, the simple scheme in equation (8.1.1) is unsatisfactory now because it does not preserve the wide sense stationary nature of an input. For instance, the scheme would impose:

$$\left.\begin{aligned} \overline{\underline{x}^2(k)} &= \overline{x^2} \qquad \text{if} \quad k = mn \qquad \text{for some integer } m \\ &= 0 \qquad \text{otherwise} \end{aligned}\right\} \qquad (8.2.1)$$

On the other hand, *randomly phased embedding* in which an ensemble of slower input samplers closes at:

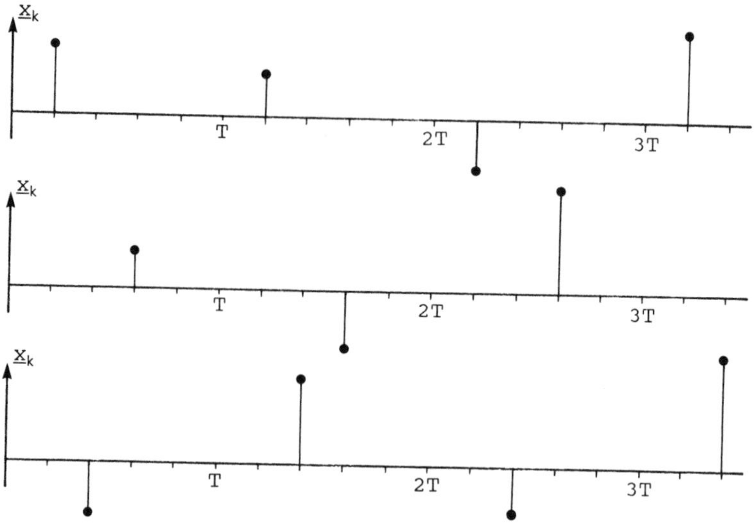

Fig 8.2.1 Three Realisations of a Randomly Phased Embedded Ensemble

$$0 \text{ or } T/n \text{ or } 2T/n \text{ or } 3T/n \text{ or } \ldots\ldots \text{ or } (n-1)T/n$$

with independent uniform probability 1/n is now shown to preserve stationary statistics. Fig 8.2.1 illustrates the artifice with three slow rate realisations embedded into a five-times faster process. Physically the scheme is seen to correspond to a form of random switch-on time for a multirate system.

The autocorrelation function of such an embedded sequence ensemble $\{\underline{X}\}$ is derived from equation (4.1.6) as:

$$\phi_{XE}(k+p,\ k) = \iint_{-\infty}^{\infty} \underline{x}_1\,\underline{x}_2\,P(\underline{x}_1,\underline{x}_2)\,d\underline{x}_1\,d\underline{x}_2 \qquad (8.2.2)$$

where for ease of nomenclature:

$$\underline{x}_1 = \underline{x}(\overline{k + pT/n}) \qquad \text{and} \qquad \underline{x}_2 = \underline{x}(kT/n) \qquad (8.2.3)$$

Expanding the above joint probability density function in terms of the conditional probability function $P(\underline{x}_1 | \underline{x}_2)$ yields:

$$\phi_{XE}(k + p,\ k) = \iint_{-\infty}^{\infty} \underline{x}_1 \underline{x}_2 P(\underline{x}_1 | \underline{x}_2) P(\underline{x}_2) d\underline{x}_1 d\underline{x}_2 \qquad (8.2.4)$$

By design the ensemble $\{\underline{x}_2\}$ takes the independent probability 1/n of being definitely non-zero at a fast rate sampling instant, so that:

$$P(\underline{x}_2) = \frac{1}{n} P(x_2) \qquad (8.2.5)$$

Furthermore, granted that $\underline{x}_2$ is not definitely zero, then:

$$\left. \begin{aligned} P(\underline{x}_1 | \underline{x}_2) &= P(x_1 | x_2) \quad \text{if} \quad p = mn \quad \text{for some integer } m \\ x_1 &= 0 \quad \text{otherwise} \end{aligned} \right\} \qquad (8.2.6)$$

Substituting equation (8.2.5) and (8.2.6) into equation (8.2.4) gives:

$$\left. \begin{aligned} \phi_{XE}(k + p,\ k) &= \frac{1}{n} \iint_{-\infty}^{\infty} x_1 x_2 P(x_1, x_2) dx_1 dx_2 \qquad \text{for} \quad p = mn \\ &= 0 \quad \text{otherwise} \end{aligned} \right\}$$

and if the slow rate sequence ensemble $\{X\}$ is wide sense stationary, the above reduces to:

$$\left. \begin{aligned} \phi_{XE}(k + p,\ k) &= \frac{1}{n} \phi_X(p) \qquad \text{if} \quad p = mn \text{ for some integer } m \\ &= 0 \qquad \text{otherwise} \end{aligned} \right\} \qquad (8.2.7)$$

Randomly phased embedding of a wide sense stationary sequence ensemble creates therefore a wide sense stationary ensemble at the faster sampling rate, and its pulse power spectrum is obtained from equation (4.4.1) as:

$$\Phi_{XE}(z_n) = \frac{1}{n}\sum_{-\infty}^{\infty} \phi_X(mT)\, z_n^{-mn} \qquad (8.2.8)$$

If the continuous input data were actually sampled at the faster rate, its pulse spectrum would be:

$$\Phi_X(z_n) = \sum_{-\infty}^{\infty} \phi_X(kT/n)\, z_n^{-k} \qquad (8.2.9)$$

so that in the notation of equations (8.1.12) and (8.1.21):

$$\Phi_{XE}(z_n) = \frac{1}{n}\mathscr{S}\left\{\Phi_X(z_n)\right\} = \frac{1}{n}\, Z\left\{\Phi_X(z_n)\right\}_{z=z_n{}^n} \qquad (8.2.10)$$

Finally applying equations (8.1.15) and (8.1.16) yields:

$$\Phi_{XE}(z_n) = (1/n)^2 \sum_{m=0}^{n-1} \Phi_X(p_m) \qquad (8.2.11)$$

with:

$$p_m = z_n \exp(j2\pi m/n) \qquad (8.2.12)$$

or in terms of real frequency:

$$\Phi^*_{XE}(\omega) = (1/n)^2 \sum_{m=0}^{n-1} \Phi^*_X(\omega + 2\pi m/T) \qquad (8.2.13)$$

where according to equation (4.5.35)

$$\Phi^*_X(\omega) = (n/T) \sum_{-\infty}^{\infty} \Phi_X(\omega - 2\pi nk/T) \qquad (8.2.14)$$

Two important conclusions stem from the above analysis. Firstly, if the pulse transfer function of a stable multirate system is $H(z_n)$ and if its continuous data has the power spectrum $\Phi_X(s)$, then the pulse power spectrum of the output is specified directly by equation (4.6.17) as:

$$\Phi_Y(z_n) = H(z_n)H(z_n^{-1}) \frac{1}{n} \mathscr{S}\left[\Phi_X(z_n)\right] \tag{8.2.15}$$

Secondly, equations (8.2.13) and (8.2.14) prove that multirate controllers still introduce ripple producing spectral sidebands which are centred about integral multiples of the lower sampling frequency $2\pi/T$. No longer can it be taken for granted therefore that a multirate compensator endows an intrinsically superior ripple performance on a DDC system.

The steady state ripple power in a stable multirate DDC system can now be analysed in the same way as for the single rate system in Section 5.3. Firstly, a feedback system is replaced by its equivalent open loop counterpart in which the plant and zero order hold are excited by the same forcing sequence. In this respect, equations (8.1.8) and (8.1.20) establish that a slow input sequence $\{x_k\}$ to the unity feedback system in Fig 8.1.2 produces the fast rate forcing sequence*:

$$F(z_n) = \left\{ \frac{\lambda D(z_n)}{1 + \lambda \mathscr{S}\left[D(z_n)H(z_n)\right]} \right\} X(z_n{}^n) \tag{8.2.16}$$

or

$$F(z_n) = \left[K(z_n)/H(z_n)\right] X(z_n{}^n) \tag{8.2.17}$$

where $K(z_n)$ denotes the overall pulse transfer function of the DDC system. Thus the feedback system and the open loop system in

---

*with a network B(s) in the feedback path $\mathscr{S}\big(D(z_n)H(z_n)\big)$ is replaced by $\mathscr{S}\big(D(z_n)HB(z_n)\big)$ in equation (8.2.16)

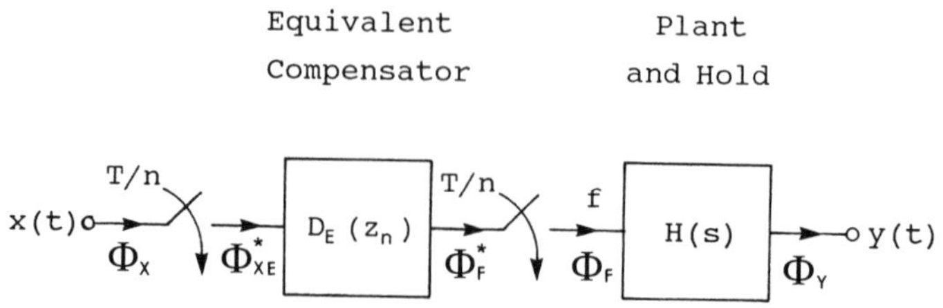

Fig 8.2.2 The Open Loop Equivalent to the Feedback System in Fig 8.1.2

Fig 8.2.2 are equivalent, in the sense of eliciting the same continuous output data, provided that:

$$D_E(z_n) = K(z_n)/H(z_n) \tag{8.2.18}$$

The pulse spectrum of the plant forcing sequence Φ^*_F is derived from that of the embedded input sequence Φ^*_{XE} by equation (8.2.15) as:

$$\Phi^*_F(\omega) = \left|K^*(\omega)/H^*(\omega)\right|^2 \Phi^*_{XE}(\omega) \tag{8.2.19}$$

and substituting equations (8.1.13) and (8.1.14) yields:

$$\Phi^*_F(\omega) = (1/nT)\left|K^*(\omega)/H^*(\omega)\right|^2 \sum_{m=0}^{n-1} \sum_{-\infty}^{\infty} \Phi_X(\omega + m\omega_0 - kn\omega_0) \tag{8.2.20}$$

where:

$$\omega_0 = 2\pi/T \tag{8.2.21}$$

Translation of the above pulse spectrum into continuous time is quite subtle. Because the plant's forcing sequence is at the fast rate, a first reaction based on equation (5.2.15) is to multiply $\Phi^*_F(\omega)$ by n/T. More circumspectly, however, the right-hand side of equation (8.2.20) is the spectrum of a slow rate sequence, because the sidebands are centred about multiples of ω_0 and not $n\omega_0$*. Therefore to avoid such misconceptions, it is recommended that equation (5.2.15) is regarded as the spectral relationship:

$$(\omega_0/2\pi) \sum_{-\infty}^{\infty} \Phi(\omega - k\omega_0) \longrightarrow (\omega_0/2\pi)^2 \sum_{-\infty}^{\infty} \Phi(\omega - k\omega_0)$$

Discrete Time *Continuous Time*

(8.2.22)

Accordingly, the spectrum of the continuous output data from the multirate DDC is derived as:

$$\Phi_Y(\omega) = (1/nT^2) \left| K^*(\omega) H(\omega)/H^*(\omega) \right|^2 \sum_{m=0}^{n-1} \sum_{-\infty}^{\infty} \Phi_X(\omega + m\omega_0 - kn\omega_0) \tag{8.2.23}$$

A similar argument to that in Section 5.3 confirms that the infinite series of functions:

$$\left\{ (1/nT^2) \left| K^*(\omega) H(\omega)/H^*(\omega) \right|^2 \Phi_X(\omega + m\omega_0 - kn\omega_0) \right\}$$

for

$$0 \le m \le n - 1 \quad \text{and} \quad -\infty \le k \le \infty$$

is normally summable over the range of real frequencies $[-\infty,\infty]$. Consequently its terms and the order of their integration and summation can be freely rearranged. For present purposes, an apposite rearrangement is symbolically represented by:

$$\sum_{k=-\infty}^{\infty} A_k = \sum_{m=0}^{n-1} \sum_{k=-\infty}^{\infty} A_{m-kn} \tag{8.2.24}$$

---

*equivalently, this represents the failure of the multirate strategy to reduce the ripple component of control error which, applied to equation (8.2.23), enables the steady state ripple component of error to be specified as:

$$\phi_r = (1/2\pi nT^2) \int_{-\infty}^{\infty} \left|Q^*(\omega)H(\omega)\right|^2 \left\{ \sum_{k=1}^{\infty} \Phi_x(\omega - k\omega_0) + \Phi_x(\omega + k\omega_0) \right\} d\omega \tag{8.2.25}$$

where by definition:

$$Q^*(\omega) = K^*(\omega)/H^*(\omega) \tag{8.2.26}$$

Because the integrands involve even functions of the real frequency variable, the above equation reduces to:

$$\phi_r = (1/\pi nT^2) \sum_{1}^{\infty} \int_{-\infty}^{\infty} \left|Q^*(\omega)H(\omega)/H^*(\omega)\right|^2 \Phi_x(\omega - k\omega_0) d\omega \tag{8.2.27}$$

and the further change of variable:

$$\omega \longrightarrow \omega + k\omega_0$$

gives:

$$\phi_r = (1/\pi nT^2) \int_{-\infty}^{\infty} \left\{ \sum_{1}^{\infty} \left|Q^*(\omega + k\omega_0)\ H(\omega + k\omega_0)\right|^2 \right\} \Phi_x(\omega) d\omega \tag{8.2.28}$$

where by definition:

$$Q^*(\omega) = K^*(\omega)/H^*(\omega) \tag{8.2.29}$$

Because pulse transfer functions are periodic with respect to the sampling frequency ($n\omega_0$ in this case), it is again convenient to rearrange the above series symbolically according to:

$$\sum_{k=1}^{\infty} A_k = \sum_{k=1}^{\infty} \sum_{m=1}^{n-1} A_{k\,n-m}$$

so that:

$$\phi_r = \left(\frac{1}{\pi n T^2}\right) \int_{-\infty}^{\infty} \left\{ \sum_{1}^{\infty} \sum_{m=0}^{n-1} \left| Q^*(\omega - m\omega_0) \right|^2 \left| H(\omega - m\omega_0 + kn\omega_0) \right|^2 \right\} \Phi_X(\omega) d\omega \tag{8.2.30}$$

Generally, the sampling frequency must be at least an order of magnitude larger than the bandwidth of the continuous input data to achieve adequately accurate data reconstruction. Consequently, the above integral has limits whose magnitudes are effectively much less than ω_0, and so with the nomenclature of equation (5.3.27):

$$\left| H(\omega - m\omega_0 + kn\omega_0) \right| \simeq 2G_\infty \left| \sin(\overline{nk - m}\ \pi/n) \right| . \left[\omega_0(nk - m)\right]^{-R-1} \tag{8.2.31}$$

Also as a closed bandwidth is usually commensurate with that of the input data, it follows from equations (8.1.22) and (8.1.36) that:

$$\left.\begin{aligned} Q^*(\omega - m\omega_0) &\simeq \lambda D^*(\omega)/\ 1 + \lambda D^*(\omega) H^*(\omega) \quad \text{for} \quad m = 0 \\ &\simeq \lambda D^*(m\omega_0) \quad \text{for} \quad 1 \le m \le n-1 \end{aligned}\right\} \tag{8.2.32}$$

The above approximation evidently simplifies the computation of the ripple error component for a multirate DDC system by exploiting data already available from the construction of its Nyquist diagram.

If the continuous input data belongs to the ensemble:

$$\left\{x(t)\right\} = \left\{A \sin(\Omega t + \psi)\right\} \tag{8.2.33}$$

whose amplitude and frequency are constant, but whose phase is uniformly distributed within $[0, 2\pi]$, then its power spectrum is specified by equation (5.2.18) as:

$$\Phi_X(\omega) = (A^2\pi/2)\left[\delta(\omega - \Omega) + \delta(\omega + \Omega)\right] \tag{8.2.34}$$

For this particular input, the ripple power in the sidebands* centred about $\pm k\omega_0$ is derived from equations (8.2.27) and (8.2.28) as:

$$\phi_r \Big|_{k\text{-sideband}} = (A^2/2nT^2)\left(\left| Q^*(\Omega + k\omega_0)H(\Omega + k\omega_0) \right|^2 + \left| Q^*(-\Omega + k\omega_0)H(-\Omega + k\omega_0) \right|^2 \right) \tag{8.2.35}$$

Equation (8.2.34) formed the basis for an experimental validation of the preceding analysis by S M Patel (118). Section 5.5 describes the experimental equipment and the study concerns a unity feedback system with zero order hold and plant:

$$G(s) = \frac{1}{s(1 + 0.05s)} \tag{8.2.36}$$

With the adopted sampling parameters of:

$$T = 40 \text{ ms} \quad ; \quad n = 3 \tag{8.2.37}$$

the multirate compensator:

$$D(z_s) = 7.5\left(z_3^{-1} - 0.45\, z_3^{-2}\right) \tag{8.2.38}$$

imposes adequate stability. Measurements of the overall real frequency response based on the continuous input and output data** were first compared against the single rate approximation in equation (8.1.36). These results presented in Table 8.2.1 confirm the simulation and they may now also be used as a tutorial exercise on the analysis in Section 8.1. Because the minimum centre frequency of the tunable bandpass filter was 133 rad/s, it was therefore possible only to isolate the third and higher order sidebands. Their measured and predicted mean square values in Fig 8.2.3 clearly validate the analysis.

---

* corresponding to the kth term in equation (8.2.27)

** on which ripple was imperceptible

Input Frequency (rad/s)	Closed Loop Gain	
	Measured	Predicted
0.63	0.925	0.927
1.26	0.789	0.777
1.88	0.637	0.635
2.51	0.550	0.524
3.14	0.438	0.443
4.40	0.308	0.330
6.28	0.231	0.237

Table 8.2.1 Closed Loop Gain Characteristic for the Experimental System

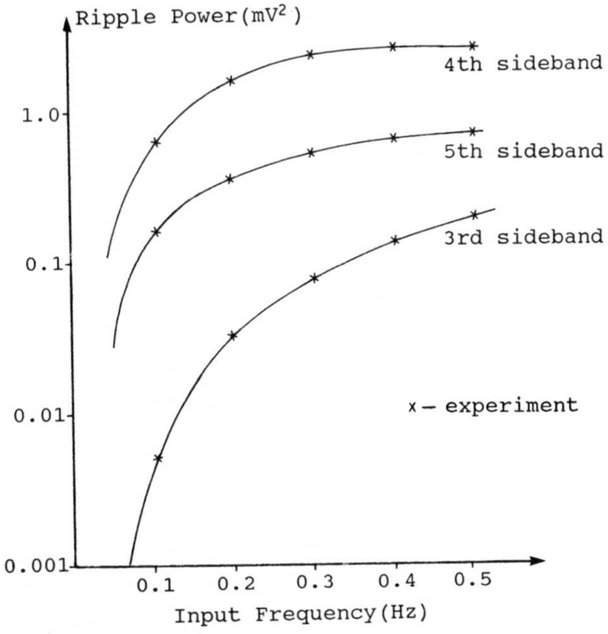

Fig 8.2.3 Measured and Calculated Ripple Powers in the Experimental System

8.3 ROUNDING ERRORS IN MULTIRATE SYSTEMS

The embedding techniques described in Sections 8.1 and 8.2 enable a multirate system to be analysed as a fast rate system with a modified input. Consequently, the statistical analysis of arithmetic rounding errors in Section 6.5 can be extended to multirate systems in a quite straightforward fashion (155). As in the single rate analysis, actual multirate variables are primed while an unprimed quantity is its 'ideal' evaluation with a computer having an arbitrary long wordlength. If the compensator of the multirate DDC system in Fig 8.1.2 has the pulse transfer function:

$$D(z_n) = A(z_n)/B(z_n) \tag{8.3.1}$$

with:

$$A(z_n) = \sum_{p=1}^{P} a_p z_n^{-k} \quad \text{and} \quad B(z_n) = 1 + \sum_{q=1}^{Q} b_q z_n^{-q} \tag{8.3.2}$$

it follows from equation (6.5.12) that its actual output is the ideal output additively contaminated with a computational error sequence $\{E_k\}$, whose z_n-transformation is:

$$\mathcal{E}(z_n) = R(z_n)/B(z_n) \tag{8.3.3}$$

With a directly programmed compensator, terms of the sequence represented by $R(z_n)$ are the total arithmetic rounding error in evaluating the right-hand side of:

$$f'_k = \sum_{p=1}^{P} a_p e'_{k-p} - \sum_{q=1}^{Q} b_q f'_{k-q} \tag{8.3.4}$$

at the sampling instants $\{kT/n\}$. Because of the embedding procedure, a control error sequence $\{e'_k\}$ contains terms that are exactly zero as illustrated by Fig 8.2.1. Thus, the number of *significant multiplications** involved in equation (8.3.4) generally varies from sample to sample time, and this function μ_k is termed the *multiplication pattern* of a multirate controller (155). It follows that terms in the total arithmetic rounding error sequence $\{r_n\}$ take the form:

$$r_k = \sum_{i=1}^{\mu(k)} \rho_i(k) \qquad (8.3.5)$$

where the quantisation error $\rho_i(k)$ is incurred when the product from the ith significant multiplication at time kT/n is truncated into the single length operational format. Although the variable upper limit of the above summation is a complication, it is now shown that the steady state finite wordlength control error component (ϕ_{fw}) can still be predicted by the power spectrum techniques in Section 6.5.

Often in practice all the P+Q coefficients in equation (8.3.4) correspond to a significant multiplication. Under these conditions, the required steady state multiplication pattern is derived in Appendix A8.3 as:

$$\left.\begin{aligned} \mu(k) &= Q + m && \text{for} \quad in + \nu + 1 \le k \le (i+1)n \\ &= Q + m + 1 && \text{for} \quad (i+1)n + 1 \le k \le (i+1)n + \nu \end{aligned}\right\} \qquad (8.3.6)$$

where by definition:

$$P = mn + \upsilon \qquad \text{with} \qquad 1 \le \upsilon < n \qquad (8.3.7)$$

and i denotes any positive or negative integer. Consonant with intuition, these equations establish that a multiplication pattern is periodic in the fast rate index (k) with period n**. The

---

*this excludes multiplication by: zero, unity or positive integral powers of 2 which incur no error

**the period is evidently T in real time

autocorrelation function of the total rounding error sequence is obtained from equation (8.3.5) as:

$$\overline{r_{k+p}r_k} = \overline{\sum_{i=1}^{\mu(k+p)} \sum_{j=1}^{\mu(k)} \rho_i(k+p)\rho_j(k)} \qquad (8.3.8)$$

Because the variable summation limits are awkward for analysis, equation (8.3.8) is re-written as:

$$\phi_r(k+p,k) = \sum_{i=1}^{P+Q} \sum_{j=1} \overline{\alpha_i(k+p)\rho_i(k+p)\alpha_j(k)\rho_j(k)} \qquad (8.3.9)$$

Here the auxiliary variables $\{\alpha_i\}$ take values of 0 or 1 depending on whether the multiplication by the corresponding numerator or denominator term actually occurs, and with this nomenclature:

$$\mu_k = \sum_{i=1}^{P+Q} \alpha_i(k) \qquad (8.3.10)$$

By introducing randomly phased embedding, the auxiliary variables assume a statistical character that is independent of the input data and therefore of the individual rounding errors $\{\rho_i\}$. Thus:

$$\phi_r(k+p,k) = \sum_{i=1}^{P+Q} \sum_{j=1} \overline{\alpha_i(k+p)\alpha_j(n)} . \overline{\rho_i(k+p)\rho_i(k)} \qquad (8.3.11)$$

and applying the established properties of quantisation errors in equations (6.1.1) to (6.1.4) yields:

$$\left.\begin{aligned} \phi_r(k+p,k) &= \sum_{i=1}^{P+Q} \overline{\alpha_i(k)} . (q^2/12) \qquad \text{for } p = 0 \\ &= 0 \quad \text{otherwise} \end{aligned}\right\} \qquad (8.3.12)$$

which equation (8.3.10) reduces to:

$$\left.\begin{aligned} \phi_r(k+p,k) &= \overline{\mu_k}\cdot(q^2/12) \quad \text{for } p=0 \\ &= 0 \quad \text{otherwise} \end{aligned}\right\} \qquad (8.3.13)$$

Though there are generally n different possible values in a multiplication pattern, equation (8.3.6) reveals that the value of Q + m is assumed n - ν times and that the value Q + m + 1 is assumed ν times in this practically important case. As a result of randomly phased embedding it follows that:

$$\left.\begin{aligned} \text{Probability } \{\mu_k = Q + m\} &= (n - \nu)/n \\ \text{Probability } \{\mu_k = Q + m + 1\} &= \nu/n \end{aligned}\right\} \qquad (8.3.14)$$

for all values of the integer index k, and therefore:

$$\overline{\mu_k} = (Q + m)(n - \nu)/n + (Q + m + 1)(\nu/n)$$

or

$$\overline{\mu_k} = (Qn + mn + \nu)/n \qquad (8.3.15)$$

Substituting equation (8.3.15) into equation (8.3.13) proves that the total rounding error is wide sense stationary with correlation sequence:

$$\phi_r(k) = (1/n)(Qn + mn + \nu)(q^2/12)\cdot\delta_{k\,0} \qquad (8.3.16)$$

and pulse power spectrum:

$$\Phi_r(z_n) = (1/n)(Qn + mn + \nu)(q^2/12) \qquad (8.3.17)$$

The above derivation of the total rounding error spectrum is actually simpler than the original in reference 155, which applied Kolmogorov's theorem (158) to an assumed ergodic error ensemble.

Straightforward manipulation of pulse transfer functions in a linearised control system model, as shown by Figs 6.5.1 and 6.5.2, enables the noise sources associated with A-D conversion and computation errors to be represented as additive inputs to the ideal system having an arbitrary long wordlength. Computation of the steady state finite wordlength error component is most conveniently implemented using the deterministic procedure in Fig 6.5.3. Now the closed loop pulse transfer function is of course $K(z_n)$, and the shaping filters for the A-D conversion and total rounding errors are:

$$\chi_1(z_n) = 1/\sqrt{6} \tag{8.3.18}$$

and

$$\chi_2(z_n) = \left\{(Qn + mn + \nu)/12n\right\}^{1/2} / A(z) \tag{8.3.19}$$

respectively. The analysis is readily extended to other programming techniques that in essence contain directly programmed factors of a pulse transfer function in parallel or cascade. Specifically, each directly programmed factor $A_p(z_n)/B_p(z_n)$ generates an additive output noise process with spectrum:

$$\left\{ \frac{\bar{\mu}_p}{B_p(z_n)B_p(z_n^{-1})} \right\} (q^2/12) \tag{8.3.20}$$

which by means of equation (4.6.19) is transferred to the input of the DDC system or filter in the manner of Fig 6.5.5a. Calculation of the steady state finite wordlength control error or output noise now involves the arrangement of shaping filters and overall pulse transfer functions in Fig 6.5.5b.

If the plant of the linearised multirate DDC system in Fig 8.1.2 is:

$$G(s) = \frac{1}{s^2(1 + s)} \tag{8.3.21}$$

with the sampling parameters:

$$T = 0.1 \text{ second} \quad ; \quad n = 3 \tag{8.3.22}$$

then adequate closed loop stability is imposed by either of the triple rate digital compensators:

$$D_1(z_3) = \frac{16.5(z_3 - 0.978)(z_3 - 0.955)}{z_3(z_3 - 0.78 - j0.15)(z_3 - 0.78 + j0.15)} \qquad (8.3.23)$$

or

$$D_2(z_3) = \frac{8.3(z_3 - 0.98)^2}{z_3(z_3 - 0.79)^2} \qquad (8.3.24)$$

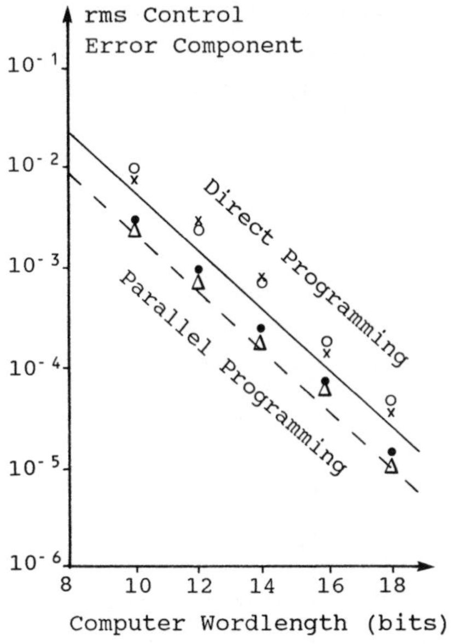

Direct Programming
x 0.01 Hz Sinusoidal input
o Drift excitation

Parallel Programming
• 0.01 Hz Sinusoidal input
Δ Drift excitation

Fig 8.3.1 Compensation by $D_1(z_3)$

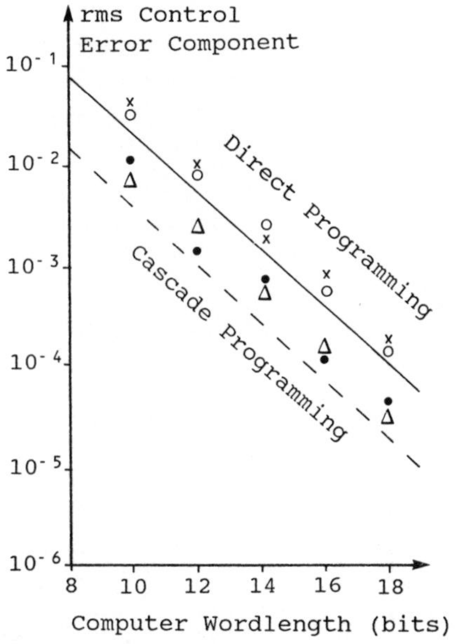

Direct Programming
x 0.1 Hz Sinusoidal input
o Drift excitation

Cascade Programming
• 0.01 Hz Sinusoidal input
Δ Drift excitation

Fig 8.3.2 Compensation by $D_2(z_3)$

Figs (8.3.1) and (8.3.2) depict the respective finite wordlength control error components as a function of controller wordlength from calculations based on the numerical scheme in Fig 6.5.3. Also shown are corresponding rms measurements derived from simulations of these systems with a sinusoidal input of 0.01 Hz or a drift excitation*. Within a predicted sample-size error bound of around 7%, these measurements are seen to confirm the analysis, and the smaller wordlength requirements achieved by the use of complex poles in phase advance type compensators.

A comprehensive comparison of the performance achievable by single- and multirate compensators is described in reference 157. There the plant in equation (8.3.21) and its zero order hold are stabilised by the single rate controller:

$$D(z) = 4.8(z - 0.975)^2 / z(z - 0.800)^2 \qquad (8.3.25)$$

with

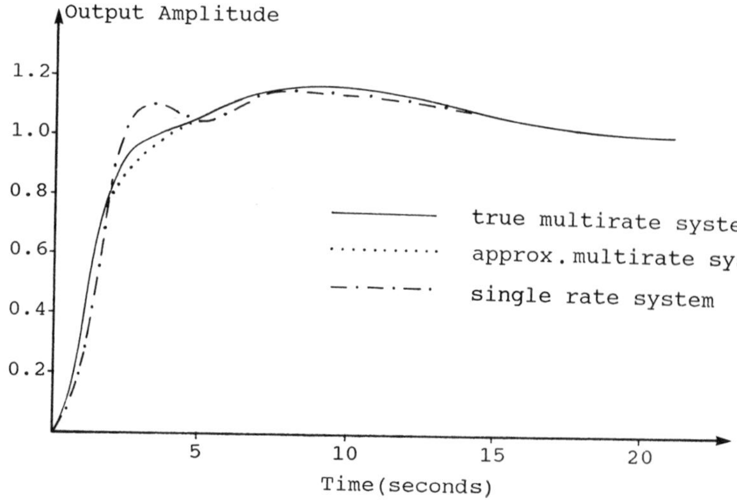

Fig 8.3.3 Comparison of Unit Step Function Responses

---

*see Section 6.5

$$T = 0.1 \text{ second} \tag{8.3.26}$$

and the multirate controller:

$$D(z_3) = 27.78(z_3 - 0.991)^2 \big/ z_3(z_3 - 0.900)^2 \tag{8.3.27}$$

with the same sampling parameters as in equation (8.3.22). As shown in Fig (8.3.3), the two controllers impose very similar transient characteristics, so that a fair basis exists for comparing their respective ripple and finite wordlength control error components. Table 8.3.1 contains the calculated rms ripple error components for the input spectra:

$$\Phi_1(\omega) = \frac{2.7}{\omega^2 + (0.4\pi)^2} \tag{8.3.28}$$

and

$$\Phi_2(\omega) = \frac{6.35 \times 10^{-7}}{\left[\omega^2 + (0.02\pi)^2\right]^2} \tag{8.3.29}$$

The finite wordlength error component for the single rate system with direct programming is computed as:

$$\sqrt{\phi_{fw}} = 18.5q \tag{8.3.30}$$

and for the multirate system as:

$$\sqrt{\phi_{fw}} = 11.3q \tag{8.3.31}$$

Though the degeneration in control accuracy due to arithmetic rounding errors is slightly greater for the single rate system, its ripple error component is seen to be about three-times smaller. A detailed argument in reference 157 establishes that, contrary to Kranc's original conjecture (154), multirate systems are generally inferior to single rate systems in this respect.

Input Spectrum	Single rate $\sqrt{\phi_r}$	Multirate $\sqrt{\phi_r}$
$\Phi_1(\omega)$	0.96×10^{-6}	2.9×10^{-6}
$\Phi_2(\omega)$	2.4×10^{-9}	10.4×10^{-9}

Table 8.3.1 Comparison of Single- and Multirate System Ripple Performance

The next Section 8.4 considers the stability, ripple performance and the effect of arithmetic rounding errors in DDC systems whose controllers operate with input data rates that are an integral number of times faster than their outputs. Various practical alternatives for the design of single-input DDC systems will then have been examined.

8.4 ASPECTS OF SUBRATE DDC SYSTEMS

A subrate DDC system has a controller whose input sampling rate is an integral multiple times faster than its output rate. Computing speeds in the 1960's were much slower than at present, and in some centralised multiloop control systems the required stabilising calculations were almost too demanding. Subrate controllers then appeared to offer the required computational economies whilst preserving fast enough monitoring of the control errors to initiate accident management. Alternatively, the envisaged release of computer time might have been used for off-line data logging or process improvement etc. The principal disadvantage of the strategy is clearly an intrinsically inferior ripple performance. Distributed control and the greatly enhanced speed of μ-processors now render the above arguments for subrate systems less relevant. Nevertheless, a brief description of subrate system analysis is presented here to demonstrate that the aspired economies in computing time are not necessarily achieved, and that the design basis is much weaker than that for single or multirate control.

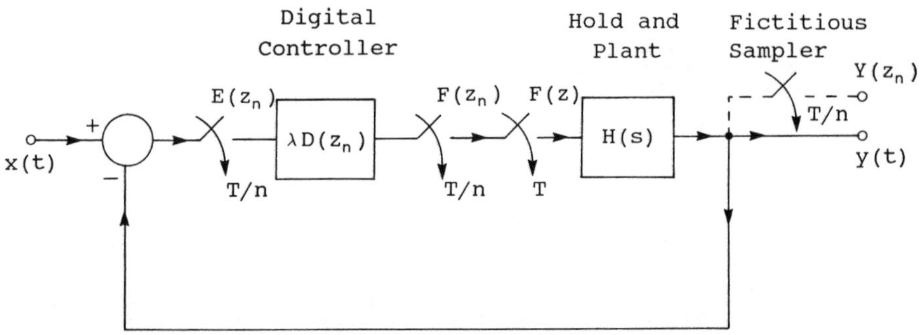

Fig 8.4.1 A Linearised Subrate DDC System

It is analytically convenient to represent the slower output sampling process in a subrate DDC system by a fast rate switch followed by the actual slow rate switch in the manner of Fig 8.4.1. According to equation (8.1.8), the fictitiously sampled output of the plant is given by:

$$Y(z_n) = H(z_n)F(z)\Big|_{z = z_n} \tag{8.4.1}$$

with

$$F(z) = Z\left[\underline{F}(z_n)\right] = \sum_{0}^{\infty} \underline{f}(kn)z^{-k} \tag{8.4.2}$$

Expressing equation (8.4.1) in terms of the control error sequence yields:

$$Y(z_n) = H(z_n).Z\left\{\lambda D(z_n)\left[X(z_n) - Y(z_n)\right]\right\}\Bigg|_{z=z_n{}^n} \tag{8.4.3}$$

and from equation (8.1.15) it follows that:

$$Y(z_n) = (\lambda/n)H(z_n).Z\left\{D(z_n)\left[X(z_n) - Y(z_n)\right]\right.$$

$$\left. + \sum_{p=1}^{n-1} D(z_n\alpha_p)\left[X(z_n\alpha_p) - Y(z_n\alpha_p)\right]\right\}$$

where (8.4.4)

$$\alpha_p = \exp(j2\pi p/n) \tag{8.4.5}$$

Thus it is not possible to derive an explicit expression for the overall pulse transfer function of a subrate DDC system as defined by the fast rate sequence ratio $Y(z_n)/X(z_n)$. Likewise, an explicit overall pulse transfer function as defined by the slow rate sequence ratio $Y(z)/X(z)$ is also unobtainable. The preceding analyses of the ripple error component (ϕ_r) in single- and multirate DDC systems invoke equivalent open loop controllers, that are seen from equations (5.1.10) and (8.2.18) to include the explicit closed loop pulse transfer functions $K(z)$ or $K(z_n)$. Consequently, these analytical techniques are not available for quantifying the ripple error component in subrate DDC systems. Nevertheless, on account of the generous stability margins normally allowed, it is possible to engineer adequate stability in closed loop subrate systems by means of the following approximation.

An approximate overall pulse transfer for a DDC system must evidently preserve its DC open loop gain. The previously successful approximation for a multirate controller, in which its slower sampler is replaced by one at the faster rate cascaded with a scalar gain of 1/n, suggests the possibility of a similar approximation for subrate controllers. It is proposed therefore to investigate the accuracy of approximating the subrate system in Fig 8.4.1 by the overall pulse transfer function:

$$K(z_n) = \frac{Y(z_n)}{X(z_n)} = \frac{(\lambda/n)D(z_n)H(z_n)}{1 + (\lambda/n)D(z_n)H(z_n)} \tag{8.4.6}$$

Equation (8.4.4) clearly takes the form:

$$\Psi(Y) = 0 \tag{8.4.7}$$

and its implicitly contained true solution (or root) represents the actual output sequence of the subrate system. By regarding an approximate solution like equation (8.4.6) as an element of the set Q, that consists of all rational functions of z, considerable insight can be acquired from functional analysis techniques (10). Specifically, provided the approximate output sequence (Y) is close enough to the true solution ($\hat{Y}$) then a Taylor series expansion yields:

$$\Psi(\hat{Y}) \simeq \Psi(Y) + \Psi'(Y)\big|_{\hat{Y}-Y} \tag{8.4.8}$$

where the total derivative $\Psi'(Y)$ is a continuous linear function on Q. 'Close enough' implies the existence of a norm on Q, and because the behaviour of Y on the unit circle in the z_n- plane is the paramount concern, a suitable norm at any real frequency ω is given by:

$$\|Y\| = |Y^{*}(\omega)| \tag{8.4.9}$$

Granted that the linear mapping $\Psi'(Y)$ has an inverse, then equation (8.4.8) provides the generalised Newton-Raphson algorithm:

$$0 = \left[\Psi'(Y)\right]^{-1} . \Psi(Y) + (\hat{Y} - Y) \tag{8.4.10}$$

or

$$Y - \hat{Y} = \left[\Psi'(Y)\right]^{-1} . \Psi(Y) \tag{8.4.11}$$

Thus the residual $\Psi(Y)$ alone cannot quantify the approximation error $(Y - \hat{Y})$, which also can depend significantly* on $[\Psi'(Y)]^{-1}$. Originally in reference 156, the ratio $\Psi(Y)/Y$ is employed as a measure of the error, but the above formulation clearly demonstrates its inadequacy in this respect.

Substituting equation (8.4.6) into equation (8.4.4) establishes that:

$$\Psi(Y) = \sum_{p=1}^{n-1} K(z_n \alpha_p) \left\{ H(z_n)/H(z_n \alpha_p) \right\} X(z_n \alpha_p) \tag{8.4.12}$$

while more erudite considerations yield:

$$\Psi'(Y) = -\left\{1 + (\lambda/n) D(z_n) H(z_n)\right\} \left\{ I + \sum_{p=1}^{n-1} K(z_n) \left\{ D(z_n \alpha_p)/D(z_n) \right\} M_p \right\} \tag{8.4.13}$$

where I denotes the identity mapping on Q, and M_p the continuous linear mapping defined by:

$$M_p(Y) = Y(\ .\alpha_p) \qquad \text{for} \qquad 1 \le p \le n - 1 \tag{8.4.14}$$

An attempt to bound the frequency domain approximation error from equation (8.4.8) produces:

$$\|\Psi(Y)\| = \left\| \Psi'(Y) \Big|_{Y-\hat{Y}} \right\| \le \|\Psi'(Y)\| . \|Y - \hat{Y}\| \tag{8.4.15}$$

which is obviously futile. It is therefore necessary to invoke equation (8.4.11) in the form:

$$\|Y - \hat{Y}\| \le \left\| \left[\Psi'(Y)\right]^{-1} \right\| . \|\Psi(Y)\| \tag{8.4.16}$$

where the inverse derivative mapping is derived from equation

---

*Consider the Newton-Raphson algorithm applied to finding the root x of the real function f(x) when f'(x) is very small indeed

(8.4.13) as:

$$\left[\Psi'(Y)\right]^{-1} =$$

$$-\left(1 + (\lambda/n)D(z_n)H(z_n)\right)^{-1} \cdot \left[I + \sum_{p=1}^{n-1} K(z_n)\left(D(z_n\alpha_p)/D(z_n)\right) M_p\right]^{-1} \qquad (8.4.17)$$

Generally for an arbitrary continuous linear mapping Θ:

$$\|\Theta\| = \text{Sup}\left\{\|\Theta(Y)\| \text{ for any } \|Y\| = 1\right\}$$

so that for elements of the set $\{M_p\}$ in particular, equations (8.4.9) and (8.4.14) give:

$$\|M_p\| = \text{Sup}\left\{\left|Y^*(\omega + p\omega_o)\right|\left|Y^*(\omega)\right|^{-1} \text{ for any } Y(\omega) \neq 0 \text{ from } Q\right\} \qquad (8.4.18)$$

where

$$\omega_o = 2\pi/T \qquad (8.4.19)$$

It can be shown that (10):

$$\left(I + \Theta\right)^{-1} = \sum_{m=o}^{\infty} (-1)^m \Theta^m \qquad \text{for } \|\Theta\| < 1 \qquad (8.4.20)$$

Unfortunately, and particularly with phase advance type compensators, the norm of the sum involving $\{M_p\}$ exceeds unity even for the restricted frequency band that largely determines closed loop stability. Denied also the opportunity to simplify or reasonably approximate the derivative $\Psi'(Y)$, the frequency domain error bound associated with equation (8.4.6) is presently not considered quantifiable; except at zero frequency where zero error is incurred. Each potential compensator of the linearised model of a subrate DDC system must therefore be compared against a time domain simulation during its iterative design. On the other hand, an adequate non-linear transient simulation for the finalised Nyquist based design is usually sufficient with single- and multirate systems on account of the previously established precise relationship between time and frequency response characteristics.

The approximation of a subrate DDC system by equation (8.4.6) and an application of the gain-phase curves in Section 7.3 result in a 'compensator' that operates on fast rate input and output sequences. As a first step in the realisation of a true subrate design from such a parent, consider the simple compensator:

$$D(z_n) = \frac{F(z_n)}{E(z_n)} = \frac{z_n^{-1}(a_o + a_1 z_n^{-1})}{1 - b_1 z_n^{-1}} \tag{8.4.21}$$

whose fast rate recurrence equation is:

$$f_k = b_1 f_{k-1} + a_o e_{k-1} + a_1 e_{k-2} \tag{8.4.22}$$

It follows therefore that:

$$f_{mn} = b_1 f_{mn-1} + a_o e_{mn-1} + a_1 e_{mn-2} \tag{8.4.23}$$

so that after repeated substitution to eliminate the fast rate output terms one obtains:

$$f_{mn} = (b_1)^n f_{mn-n} + a_o e_{mn-1} + a_1 (b_1)^{n-1} e_{mn-n-1} + (a_o b_1 + a_1) \sum_{p=1}^{n-1} (b_1)^{p-1} e_{mn-p-1} \tag{8.4.24}$$

By setting:

$$\left.\begin{aligned} &\beta_1 = (b_1)^n \quad ; \quad c_o = a_o \quad ; \quad c_n = a_1 (b_1)^{n-1} \\ &c_p = (a_o b_1 + a_1)(b_1)^{p-1} \qquad \text{for } 1 \le p \le n-1 \end{aligned}\right\} \tag{8.4.25}$$

in equation (8.4.24), the slow output rate recurrence relation that defines the required subrate controller is derived as:

$$f_{mn} = \beta_1 f_{mn-n} + \sum_{p=o}^{n} c_p e_{mn-p-1} \qquad (8.4.26)$$

and its z_n-pulse transfer function is evidently:

$$D_s(z_n) = \frac{z_n^{-1} C(z_n)}{1 - \beta_1 z_n^{-n}} \qquad (8.4.27)$$

where

$$C(z_n) = \sum_{p=o}^{n} c_p z_n^{-p} \qquad (8.4.28)$$

More complex controllers designed by the single rate approximation can be translated into the actual subrate form by first expanding them as for a parallel realisation:

$$D(z_n) = \sum_{i} \left\{ \frac{z_n^{-1}(a_{0i} + a_{1i} z_n^{-1})}{1 - b_i z_n^{-1}} \right\} \qquad (8.4.29)$$

and then treating each factor as above to yield:

$$D_s(z_n) = \sum_{i} \left\{ \frac{z_n^{-1} C_i(z_n)}{1 - \beta_i z_n^{-n}} \right\} \qquad (8.4.30)$$

Equation (8.4.30) applies even when some of the poles $\{b_i\}$ are complex, but because these can exist only as conjugate pairs, real arithmetic operations can be preserved by combining any two such conjugates β_i and $\overline{\beta_i}$ into the pulse transfer function component:

$$\frac{z_n^{-n} \Psi(z_n)}{1 - (2\mathbf{R}\beta_i) z_n^{-n} + |\beta_i|^2 z_n^{-2n}} \qquad (8.4.31)$$

where:

$$\Psi(z_n) = 2 \sum_{p=o}^{n} \left[\mathbf{R}\, c_{p\,i} \right] z_n^{-p} + 2 z_n^{-n} \sum_{p=o}^{n} \left[\mathbf{R}\, c_{p\,i}\, \overline{\beta_i} \right] z_n^{-p} \qquad (8.4.32)$$

A directly programmed controller devised by the single rate approximation would evidently satisfy a recurrence relation of the general form:

$$f_k = \sum_{i=1}^{I} b_i\, f_{k-i} + \sum_{p=1}^{P} a_p\, e_{k-p} \qquad (8.4.33)$$

and it can be inferred from equations (8.4.26) and (8.4.32) that the corresponding subrate realisation is given by:

$$f_{mn} = \sum_{i=1}^{I} \beta_i\, f_{mn-in} + \sum_{p=1}^{nP} c_p\, e_{mn-p-1} \qquad (8.4.34)$$

The extra number of coefficients $\{c_p\}$ compared to $\{a_p\}$ evidently confounds aspirations that subrate controllers involve fewer multiplications than single rate systems; at least when they are designed in the above manner.

Equation (8.4.34) specifies the ideal recurrence equation performed by a directly programmed subrate compensator with an arbitrary long wordlength. Due to the finite wordlength of practical compensators, the computed output is actually:

$$f'_{mn} = \sum_{i=1}^{I} \beta_i\, f'_{mn-in} + \sum_{p=1}^{nP} c_p\, e'_{mn-p-1} + r_D(mn) \qquad (8.4.35)$$

Here, $\{e'_k\}$ is the fast rate input sequence from the A-D converter(s) which introduces the quantisation errors defined in equation (6.5.3), and $r_D(mn)$ denotes the total arithmetic rounding error in computing:

$$\sum_{i=1}^{I} \beta_i f'_{mn-in} + \sum_{p=1}^{nP} c_p e'_{mn-p-1} \tag{8.4.36}$$

If the above calculation involves a total of μ multiplications excluding zero, unity and integral powers of 2, then:

$$r_D(mn) = \sum_{1}^{\mu} \rho_k(mn) \tag{8.4.37}$$

where ρ_k represents the rounding error on one such multiplication. Defining the computational error quantity:

$$f'_{mn} = f_{mn} + E_D(mn) \tag{8.4.38}$$

then as in Section 6.5, its z_n-transformation is derived as:

$$\mathcal{E}(z_n) = R_D(z_n)/B(z_n) \tag{8.4.39}$$

where:

$$\left.\begin{aligned} R_D(z_n) &= \sum_{mn=1}^{\infty} r_D(mn) z_n^{-mn} \\ B(z_n) &= 1 + \sum_{i=1}^{I} \beta_i z_n^{-in} \end{aligned}\right\} \tag{8.4.40}$$

This computational error sequence can be referred to the faster rate input of the subrate controller to give:

$$\mathcal{E}_I(z_n) = R_D(z_n)/C(z_n) \tag{8.4.41}$$

where:

$$C(z_n) = \sum_{p=1}^{nP} c_p z_n^{-p} \qquad (8.4.42)$$

By applying the artifice of randomly phased embedding described in Section 8.2, the power spectrum of the computational error sequence to be superimposed on the input of an idealised closed loop system is determined as:

$$\Phi_{EI}(z_n) = (\mu q^2/12n)\left[C(z_n)C(z_n^{-1})\right]^{-1} \qquad (8.4.43)$$

With phase advance type compensators, the bandwidth of the pulse transfer function $1/C(z_n)$ is markedly smaller than that of the control system as a result of negative feedback. Consequently, wordlength requirements for subrate DDC systems can be estimated using the computational scheme in Fig 6.5.3 with the overall pulse transfer function approximated by equation (8.4.6)*. The shaping filter for A-D conversion errors is still:

$$\chi_1(z_n) = 1/\sqrt{6} \qquad (8.4.44)$$

but that for multiplicative rounding errors now becomes:

$$\chi_2(z_n) = \sqrt{\mu/12n}/C(z_n) \qquad (8.4.45)$$

The above analysis of the finite wordlength error component is readily extended (156) to other programming techniques, which in essence contain directly programmed factors of a pulse transfer function in parallel or cascade.

Although the single rate approximation in equation (8.4.6) allows the formulation of an equivalent open loop controller like that in equation (8.2.18), its conversion into subrate form is considered a generally inappropriate basis for calculating the ripple component of control error. Specifically, the discussion of Table 5.3.1 establishes that a ripple error component is

---

*at least it is better than nothing!

largely determined by the first sideband of the sampled input data, so that the 'beyond bandwidth' response of a subrate DDC system is important in this respect. Tables 5.4.1 and 8.3.1 confirm this deduction by demonstrating the importance in particular cases of the input data's spectral bandwidth and cut-off rate. Because the accuracy of equation (8.4.6) can be guaranteed only close to zero frequency, its use in ripple power calculations would therefore be suspect.

As an example of designing with the approximate pulse transfer function relationship in equation (8.4.6),consider a subrate DDC system with the linearised plant model:

$$G(s) = \frac{1}{s^2(1+s)} \tag{8.4.46}$$

and sampling parameters:

$$T = 0.1 \text{ second} \quad ; \quad n = 3 \tag{8.4.47}$$

Though its single rate approximation appears at first sight very similar to the multirate example in Section 8.1, the Laplace transfer functions within the respective feedback loops are subtly different:

$$\left.\begin{aligned} H(s) &= \exp(-sT/n)\left[\frac{1-\exp(-sT/n)}{s}\right]G(s) \quad &\text{multirate system} \\ H(s) &= \exp(-sT)\left[\frac{1-\exp(-sT)}{s}\right]G(s) \quad &\text{subrate system} \end{aligned}\right\} \tag{8.4.48}$$

Attention is drawn in particular to the dead-time lag exp(-sT), which is associated with the subrate controller because it can only initiate corrective action after one longer sampling period. As described in Section 7.3, it is convenient to segregate inherent computational delays into the plant transfer function when iteratively designing a digital compensator. Despite the somewhat larger phase lag of the open loop subrate system, the

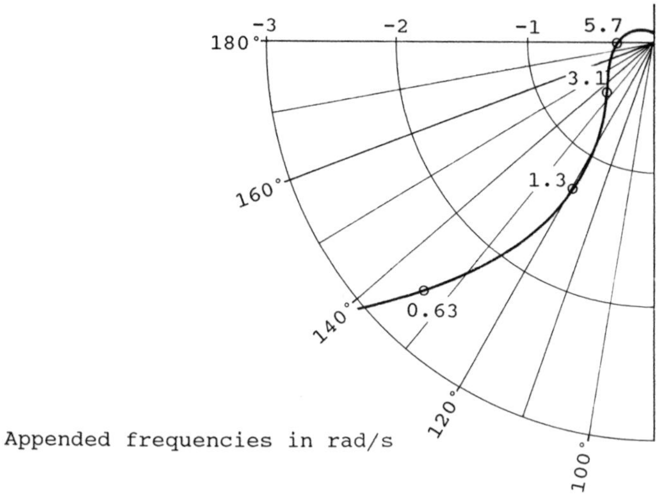

Fig 8.4.2 Approximate Nyquist Diagram of the Subrate DDC System

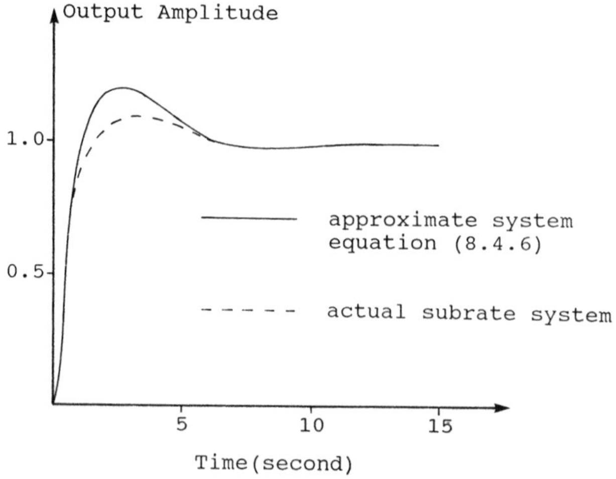

Fig 8.4.3 Step Responses of the actual and approximate Subrate System

same single rate controller*:

$$\lambda D(z_3) = \frac{16.5(z_3 - 0.978)(z_3 - 0.955)}{z_3(z_3 - 0.78 - j0.15)(z_3 - 0.78 + j0.15)} \tag{8.4.49}$$

imposes adequate stability margins on its Nyquist diagram in Fig (8.4.2). It may appear paradoxical that an additional delay of T/3 is included in the above controller, when the dead time lag exp(-sT) is already included in the plant to account for computational delays. However, reference to equation (8.4.24) for instance shows that this factor z_3 is necessary so that control error samples no earlier than (mn - 1)T are involved in producing the output at time mnT. Although the closed loop step responses in Fig 8.4.3 demonstrate that the approximate pulse transfer function is satisfactory in this case, its unquantifiable error in the frequency domain remains a source of concern.

To validate the above analysis of the finite wordlength control error component, the subrate compensator corresponding to equation (8.4.49) was realised by parallel programming (156). For this purpose, the appropriate real and imaginary parts of equation (8.4.26) are computed using real arithmetic operations before being combined to provide the real output sequence. (Details of the procedure should be by now self-evident to the reader.) Simulations of the actual subrate DDC system with various practical wordlengths were compared against a virtual ideal (having main frame accuracy) for sinusoidal and drift excitation inputs to isolate the finite wordlength control error component (156). These results are compared in Fig 8.4.4 against predictions based on the approximate overall pulse transfer function in the computational scheme of Fig 6.5.3. Though good agreement obtains within the estimated finite sample-size error

---

*Note its scalar gain of 16.5 is reduced by a factor 3 when included in the approximate Nyquist diagram and the approximate simulation

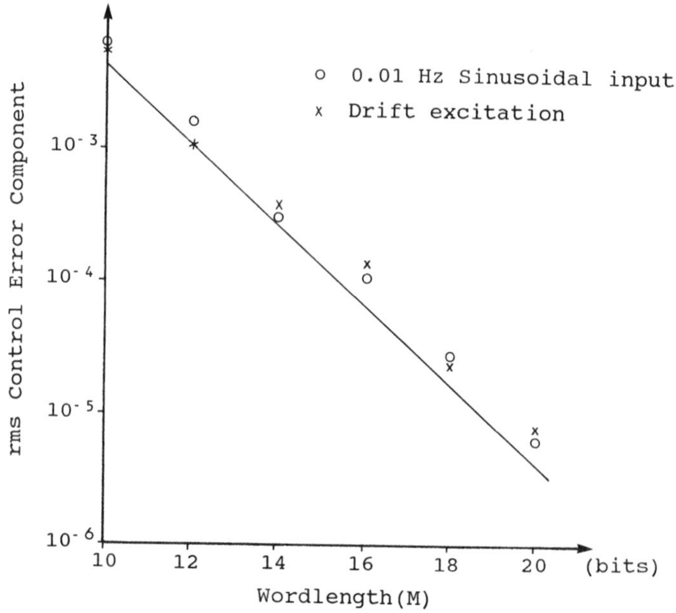

Fig 8.4.4 Predicted and Simulated rms Values of the Finite Wordlength Control Error component for the Subrate DDC System Example

bound of ±7%, satisfaction must again be tempered by recalling the unquantifiable uncertainty on the single rate approximation.

This last chapter may well be considered too abbreviated. Against such an argument it is suggested that multirate and subrate systems have been shown not to fulfil the aspirations of their innovators. Specifically, multirate systems can have a larger ripple component of control error than the single rate system (157), and subrate systems do not always release computing time for other plant control functions. Moreover, the advances in large scale integrated circuit technology have made these strategies less relevant. Nevertheless, their inclusion here is merited if only to exemplify the general danger of implementing an engineering design based solely on plausible intuitive arguments.

CHAPTER 9

Principal Conclusions and Final Remarks

Alpha and Omega, the beginning and the end - Revelation of St John

Digital systems can be engineered with constructional and operational advantages over their counterparts operating with continuous data provided that certain fundamentals are well-understood. Chapter 1 introduces those aspects of data sampling, amplitude quantisation and binary arithmetic which are crucial to a sound design. Firstly, periodic sampling of continuous data is shown to create an infinite series of perverse spectral sidebands centred on integral multiples of the sampling frequency. Imperfect rejection of these sidebands by a plant or engineering process leads to irreversible distortion (ripple) which diminishes the potentially available control accuracy. However, it can be inferred even from so early on that this loss can be contained by the deployment of a high enough sampling frequency. Amplitude quantisation of data in digital computer systems and the amazing property of two's complement arithmetic with respect to overflow are also outlined prior to more detailed consideration later on. Descriptions of some available hardware for digital to analogue conversion and its mathematical model complete Chapter 1.

If the continuous output of a linearised DDC system model is fictitiously sampled, its dynamics at the sampling instants can be determined using the z- Transformation. This application of mathematical complex variable theory in Chapter 2 is shown to produce pulse transfer functions with identical combinational properties to Laplace transfer functions, as well as for instance an initial value theorem and inversion integral. Similarities between the z- and Laplace Transformations are deliberately emphasised throughout the book, and readers familiar with the concepts of continuous data control systems are intentionally made to feel 'quite at home'. Section 2.2 demonstrates the correspondence between linear recurrence relations and pulse transfer functions whose various factorisations lead to direct, parallel, cascade and canonical programming techniques. Though this and many other of the theoretical developments apply equally well to digital filters, the engineering objectives and constraints are radically different. Accordingly, this book can only really focus on the specific practical design aspects of DDC systems. Electronic data sampling and transfers are usually executed in such small fractions of a sampling period that these periodic operations in a DDC system can be regarded as impulse modulation. Each number in a time domain sequence is therefore represented, as appropriate, by a δ-function of the same area and this approximation simplifies the analysis later in Chapter 5 of control errors due to ripple.

Dynamic stability intuitively requires that perturbed system variables do not wander too far away from, or eventually converge to, a particular equilibrium state designated as the stationary or steady state solution. Although non-linear differential equations behave in a variety of fascinating ways that lead to stimulating theoretical investigations, the complexity of industrial processes as yet generally obstructs the practical application of these techniques. Moreover, control engineering is concerned not so much with ascertaining stability as with its imposition by the introduction of additional components or modifications to system topology. Chapter 3 first establishes necessary and sufficient conditions for the intrinsically global stability of linear discrete data systems in terms of their pole locations lying strictly within the unit circle of the z- plane. While at first sight these conditions contravene the asserted parity between the z- and Laplace Transformations, the paradox can be resolved by properly comparing both in the s- plane. Mathematical similarities between discrete and continuous data control systems are further re-enforced by demonstrating in particular the existence of an identical Nyquist Diagram technique. For continuous data systems, a Nyquist diagram offers far more than a bald statement of closed loop stability or instability on account of a firm relationship between time and frequency domain behaviour. Granted data sampling rates complying with the common sense algorithm that 'Around ten points are need to define a sinusoid', it is shown that an essentially identical relationship exists for discrete data systems; thereby releasing to DDC systems the more familiar frequency domain design schemes for continuous data systems.

The effect of noise associated with the discreteness of electronic charge is largely eliminated in digital systems, but much larger unwanted disturbances can result from amplitude quantisation errors in A-D conversions or in rounding arithmetic products. Also the continuous input or output data is sometimes contaminated with a significant noise component. It follows that the influence of such noise processes on the accuracy of a control system merits consideration. Chapter 4 presents a succinct introduction to the statistical analysis of so-called wide sense stationary stochastic processes. Though subtle pitfalls await the non-expert who applies correlation techniques to epidemiology, easily interpreted spectral input-output relationships are derived for continuous and discrete linear systems. As with the deterministic signals studied in Chapter 2, contour integration enables a straightforward determination of a z- transformation from the Laplace transformation of the corresponding continuous data. Frequency response measurements on a real system can involve its excitation with sinusoids or single pulses. However, the necessity that the induced perturbations stand-out against ambient noise amplitudes can render such deterministic measurements unacceptable in some cases. Good reasons obtain therefore for sometimes using the low amplitude stochastic measurement technique described in Section 4.7. On the other hand, statistical estimates converge no faster than the reciprocal of the square root of measurement time, and they require carefully constructed stochastic signal sources. Consequently, the recommended deterministic procedure for computing the ensemble mean square value of a stochastic variable has definite advantages; especially later in Chapter 6 where it is deployed to predict the wordlength requirement of a DDC system.

The performance specification of a DDC system generally relates in part to its control error as a function of continuous time. Under transient conditions, control errors are dominated by conventional plant dynamics, and they can be adequately assessed by the z- Transformation. Under steady state conditions, when control errors due to conventional plant dynamics are much smaller, ripple and finite wordlength effects can provide significant separate contributions. Because the amplitudes of high frequency ripple errors vary markedly over a sample period, a control accuracy assessment by the z- Transformation is usually insufficient on its own. The Modified z- Transformation discussed in Section 2.5 represents an early attempt to evaluate such intersample behaviour. However, because predicted ripple errors do not evolve in a convenient form for design purposes (eg as a single number), it has now become largely obsolete. Chapter 5 derives the mean square value of the ripple error component in DDC systems by means of an equivalent open loop controller and a surprisingly simple relationship between power spectra in discrete and continuous time (ie between number sequences and corresponding pulse trains). Both z- and Laplace Transformation are involved in this analysis, so there are more than pedagogical reasons for establishing their similarities in previous Chapters. Because ripple power depends on the controller's pulse transfer function, which itself is an implicit function of sampling frequency, an iterative design procedure for DDC systems is strictly necessary. A reasonable initial estimate of the sampling frequency for this purpose is established in Section 5.4 on the basis of the rms ripple error produced by the maximum amplitude sinusoidal input at the frequency corresponding to the closed loop bandwidth specification. Though demonstrably not an actual 'worst case' situation, it is in practice close enough, and certainly better than previous guesses. Furthermore, existing electronic hardware admits on occasion somewhat higher sampling rates at no extra cost, and then the inherently dramatic reduction of ripple power with sampling frequency can be exploited to obviate the tedium of an iterative design. Early experiments described in Section 5.5 fully validate the above analysis.

Amplitude quantisation errors after processing by the recurrence equation of a digital controller produce the finite wordlength control errors. Chapter 6 begins with earlier derivations of the idealised quantisation error statistics which apply if the characteristic function of the unquantised data is bandlimited to within half the amplitude quantisation frequency. Despite the fact that perfectly bandlimited characteristic functions appear exceptional, the original authors leave the utility of their analyses as a largely open question. The possibly novel development in Section 6.3 reveals that these idealised statistics are generally accurate enough approximations for practical computer wordlengths ($\gtrsim$ 8 bits) and sampling rates. A 'sprinkling' of wide bandwidth noise on the unquantised data is helpful in this respect. The derivation of an upper bound on the finite wordlength control error component in Section 6.4 elicits a clear mental picture of the aggravation of amplitude quantisation errors by the particular recurrence equation of a digital controller. Specifically, the error component increases as its real zeros are moved closer to the $1\underline{/0}$ point, which corresponds to a stronger derivative action. In this same context, it is shown that wordlength requirements can be markedly reduced by engineering digital compensators with complex poles (whenever possible). A spectral relationship between quantisation errors as inputs and the induced control errors as outputs is deduced in Section 6.5. It enables the steady state mean square value of this control error component to be defined as a function of wordlength by a single deterministic calculation. Different programming techniques: direct, parallel, cascade etc can all be assessed in the same straightforward manner, and simulation results confirm the analysis within the uncertainties dictated by the finite sample sizes. Finite wordlength control errors stem from a linear system's convolution of its quantisation errors, and these fulfil the conditions of the Central Limit theorem apart from possibly statistical independence. A statistical test applied to some simulated control errors supports the hypothesis of their Normality, so that a mean square value computed as above can be translated into precise confidence limits.

Chapter 7 combines the preceding analyses into a comprehensive, yet computationally undemanding, Nyquist Diagram based design technique. The complete procedure is illustrated in Fig 7.1.2 for a unity feedback configuration, and it first addresses the possibility of significant wide bandwidth noise on the continuous input or output data. Where necessary and if practicable, suitable anti-aliasing filters are recommended. If these cannot be deployed, then a control accuracy specification can be satisfied only by a high enough sampling frequency to restrict the intrusion of noise sidebands into the signal baseband. After an initial selection of sampling frequency based on the proposal in Section 5.4, a digital phase advance compensator for plants with one or more integrations is devised with the aid of two simple empirical rules and some standard curves which are easily computed once and for all time. Destabilising structural resonances in servo systems are countered in practice by divided reset, feedforward*, or notch networks. A simple form of digital notch filter is examined in terms of its resonant frequency, bandwidth, and wordlength requirements for a satisfactory implementation. Experience recommends the compensation of a Class-0 system with an RC phase retard network which is realised by placing a suitable capacitor across the feedback resistor of a D-A converter. Such analogue networks evolve as superior to their digital counterparts because a required attenuation is achieved with a similar phase lag over a narrower frequency band. At this juncture the ripple and finite wordlength error components are evaluated by the computational techniques described in Chapters 5 and 6. Depending on the performance specification and the limitations of available hardware, these results could dictate changes to the estimated sampling frequency, so that the design might have to continue iteratively to meet these particular constraints. Amplitude

---

*allowing in some circumstances a reduction in the added scalar gain

saturation, or overflow with a digital controller, is shown in Section 7.7 to be highly destabilising. However, by deploying two's complement arithmetic and repeated addition operations in preference to left-shifts, its onset is readily predictable so that appropriate coarse-fine control schemes can be readily engineered. The linearisation of a plant model or the introduction of non-linear control schemes evidently renders vindication of a final design by an adequate simulation absolutely imperative.

Time domain syntheses of digital controllers are shown to be generally unsatisfactory. Although the transient response to the specific design input appears at first sight to be exceptionally good, rate constraints are almost always contravened and the transient response to other inputs leaves much to be desired. Though Smelka's synthesis uniquely accommodates reasonable rate constraints, it can involve differentiating controllers as well as inconsistent specifications of transient rise time and overshoot. On the other hand, the recommended frequency domain procedure is consistent and it demonstrably produces 'even-tempered' designs.

A multirate DDC system is defined in Chapter 8 as one having a controller whose output sampler operates at an integral number of times faster than its input unit. Because the output of a multirate controller appears as a closer approximation to continuous data, Kranc suggested that ripple errors would tend to be less than for a single rate controller sampling at the slower input rate. This argument overlooks the lower frequency sidebands generated by the input sampler which cannot be eliminated by faster output sampling. To enable a closer scrutiny of the conjecture, Nyquist Diagram theory, ripple and finite wordlength control error components are derived for multirate systems and then validated by simulation. Using the Class-2 plant in Section 7.3 as an example, it is then shown that a triple rate controller giving essentially the same closed loop transient performance as the corresponding single rate unit actually results in rms ripple errors that are three times larger. Reference 157 establishes that a multirate system has generally an inferior ripple performance in practice.

A subrate DDC system has a controller whose input sampler operates at an integral number of times faster than its output unit. Computing speeds in the 1960's were very much slower than at present, so the introduction of subrate controllers with a centralised computer was believed to release time for off-line data logging, process optimisation etc. Section 8.4 first establishes that an explicit overall pulse transfer function for a subrate DDC system does not exist. A single rate approximation, which preserves the same zero frequency gain through the controller, is then proposed. Though the uncertainties this introduces on a Nyquist diagram could not be quantified, it does enable a reasonably successful compensation of the above Class-2 plant. However, the subrate recurrence relation involves virtually as many multiplications as a single rate design. It is therefore concluded that neither multirate nor subrate controllers fulfil original aspirations. Indeed distributed control and the greatly enhanced speed of contemporary digital controllers now often enable the design of quite adequate single rate DDC systems, so that such (unsuccessful) strategies for reducing ripple or computing time have become less relevant.

Appendices

A2.1 The Limit in Equation (2.1.12)

Suppose there exists a real constant $0 \leq G < 1$ such that for some integer N the sequence $\{u_n\}$ of complex numbers satisfies:

$$|u_{n+1}| \leq G\, |u_n| \qquad \text{for all } n \geq N \tag{A2.1.1}$$

then by mathematical induction it follows that:

$$|u_{N+k}| \leq G^k\, |u_N| \qquad \text{for all integer } k$$

Application of the extension of inequalities (10) yields:

$$\lim_{k\to\infty} |u_{N+k}| \leq |u_N| \lim_{k\to\infty} G^k = 0$$

so that:

$$\lim_{n\to\infty} |u_n| = 0$$

and because the modulus function is continuous over the set of complex numbers:

$$\lim_{n\to\infty} u_n = 0 \tag{A2.1.2}$$

It is taken as self-evident that:

$$\lim_{n \to \infty} \left(\frac{n}{n + 1} \right) = 1$$

and also that if $|p_m| < 1$, then:

$$1 - |p_m|^{1/q} > 0 \qquad \text{for integer } q \geq 1$$

By definition of a limit, there exists an integer N for which:

$$0 \leq |p_m|^{1/q} < \frac{N}{N + 1}$$

or

$$0 \leq |p_m| \left(\frac{N}{N + 1} \right)^q < 1 \qquad \text{(A2.1.3)}$$

Terms of the sequence in equation (2.1.2) satisfy:

$$|u_{n+1}| = |p_m| \left(\frac{n + 1}{n} \right) |u_n| \qquad \text{for all } n$$

so setting the constant G to:

$$G = |p_m| \left(\frac{N + 1}{N} \right)$$

ensures that terms of this sequence satisfy equation (A2.1.1). Hence by equation (A2.1.2):

$$\lim_{n \to \infty} n^q p_m = 0$$

as stated in the main text.

A2.4 The Normal Summability of $\left\{H(s)(z^{-1}e^{sT})^n\right\}$ over [c - j∞, c + j∞]

Stray capacitance, inertia etc ensure that in practice Laplace transformations of interest satisfy:

$$\lim_{s \to \infty} \left| sH(s) \right| = \alpha \qquad \text{a constant} \qquad \text{(A2.4.1)}$$

and because:

$$H(s) = \frac{1}{s}\left(sH(s)\right) \qquad \text{for } s \neq 0$$

then:

$$\lim_{s \to \infty} \left| H(s) \right| = \lim_{s \to \infty} \left| \frac{1}{s} \right| . \lim_{s \to \infty} \left| sH(s) \right| = 0$$

Hence it follows that there exists a real number R such that:

$$\left| H(s) \right| \leq 1 \qquad \text{for all } \left| s \right| > R \qquad \text{(A2.4.2)}$$

In general, the Laplace transformation is also meromorphic*, so that it is analytic and therefore continuous along the line [c - j∞ ; c + j∞]. Further, the portion (D) of this line that

---

*possesses just finite order isolated singularities (poles) in all finite regions of the s- plane

forms a subset of the disc $|z| \leq R$ is compact by the Borel-Lebesque theorem (10). Because the image of a compact set by a continuous mapping is also compact (10), then for some real number b:

$$\left|H(s)\right| \leq b \qquad \text{for } s \in D \tag{A2.4.3}$$

Setting:

$$B = \text{Max}\{1, b\}$$

then from equations (A2.4.2) and (A2.4.3), it follows that H(s) is bounded by B along the whole line [c - j∞ ; c + j∞], and consequently:

$$\left|H(s)\left(z^{-1}e^{sT}\right)^n\right| \leq B\left(\left|z\right|^{-1}e^{cT}\right)^n$$

For any function F with domain [c - j∞ ; c + j∞], a convenient norm is specified by:

$$\left\|F\right\| = \text{Sup}\left\{\left|F(s)\right| \,\middle|\, s \in [c - j\infty \; ; \; c + j\infty]\right\}$$

so that:

$$\left\|H.\left(z^{-1}\exp T.\right)^n\right\| \leq B\left(\left|z\right|^{-1}e^{cT}\right)^n$$

Provided that:

$$\left|z\right|^{-1}e^{cT} < 1 \qquad \text{or} \qquad \left|z\right| > e^{cT}$$

the series of positive terms $\{B.(z^{-1}e^{cT})^n\}$ is convergent, and therefore the series $\{H(s)(z^{-1}e^{sT})^n\}$ is normally summable as stated in the main text.

B2.4 Convergence around Infinite Semicircles

Denote a semicircle of radius R in the left-half s- plane by:

$$S_L = \left\{ s = Re^{j\Theta} \,\middle|\, \frac{\pi}{2} \leq \Theta \leq \frac{3\pi}{2} \right\}$$

and a similar semicircle in the right-half s- plane by:

$$S_R = \left\{ s = Re^{j\Theta} \,\middle|\, -\frac{\pi}{2} \leq \Theta \leq \frac{\pi}{2} \right\}$$

To evaluate equation (2.4.8) by the calculus of residues, it is required to prove that:

$$\lim_{R \to \infty} \int_{S_L} \frac{H(s)ds}{1 - e^{sT}z^{-1}} = \lim_{R \to \infty} \int_{S_R} \frac{H(s)ds}{1 - e^{sT}z^{-1}} = 0 \qquad \text{(B2.4.1)}$$

where H(s) is a meromorphic function.

A similar argument to that in Appendix A2.4 shows that the series of functions $\{H(s)(e^{sT}z^{-1})^n\}$ is normally summable on S_L and S_R, provided that a semicircle is large enough and provided that:

$$|z| > e^{RT} \qquad \text{(B2.4.2)}$$

Under these conditions, the order of integration and summation for the series can be interchanged to give:

$$\int_{S_L, S_R} \frac{H(s)ds}{1 - e^{sT}z^{-1}} = \int_{S_L, S_R} \left\{ \sum_{0}^{\infty} H(s) \left[e^{sT} z^{-1} \right]^n ds = \right.$$

$$= \sum_{0}^{\infty} \left\{ \int_{S_L, S_R} z^{-n} H(s) e^{snT} ds \right\} \qquad \text{(B2.4.3)}$$

In connection with equation (B2.4.3), the integrals on the extreme right relate to samples of the impulse response taken at integral multiples of the sampling period (T). Samples of continuous data are generally created in a DDC system from mean values taken over the acquisition period of an electronic A-D converter. Thus the sample h(o) can be replaced by $h(o^+)$ where o^+ denotes an arbitrary small positive real number, which becomes the lower limit of summation in equation (B2.4.3). Although this modification is obviously of no practical significance, it evolves as the crucial factor in establishing the required convergence.

The treatment of left- and right semicircles now differs in detail, though not in principle, so consider a left semicircle and the integral:

$$I_n = \int_{S_L} z^{-n} H(s) e^{snT} ds \qquad \text{for } n \geq o^+ \qquad \text{(B2.4.4)}$$

After expressing the complex variable in polar form and deploying the inequality (B2.4.2), the integral mean value theorem (10) yields:

$$\left| I_n \right| \leq e^{-RnT} \int_{\pi/2}^{3\pi/2} \left| H(Re^{j\Theta}) Re^{j\Theta} \right| \exp\left(RnT.\cos\Theta \right) d\Theta \qquad \text{(B2.4.5)}$$

Stray capacitance, inertia etc in practical systems ensure the existence of a real number (α) such that their Laplace transfer functions satisfy:

$$\lim_{s \to \infty} \left| sH(s) \right| = \alpha$$

so that for large enough semicircles:

$$\left|H(Re^{j\Theta})Re^{j\Theta}\right| \leq (\alpha + 1)$$

Thus the inequality in (B2.4.5) reduces to:

$$\left|I_n\right| \leq (\alpha + 1)e^{-RnT} \int_{\pi/2}^{3\pi/2} \exp\left[RnT.\cos\Theta\right] d\Theta \qquad \text{for } n \geq o^+ \qquad \text{(B2.4.6)}$$

Setting:

$$\Theta' = \Theta - \pi/2$$

and noting that:

$$\sin\Theta \geq 2\Theta/\pi \qquad \text{for } 0 \leq \Theta \leq \pi/2$$

simplifies statement (B2.4.6) to:

$$\left|I_n\right| \leq \frac{\pi(\alpha + 1)\exp(-RnT)}{RnT} \qquad \text{for } n \geq o^+ \qquad \text{(B2.4.7)}$$

It follows from the above inequality that:

$$\sum_{o^+}^{N} \left|I_n\right| \leq \frac{\pi(\alpha + 1)}{Ro^+ T}$$

All finite subseries of $\{\left|I_n\right|\}$ are therefore bounded, so the series $\{\left|I_n\right|\}$ and $\{I_n\}$ are convergent with (10):

$$0 \leq \left|\sum_{o^+}^{\infty} I_n\right| \leq \sum_{o^+}^{\infty} \left|I_n\right| \leq \frac{\pi(\alpha + 1)}{Ro^+ T}$$

Substituting equation (B2.4.4) into the above result gives:

$$0 \le \left| \sum_{0^+}^{\infty} \int_{S_L} z^{-n} H(s) e^{snT} ds \right| \le \frac{\pi(\alpha + 1)}{Ro^+ T}$$

and from equation (B2.4.3):

$$0 \le \left| \int_{S_L} \frac{H(s)ds}{1 - e^{sT} z^{-1}} \right| \le \frac{\pi(\alpha + 1)}{Ro^+ T}$$

Because the modulus function is continuous over the field of complex numbers, the extension of inequalities principle (10) implies:

$$0 \le \left| \lim_{R \to \infty} \int_{S_L} \frac{H(s)ds}{1 - e^{sT} z^{-1}} \right| \le \lim_{R \to \infty} \frac{\pi(\alpha + 1)}{Ro^+ T}$$

so that as required:

$$\lim_{R \to \infty} \int_{S_L} \frac{H(s)ds}{1 - e^{sT} z^{-1}} = 0$$

The corresponding result for the infinite right semicircle is similarly established.

A3.3 The Spectrum of a Sequence

If all poles of a meromorphic z- Transformation have a magnitude less than unity, the function is analytic everywhere outside a circular contour (*C*) within the unit circle. Similar analytical techniques to those deployed in Section 2.4 prove that the series $\{X(z)x_k z^{k-1}\}$ is normally summable on the contour *C*, and therefore:

$$\sum_{0}^{\infty} \frac{1}{2\pi j} \int_{C} X(z) x_k z^{k-1} dz = \frac{1}{2\pi j} \int_{C} X(z) \left(\sum_{0}^{\infty} x_k z^k \right) z^{-1} dz \tag{A3.3.1}$$

Now:

$$X(z^{-1}) = \sum_{0}^{\infty} x_k z^k$$

and by the Inversion Integral (2.1.9):

$$x_k = \frac{1}{2\pi j} \int_{C} X(z) z^{k-1} dz$$

so that equation (A3.3.1) becomes:

$$\sum_{0}^{\infty} x_k^2 = \frac{1}{2\pi j} \int_{C} X(z) X(z^{-1}) z^{-1} dz \tag{A3.3.2}$$

Because there are no poles of X(z) and $X(z^{-1})$ in the region between *C* and the unit circle of the z- plane, Cauchy's Theorem (10, 12) allows the contour of integration to be enlarged to the unit circle(Γ). Consequently:

$$\sum_{0}^{\infty} x_k = \frac{1}{2\pi j} \int_{\Gamma} X(z)X(z^{-1})z^{-1}dz \qquad (A3.3.3)$$

and because:

$$X^*(\omega)X^*(-\omega) = \left|X^*(\omega)\right|^2$$

it follows that:

$$\sum_{0}^{\infty} x_k^2 = \frac{T}{2\pi} \int_{-\pi/T}^{\pi/T} \left|X^*(\omega)\right|^2 d\omega \qquad (A3.3.4)$$

Multiplication of the left-hand side of the above equation by the sampling period (T) evidently provides an approximation to the energy of the unsampled signal x(t). Because the function $\left|X^*(\omega)\right|^2$ evidently represents the distribution of this energy with frequency, it is referred to in the main text as the 'energy' spectrum of the sequence $\{x_k\}$.

A4.6 Poles, Analytic Regions and Power Series

Suppose a function F of a complex variable u has no poles in a region S specified by:

$$|u| \leq r \tag{A4.6.1}$$

The integral of F(u) around any closed contour in S is therefore zero by Cauchy's theory of residues (10, 12). By Morera's theorem (10, 12), the function is analytic over S. Consequently, it can be represented by an absolutely convergent power series about any interior point u_o of S:

$$F(u) = \sum_{0}^{\infty} c_k (u - u_o) k \tag{A4.6.2}$$

provided that u is sufficiently close to u_o. Because the limit of an absolutely convergent series is independent of any grouping effected in its summation, the above series can be simplified to:

$$F(u) = \sum_{0}^{\infty} c'_k u^k \tag{A4.6.3}$$

In particular for:

$$u_o = R \underline{/0} \tag{A4.6.4}$$

this series is absolutely convergent so that:

$$\left| c'_k R^k \right| \leq \left\{ \sum_{0}^{\infty} \left| c_k R^k \right| \right\} \qquad \text{for all } k$$

Hence by Abel's lemma (10, 12, 38) the series $\{c'_k u^k\}$ is normally summable over the region $|u| < R$. Because u_0 is an arbitrary interior point of S, it follows that the function F can be expressed as a normally summable power series over any circular region (disk) within S (eg $|u| < r$).

Pertinent results for the inverse power series involved with the z- Transformation can be derived by substituting z^{-1} for u in the above arguments. In particular, a z- transformation H(z) can be expanded as a normally summable power series:

$$H(z) = \sum_{0}^{\infty} d_k z^{-k} \tag{A4.6.5}$$

in a region:

$$|z| > r \tag{A4.6.6}$$

in which there are no poles. When a z- transformation is derived from a continuous time function having a Laplace transformation, Section 2.4 establishes its existence as the normally summable series:

$$H(z) = \sum_{0}^{\infty} h_k z^{-k} \tag{A4.6.7}$$

over some region:

$$|z| > r'$$

Power series which equate over a partly shared region of normal summability have identical coefficients (10, 38). Accordingly, the series $\{h_k z^{-k}\}$ is normally convergent to H(z) over the region $|z| > r$, which is barren of poles. Practical systems have meromorphic (10, 12) pulse transfer functions, so just a finite number of poles lie within the unit circle. If a system is stable, all its poles lie strictly within the unit circle which lies therefore in the region of normal summability for $\{h_k z^{-k}\}$, and so:

$$\{h_k\} \text{ absolutely summable} \tag{A4.6.8}$$

as required.

A4.8 The Sum of the Squares of a Weighting Sequence

A stable discrete data system has a meromorphic pulse transfer function G(z). Because all its poles lie strictly within the unit circle (Γ) of the z- plane, it follows from Appendix A4.6 that the series:

$$G(z) = \sum_{0}^{\infty} g_k z^{-k} \qquad \text{(A4.8.1)}$$

is normally summable for:

$$|z| > r \qquad \text{with} \quad r < 1$$

Likewise, the series

$$G(z^{-1}) = \sum_{0}^{\infty} g_i z^{i} \qquad \text{(A4.8.2)}$$

is normally summable for

$$|z| < 1/r$$

Both series $\{g_k z^{-k}\}$ and $\{g_i z^{i}\}$ are therefore normally summable over a closed annular region $R_I \le |z| \le R_o$ where:

$$R_I < 1 \qquad \text{and} \qquad R_o > 1$$

Finite subsums of the 'double' series $\left\{\left\| g_k g_i z^{i-k-1} \right\|\right\}$ are evidently bounded by:

$$\left(\sum_{0}^{\infty} |g_k| R_I^{-k}\right)\left(\sum_{0}^{\infty} |g_i| R_o^{i}\right) R_I^{-1}$$

so that the series itself is normally summable over this annulus. Consequently, the order of its integration and summation can be interchanged to give:

$$\frac{1}{2\pi j} \int_{\Gamma} G(z^{-1})G(z)z^{-1}dz = \sum_{0}^{\infty}\sum g_k g_i \frac{1}{2\pi j} \int_{\Gamma} z^{i-k-1}dz$$

and because:

$$\left.\begin{aligned} \frac{1}{2\pi j} \int_{\Gamma} z^{i-k-1}dz &= 1 \qquad \text{for } i = k \\ &= 0 \qquad \text{otherwise} \end{aligned}\right\}$$

it follows that:

$$\frac{1}{2\pi j} \int_{\Gamma} G(z^{-1})G(z)z^{-1}dz = \sum_{0}^{\infty} g_k^2 \tag{A4.8.3}$$

as required.

A6.2 The First Order Characteristic Function for Quantisation Errors

The probability density function $P_E(\epsilon)$ for quantisation errors (ϵ) on a stochastic process with probability density function $P_X(x)$ is derived in Section 6.2 as:

$$\left.\begin{aligned} P_E(\epsilon) &= \sum_{-\infty}^{\infty} P_X(nq - \epsilon) \qquad \text{for } -q < \epsilon \le q/2 \\ &= 0 \qquad \text{otherwise} \end{aligned}\right\} \qquad \text{(A6.2.1)}$$

where

q - width of quantisation

It is assumed that the meromorphic two-sided Laplace transformation:

$$F_X(s) == \int_{-\infty}^{\infty} P_X(s)\ \exp(-sx)\,dx \qquad \text{(A6.2.2)}$$

converges asymptotically no slower than as s^{-2}, and also that it has no poles in a semi-infinite strip about the imaginary axis of the s- plane. Following the argument in Appendix A4.6, the series $\{P_X(nq - \epsilon).\exp(-jv\epsilon)\}$ is normally summable so that the order of its integration and summation can be interchanged to yield:

$$F_E(v) = \int_{-q/2}^{q/2} P_E(\epsilon)\exp(-jv\epsilon)d\epsilon = T_0 + T_+ + T_- \qquad (A6.2.3)$$

where:

$$\left.\begin{aligned} T_0 &= \int_{-q/2}^{q/2} P_X(-\epsilon)\exp(-jv\epsilon)d\epsilon \; ; \; T_+ = \sum_{1}^{\infty}\int_{-q/2}^{q/2} P_X(nq-\epsilon)\exp(-jv\epsilon)d\epsilon \\ T_- &= \sum_{1}^{\infty}\int_{-q/2}^{q/2} P_X(-nq-\epsilon)\exp(-jv\epsilon)d\epsilon \end{aligned}\right\} \qquad (A6.2.4)$$

Defining the Fourier transformable function:

$$\left.\begin{aligned} Q(\epsilon) &= 1 \quad \text{for } |\epsilon| < q/2 \\ &= 0 \quad \text{otherwise} \end{aligned}\right\} \qquad (A6.2.5)$$

enables equations (A6.2.4) to be written as:

$$\left.\begin{aligned} T_0 &= \int_{-\infty}^{\infty} P_X(-\epsilon)Q(\epsilon)\exp(-jv\epsilon)d\epsilon \\ T_+ &= \sum_{1}^{\infty}\int_{-\infty}^{\infty} P_X(nq-\epsilon)Q(\epsilon)\exp(-jv\epsilon)d\epsilon \\ T_- &= \sum_{1}^{\infty}\int_{-\infty}^{\infty} P_X(-nq-\epsilon)Q(\epsilon)\exp(-jv\epsilon)d\epsilon \end{aligned}\right\} \qquad (A6.2.6)$$

Each of the above integrals is the Fourier transformation of the product of two Fourier transformable functions, so by the convolution integral (1.2.25):

$$
\left.
\begin{aligned}
T_{0} &= (q/2\pi) \int_{-\infty}^{\infty} F_X(u)\ S(v+u)\,du \\
T_{+} &= (q/2\pi) \sum_{1}^{\infty} \int_{-\infty}^{\infty} F_X(u)\ S(v+u)\ \exp(junq)\,du \\
T_{-} &= (q/2\pi) \sum_{1}^{\infty} \int_{-\infty}^{\infty} F_X(u)\ S(v+u)\ \exp(-junq)\,du
\end{aligned}
\right\}
\tag{A6.2.7}
$$

where:

$$
S(w) = \frac{\sin(wq/2)}{wq/2} \tag{A6.2.8}
$$

To evaluate the above integrals and to effect their summation, apply the principle of analytic continuation (10, 12) by setting:

$$
s = j\omega
$$

so that:

$$
\left.
\begin{aligned}
T_{0} &= (q/2\pi j) \int_{-j\infty}^{j\infty} F_X(s)\ S(v-js)\,ds \\
T_{+} &= (q/2\pi j) \sum_{1}^{\infty} \int_{-j\infty}^{j\infty} F_X(s)\ S(v-js)\ \exp(snq)\,ds \\
T_{-} &= (q/2\pi j) \sum_{1}^{\infty} \int_{-j\infty}^{j\infty} F_X(s)\ S(v-js)\ \exp(-snq)\,ds
\end{aligned}
\right\}
\tag{A6.2.9}
$$

The series of functions $\{F_X(s).S(v-js).\exp(snq)\}$ for example is normally summable provided:

$$
R_s < 0 \tag{A6.2.10}
$$

because Appendix A2.4 shows that $F_X(s)$ is bounded in the

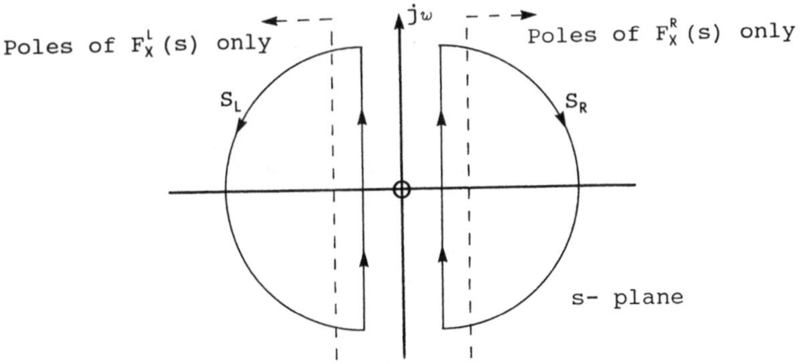

Fig A6.2.1 Illustrating the Discussion of Equation (A6.2.9)

'poleless' region about the imaginary axis. By Cauchy's theorem, the contour of integration can be displaced somewhat left in this region and closed by an infinite semicircle in the left-half s- plane. Fig A6.2.1 illustrates this contour S_L and the pole locations of $F_X(s)$ that allow its factorisation into:

$$F_X(s) = F_X^L(s) + F_X^R(s) \tag{A6.2.11}$$

where

$$\left.\begin{aligned} &F_X^L(s) \text{ - poles in left-half s- plane only} \\ &F_X^R(s) \text{ - poles in right-half s- plane only} \end{aligned}\right. \tag{A6.2.12}$$

The assumed asymptotic behaviour of $F_X(s)$ implies that $F_X^L(s)$ and $F_X^R(s)$ converge asymptotically no slower than s^{-1}, so that from Appendix B2.4 the integrals around the semicircular portion of S_L are zero. Similar arguments apply to the other series of functions $\{F_X(s).S(v - js).\exp(-snq)\}$, and consequently equations (A6.2.9) can be written as:

$$T_o = (q/2\pi j) \int_{S_L} F_X^L(s)\ S(v - js)ds + (q/2\pi j) \int_{S_R} F_X^R(s)\ S(v - js)ds$$

$$T_+ = (q/2\pi j) \int_{S_L} F_X^L(s)\ S(v - js)\ \Big[C(s) - 1\Big]ds$$

$$T_- = (q/2\pi j) \int_{S_R} F_X^R(s)\; S(v - js) \left[C(-s) - 1\right] ds \Bigg] \qquad \text{(A6.2.13)}$$

where:

$$C(s) = \left[1 - \exp(sq)\right]^{-1}$$

Hence from equation (A6.2.3) it follows that

$$F_E(v) = (q/2\pi j) \int_{S_L} F_X^L(s)\; S(v - js)C(s)ds$$
$$+ (q/2\pi j) \int_{S_R} F_X^R(s)\; S(v - js)C(-s)ds \qquad \text{(A6.2.14)}$$

As described in Section 4.4, the contours in the above integrals can be interchanged so as to enclose the simple poles of $(1 - \exp\pm sq)^{-1}$ at $\pm j2\pi k/q$ rather than those of F_X^L and F_X^R. Noting that:

$$\lim_{s \to j2\pi k/q} \left[(s - j2\pi k/q)C(\pm s)\right] = \mp 1/q \quad \text{for all integer k} \qquad \text{(A6.2.15)}$$

and that the contour S_R has negative (clockwise) orientation enables the reduction of equation (A6.2.14) to:

$$F_E(v) = \sum_{-\infty}^{\infty} F_X^L(2\pi k/q).S(v + 2\pi k/q) + \sum_{-\infty}^{\infty} F_X^R(2\pi k/q).S(v + 2\pi k/q) \qquad \text{(A6.2.16)}$$

Finally because each series is convergent, they can be combined to give:

$$F_E(v) = \sum_{-\infty}^{\infty} F_X(2\pi k/q).S(v + 2\pi k/q) \qquad \text{(A6.2.17)}$$

which is equation (6.2.6) of the main text.

A6.3 The Bandwidth of a Second Order Characteristic Function

Far more care is required in assessing the bandwidth of a second order Characteristic function than for a first order function. With a wide sense stationary input to a rounded quantiser, the samples:

$$x_1 = x(t_1) \qquad \text{and} \qquad x_2 = x(t_2) \tag{A6.3.1}$$

have the same variance σ_x^2. If the time instants t_1 and t_2 are sufficiently well separated for these variates to be essentially uncorrelated, contours of equal joint probability density in the x_1 x_2- plane resemble those of a symmetrically round hill. As the variates become more correlated, they change to those of a progressively steeper sided ridge orientated along the optimum linear regression in equation (4.2.4):

$$x_2 = \rho x_1 \tag{A6.3.2}$$

where

ρ - correlation coefficient

Though the dispersion of x_1 (regarded as the independent variable) remains at σ_x as the correlation with x_2 increases ($|\rho| \rightarrow 1$), the dispersion of the dependent variable x_2 about the optimum regression line decreases significantly according to equation (4.2.5) as:

$$\sigma_X \left(1 - \rho^2\right)^{1/2}$$

This contraction is accounted for in estimates of the double integrals in equation (6.3.7) by:

$$\int_{-\infty}^{\infty}\!\!\int \left[(\partial P/\partial x_1)^2 + (\partial P/\partial x_2)^2\right] dx_1\, dx_2$$

$$\simeq P^2(0,0)\,\sigma_X^{-2}\left[1 + (1 - \rho^2)\right] 4\sqrt{2}\,\sigma_X^2 (1 - \rho^2)^{1/2} \qquad \text{(A6.3.3)}$$

and

$$\int_{-\infty}^{\infty}\!\!\int P^2(x_1 x_2)\, dx_1\, dx_2 \simeq P^2(0,0)\, 4\sqrt{2}\,\sigma_X^2 (1 - \rho^2)^{1/2} \qquad \text{(A6.3.4)}$$

Thus the bandwidth of the second order Characteristic function is derived from equation (6.3.7) as

$$B_2 = (1/\sigma_X)\left[1 + (1 - \rho^2)^{-1}\right]^{1/2} \qquad \text{(A6.3.5)}$$

which is quoted in the main text as equation (6.3.9).

A7.2 The Non-Filterable Input or Output Situation

Sometimes it is not physically possible to prefilter the continuous input or output data of a DDC system even though they are significantly contaminated with wideband noise. Under these quite special circumstances, noise sideband power introduced into the signal baseband by sampling can be adequately attenuated only by deploying a high enough sampling frequency (ω_o). A suitable value can be engineered by means of the following calculations based on Fig A7.2.1. Here the Laplace transfer function F(s) broadly characterises the frequency domain characteristics of the particular closed loop system, so it could be, say, a low pass Butterworth filter with the same bandwidth and asymptotic cut-off rate (rank). Periodic data sampling is represented by pulse amplitude modulation with a pulse width γ and a delay (switch-on) time that is uniformly distributed over [0,T]. The scalar gain factor T/γ offsets the DC attenuation caused by sampling, and this is intrinsically compensated for by a zero order hold in an actual

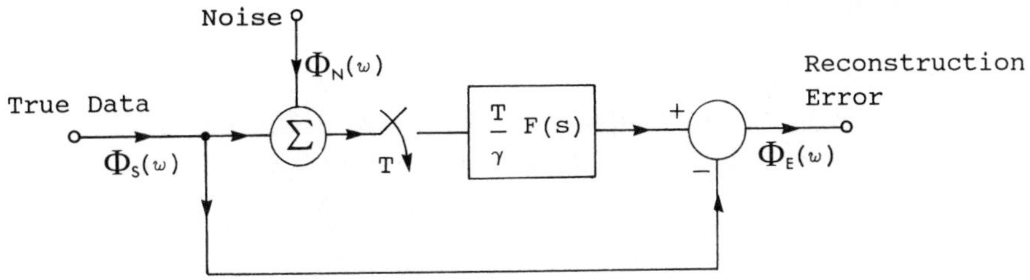

Fig A7.2.1 A Computation to engineer an appropriate Sampling Frequency

system. If the true data and noise ensembles are uncorrelated, the mean square reconstruction error equals the sum of their separate contributions as proved in Section 4.7. Furthermore, by means of the convolution integral and a Fourier Series expansion of the sampling pulse train like in Section 5.2, the power spectrum of the data reconstruction error is derived as:

$$\Phi_E(\omega) = \left|1 - F(\omega)\right|^2 \Phi_s(\omega) + \left|F(\omega)\right|^2 \Phi_N(\omega) + \left|F(\omega)\right|^2 \sum_{\substack{-\infty \\ \neq 0}}^{\infty} \left[\Phi_s(\omega - n\omega_0) + \Phi_N(\omega - n\omega_0)\right] \tag{A7.2.1}$$

Physically, the first two of the above terms represent error components incurred due to imperfect signal transmission and imperfect noise rejection in the baseband, while the infinite summation corresponds to signal and noise sideband power appearing in the baseband.

If a transient response time or equivalently a control system bandwidth is prescribed, the contributions of the first two baseband terms in equation (A7.2.1) are effectively defined. The problem then becomes the straightforward choice of sampling frequency to ensure that the sideband power arising from:

$$\left|F(\omega)\right|^2 \sum_{\substack{-\infty \\ \neq 0}}^{\infty} \left[\Phi_s(\omega - n\omega_0) + \Phi_N(\omega - n\omega_0)\right] \tag{A7.2.2}$$

is compatible with the steady state accuracy specification. To implement a numerical integration of the above spectrum, it is conveniently expressed in closed form using equation (4.5.35):

$$T\Phi^*(\omega) = \sum_{-\infty}^{\infty} \Phi(\omega - n\omega_0) \tag{A7.2.3}$$

where a pulse power spectrum $\Phi^*(\omega)$ is derived from its continuous counterpart $\Phi(\omega)$ by means of equation (4.5.26). Having made a reasonable initial choice of sampling frequency, the design proceeds according to Fig 7.1.2 with possible iterative refinements to achieve the specified bandwidth or steady state accuracy.

In another similar situation, the performance specification requires that the DDC system itself effects some steady state optimal rejection of the noise baseband spectrum (114) so the baseband terms of equation (A7.2.1) must then also be considered. A reasonably appropriate initial choice of sampling frequency and overall system bandwidth can be engineered in these circumstances by computing the mean square reconstruction error:

$$\phi_E(o) = \frac{1}{2\pi}\int_{-\infty}^{\infty} \Phi_E(\omega)\,d\omega \qquad \text{(A7.2.4)}$$

for a range of likely bandwidths and sampling frequencies in the manner of Fig A7.2.2. Afterwards the design procedure continues as in Fig 7.1.2 except that the digital compensator is devised to achieve the optimised closed loop bandwidth.

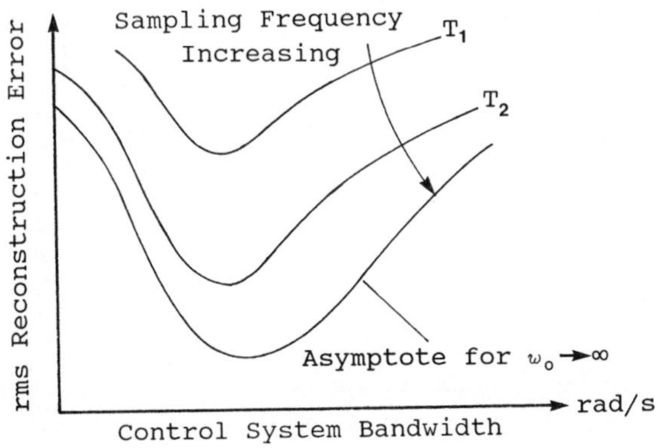

Fig A7.2.2 Appertaining to Closed Loop Bandwidth and Sampling Frequency selection with unfilterable noisy Input or Output Data

A7.3 Gain-Phase Curves for Digital Compensator Design

It is proposed in Section 7.3 that the pulse transfer function of a digital controller is a simple rational function with real coefficients and an independently variable scalar gain constant. Expressed in terms of poles and zeros, its numerator and denominator polynomials involve therefore just the two types of factor:

$$(z - a) \qquad \text{or} \qquad (z - a - jb)(z - a + jb)$$

where a and b are real numbers. In the manner specified by equation (3.3.14), each of the above terms provides the following gain and phase contributions:

$$G_a = \left|e^{j\omega T} - a\right| \qquad \text{and} \qquad \phi_a = \text{Arg}\left(e^{j\omega T} - a\right) \tag{A7.3.1}$$

or

$$G_{ab} = \left|e^{j\omega T} - a - jb\right|\left|e^{j\omega T} - a + jb\right|$$
$$\phi_{ab} = \text{Arg}\left(e^{j\omega T} - a - jb\right) + \text{Arg}\left(e^{j\omega T} - a + jb\right) \tag{A7.3.2}$$

After normalising the angular frequency variable (ω) with respect to the sampling frequency (ω_0) and applying de Moivre's theorem, equation (A7.3.1) yields:

$$G_a = \sqrt{1 - 2a\cos(2\pi\omega_N) + a^2}$$

$$\phi_a = \tan^{-1}\left\{\frac{\sin(2\pi\omega_N)}{\cos(2\pi\omega_N) - a}\right\} \tag{A7.3.3}$$

where

$$\omega_N = \omega/\omega_o \tag{A7.3.4}$$

Similar manipulation of equation (A7.3.2) gives (114, 142):

$$G_{ab} = \sqrt{\left[b^2 + G_a\cos(2\phi_a)\right]^2 + G_a\sin^2(2\phi_a)}$$

$$\phi_{ab} = \tan^{-1}\left\{\frac{G_a\sin(2\phi_a)}{b^2 + G_a\cos(2\phi_a)}\right\} \tag{A7.3.5}$$

Relating the gain and phase contributions for a complex conjugate pair of poles or zeros to those of the corresponding real entity, as shown above, simplifies their computation. Graphs of typical and useful results complete this Appendix.

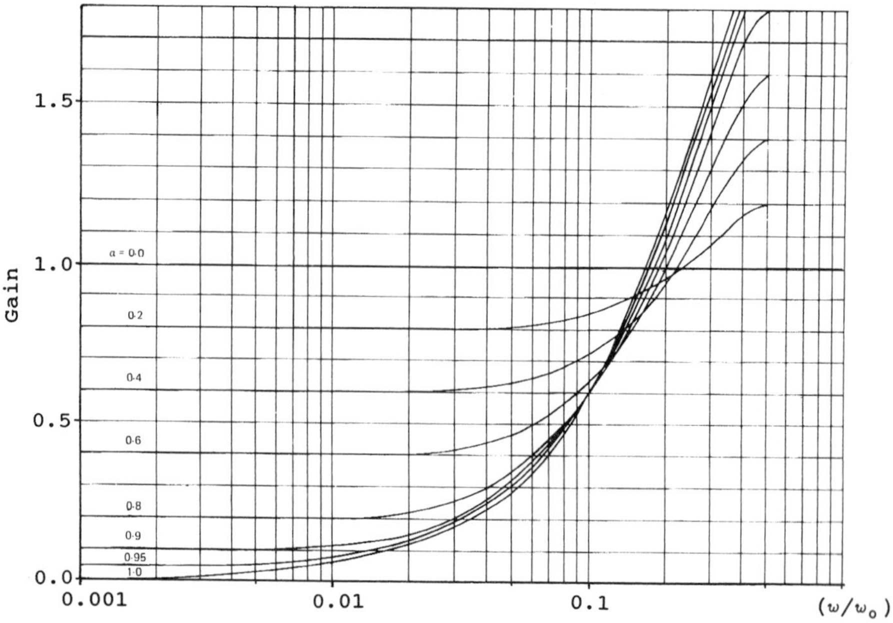

Fig A7.3.1 Gain-Frequency Graph $|\exp(j2\pi\omega/\omega_0)-a|$

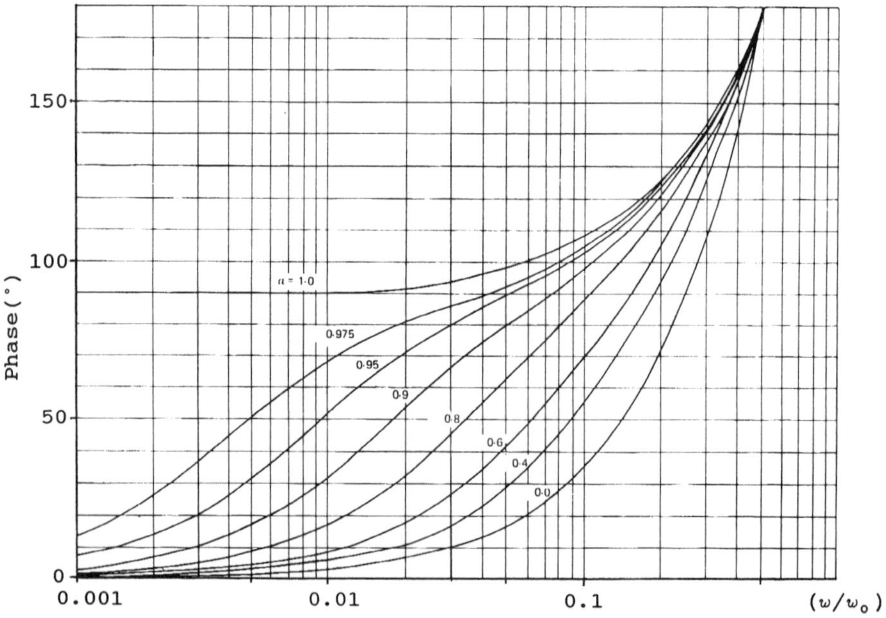

Fig A7.3.2 Phase-Frequency Graph Arg $\left[\exp(j2\pi\omega/\omega_0)-a\right]$

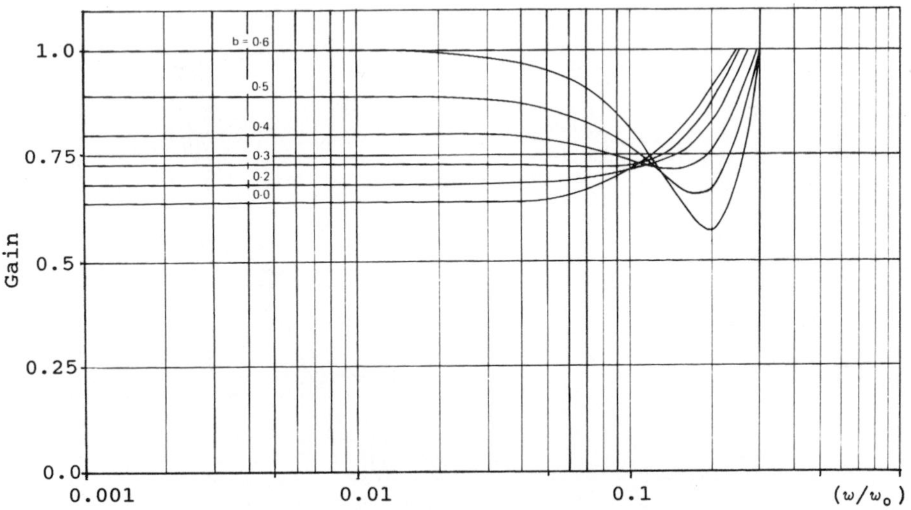

Fig A7.3.3 Gain-Frequency Graph

$$\left|\exp(j2\pi\omega/\omega)-0.2-jb\right| \left|\exp(j2\pi\omega/\omega_0)-0.2+jb\right|$$

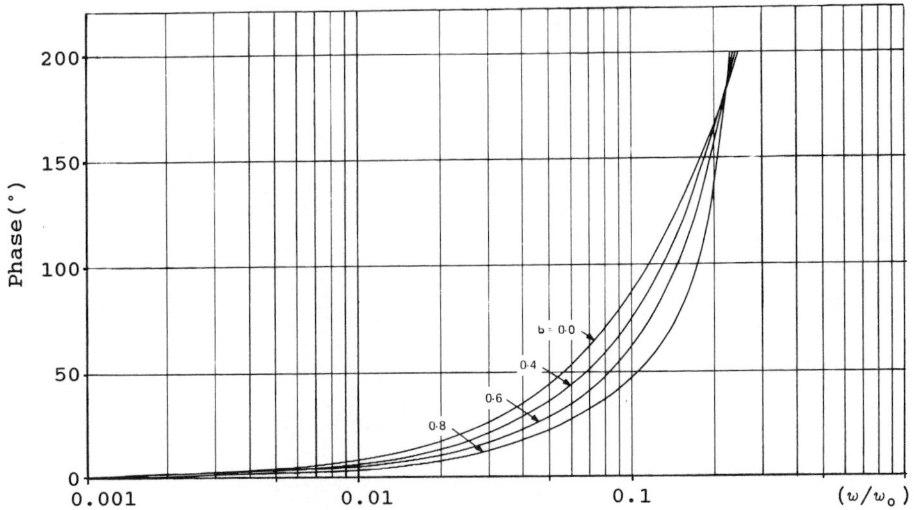

Fig A7.3.4 Phase-Frequency Graph

$$\text{Arg}\left[\exp(j2\pi\omega/\omega_0)-0.2-jb\right]\left[\exp(j2\pi\omega/\omega_0)-0.2+jb\right]$$

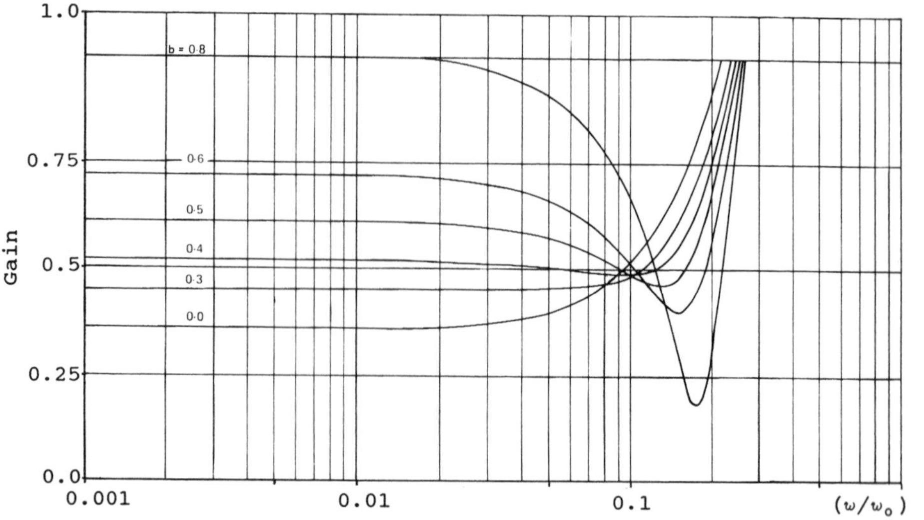

Fig A7.3.5 Gain-Frequency Graph

$$\left|\exp(j2\pi\omega/\omega_0)-0.4-jb\right|\left|\exp(j2\pi\omega/\omega_0)-0.4+jb\right|$$

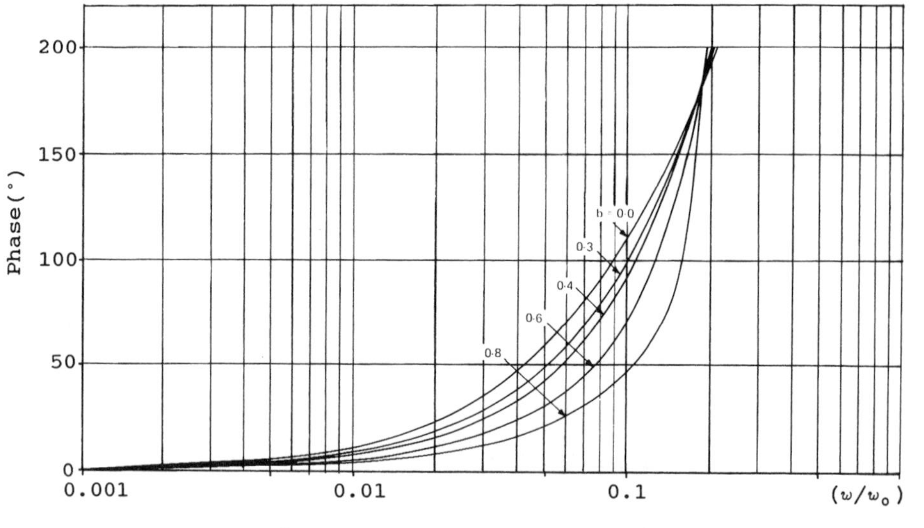

Fig A7.3.6 Phase-Frequency Graph

$$\text{Arg}\left[\exp(j2\pi\omega/\omega_0)-0.4-jb\right]\left[\exp(j2\pi\omega/\omega_0)-0.4+jb\right]$$

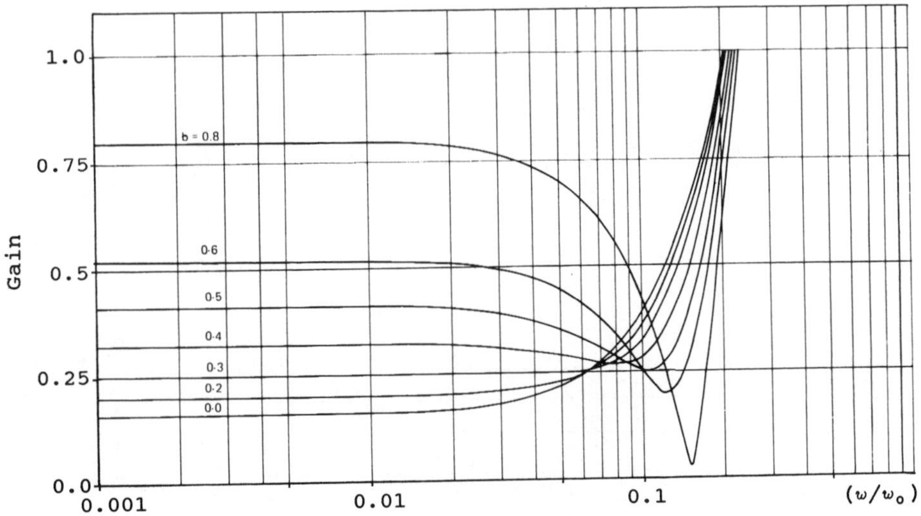

Fig A7.3.7 Gain-Frequency Graph

$$\left|\exp(j2\pi\omega/\omega_0)-0.6-jb\right|\left|\exp(j2\pi\omega/\omega_0)-0.6+jb\right|$$

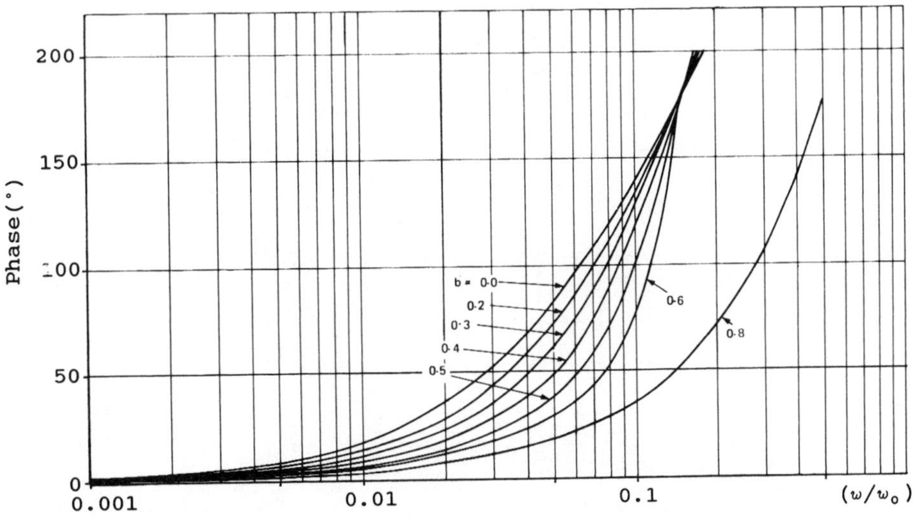

Fig A7.3.8 Phase-Frequency Graph

$$\mathrm{Arg}\left[\exp(j2\pi\omega/\omega_0)-0.6-jb\right]\left[\exp(j2\pi\omega/\omega_0)-0.6+jb\right]$$

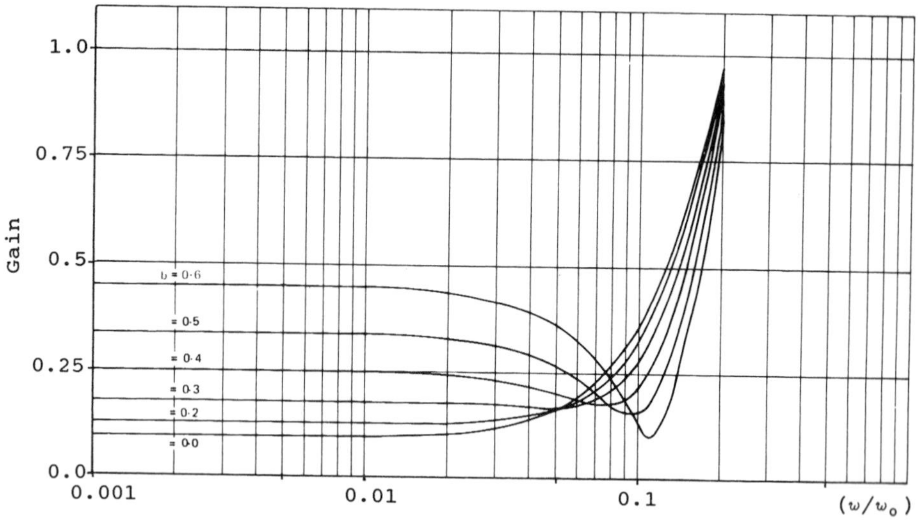

Fig A7.3.9 Gain-Frequency Graph

$$\left|\exp(j2\pi\omega/\omega_0)-0.7-jb\right|\left|\exp(j2\pi\omega/\omega_0)-0.7+jb\right|$$

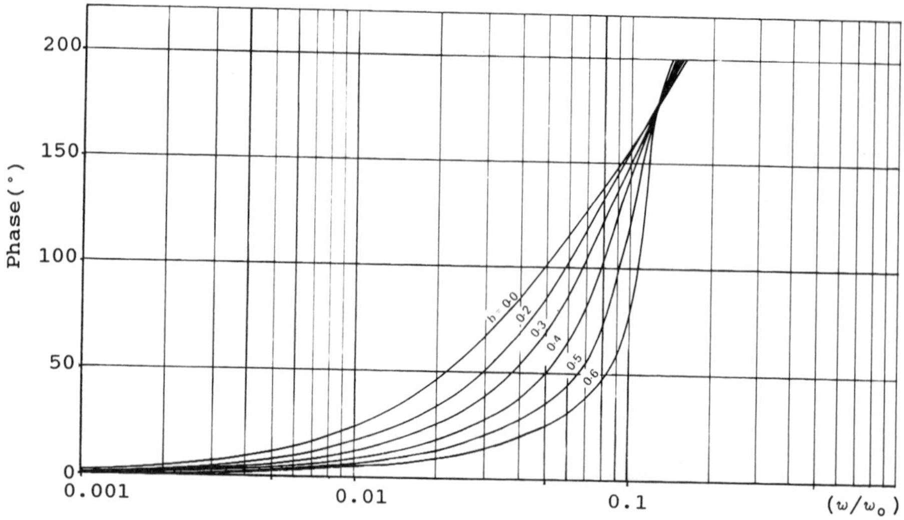

Fig A7.3.10 Phase-Frequency Graph

$$\text{Arg}\left[\exp(j2\pi\omega/\omega_0)-0.7-jb\right]\left[\exp(j2\pi\omega/\omega_0)-0.7+jb\right]$$

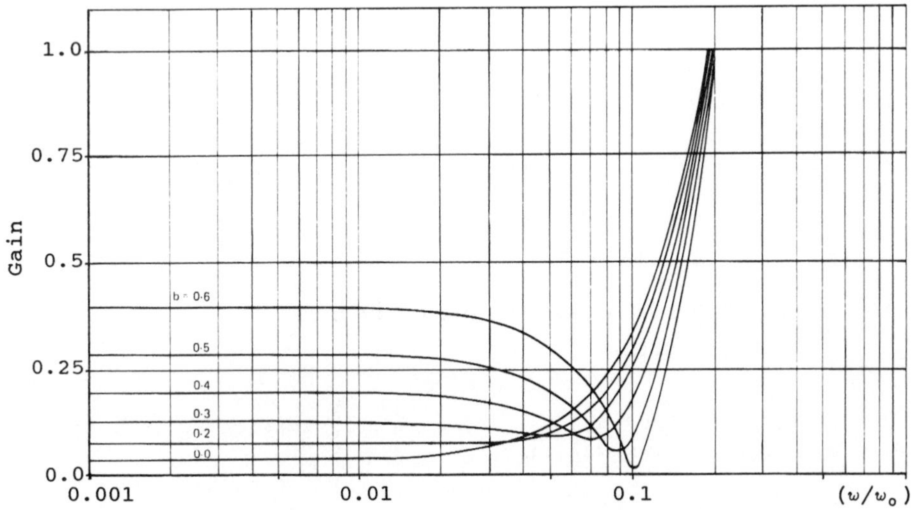

Fig A7.3.11 Gain-Frequency Graph

$$\left|\exp(j2\pi\omega/\omega_o)-0.8-jb\right|\left|\exp(j2\pi\omega/\omega_o)-0.8+jb\right|$$

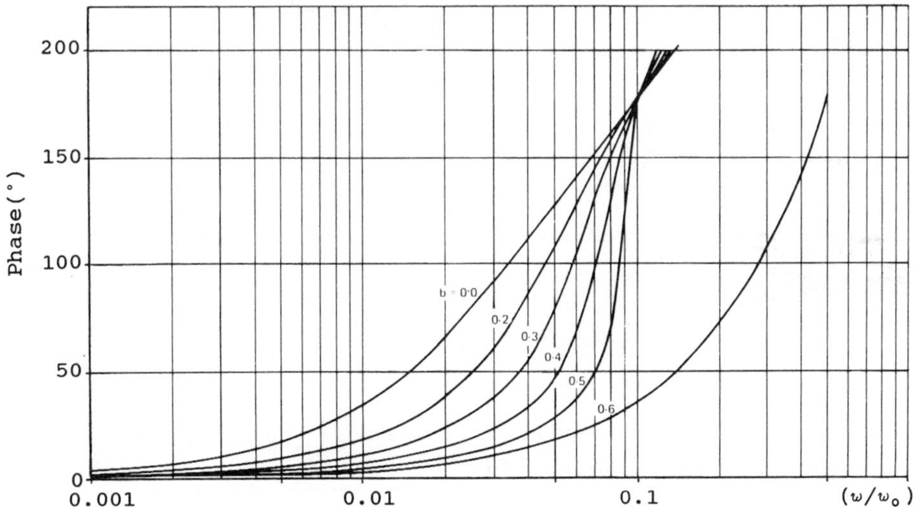

Fig A7.3.12 Phase-Frequency Graph

$$\text{Arg}\left[\exp(j2\pi\omega/\omega_o)-0.8-jb\right]\left[\exp(j2\pi\omega/\omega_o)-0.8+jb\right]$$

A8.1 Extracting a Deterministically Embedded z-Transformation

If $F(z_n)$ is the single sided z_n-transformation of an arbitrary sequence $\{f_k\}$ then by definition:

$$F(z_n) = \sum_{p=0}^{\infty} f(pT/n) z_n^{-p} \qquad (A8.1.1)$$

from which it is required to extract the embedded z-transformation:

$$F(z) = \sum_{k=0}^{\infty} f(kT) z^{-k} \qquad (A8.1.2)$$

Application of the Inversion Integral (2.1.9) yields:

$$F(z) = \sum_{k=0}^{\infty} \left(\frac{1}{2\pi j} \int_C F(z_n) z_n^{kn-1} z^{-k} dz_n \right) \qquad (A8.1.3)$$

where the contour C encompasses all the poles of $F(z_n)$. Provided that over C:

$$\left| z_n^{n} z^{-1} \right| < 1 \qquad (A8.1.4)$$

the infinite series of functions:

$$\left\{F(z_n) z_n^{-1} (z_n z^{-1})^k\right\}$$

is normally summable over C, and its order of integration and summation can be interchanged to give:

$$F(z) = \frac{1}{2\pi j} \int_C F(z_n) \left(\frac{z_n^{-1}}{1 - z_n{}^n z^{-1}} \right) dz_n \qquad \text{(A8.1.5)}$$

In the main text the above contour integral is written symbolically in equation (8.1.13) as:

$$F(z) = Z\left[F(z_n)\right] \qquad \text{(A8.1.6)}$$

and for present purposes the appropriate evaluation is according to:

$$F(z) = -\sum \text{Residues of } F(z_n) \left(\frac{z_n^{-1}}{1 - z_n{}^n z^{-1}} \right) \text{ at the poles of } \left[1 - z_n{}^n z^{-1}\right]^{-1} \qquad \text{(A8.1.7)}$$

The poles referred to above are clearly located at:

$$p_m = z^{1/n} \exp(j2\pi m/n) \qquad \text{for} \qquad 0 \le m \le n-1 \qquad \text{(A8.1.8)}$$

by which equation (A8.1.7) is straightforwardly reduced to the required result:

$$F(z) = \left(\frac{1}{n} \right) \sum_{m=o}^{n-1} F(p_m) \qquad \text{(A8.1.9)}$$

If a single sided z_n-transformation $F(z_n)$ is the product of two functions:

$$F(z_n) = F_1(z_n) F_2(z_n) \qquad \text{(A8.1.10)}$$

its corresponding sequence is specified by the convolution:

$$f(kT/n) = \sum_{p=o}^{k} f_1(pT/n)\, f_2(\overline{k - p}T/n) \qquad (A8.1.11)$$

In the special case when $\{f_1(n)\}$ is an embedded slow rate sequence with:

$$f_1(pT/n) = 0 \qquad \text{for} \qquad p = mn \qquad (A8.1.12)$$

equation (A8.1.11) becomes:

$$f(kT/n) = \sum_{p=o}^{P} f_1(pT)\, f_2(\overline{k/n - p}T) \qquad (A8.1.13)$$

where:

$$P = \text{Integral part of } k/n \qquad (A8.1.14)$$

It follows therefore that:

$$f(mT) = \sum_{p=o}^{k} f_1(pT)\, f_2(\overline{m - p}T) \qquad (A8.1.15)$$

and because the above summation is a 'slow rate convolution', then:

$$Z\left[F(z_n)\right] = F_1(z)\; Z\left[F_2(z_n)\right] \qquad (A8.1.16)$$

which is quoted in the main text.

A8.3 Derivation of a Multiplication Pattern

It is required to evaluate the multiplication pattern for the multirate recurrence relation:

$$f_k = \sum_{p=1}^{P} a_p \underline{e}_{k-p} - \sum_{q=1}^{Q} b_q f_{k-q} \qquad \text{for all integer } k \qquad \text{(A8.3.1)}$$

when all the coefficients $\{a_p\}$ and $\{b_q\}$ are non-zero, non-integral positive powers of two, and other than unity. Because $\{f_k\}$ is a fast rate sequence the last summation generally involves Q significant multiplications, so that the finite convolution:

$$C = \sum_{p=1}^{P} a_p \underline{e}_{k-p} \qquad \text{(A8.3.2)}$$

alone requires detailed consideration. For this purpose set:

$$P = mn + \nu \qquad \text{(A8.3.3)}$$

where m and ν are integers with:

$$0 \leq \nu \leq n - 1 \qquad \text{(A8.3.4)}$$

If m = 0, then at time instants corresponding to:

$$\nu + 1 - n \leq k \leq 0 \qquad \text{(A8.3.5)}$$

the correlation is zero because terms of the embedded sequence $\{\underline{e}_{k-p} | 1 \le p \le \nu\}$ are all zero.

However, at the time instants:

$$1 \le k \le \nu \qquad \text{(A8.3.6)}$$

just one of the coefficients $\{a_p\}$ multiplies the generally non-zero input sample $\underline{e}_0$, so that:

$$\left.\begin{aligned} \mu_k &= Q + 0 \qquad \text{for} \quad \nu + 1 - n \le k \le 0 \\ &= Q + 1 \qquad \text{for} \quad 1 \le k \le \nu \end{aligned}\right\} \qquad \text{(A8.3.7)}$$

For an arbitrary value of k, there exists an integer i such that:

$$k = in + k' \qquad \text{(A8.3.8)}$$

with

$$\nu + 1 - n \le k' \le \nu \qquad \text{(A8.3.9)}$$

The same argument as above with k replaced by k' and $\underline{e}_0$ replaced by $\underline{e}_{in}$ yields the same values as in equation (A8.3.7). Hence this multiplication pattern is periodic with period n.

If m=1 and ν=0, then when k=1, the convolution involves just the one multiplication of $\underline{e}_0$ by a_1. If the convolution (C) is assumed to contain just one significant multiplication for any k>1, then there exists a unique term in the set $\{k-p | 1 \le p \le n\}$ which is exactly divisible by n. Thus there are unique integers p'_k and i_k for which:

$$k - p'_k = i_k n \qquad \text{for} \quad 1 \le p'_k \le n \qquad \text{(A8.3.10)}$$

For the fast rate index k+1, it is readily verified that these unique integers then become:

$$p'_{k+1} = p'_k + 1 \quad \text{and} \quad i_{k+1} = i_k \quad \text{for} \quad 1 \le p'_k < n$$

$$p'_{k+1} = 1 \quad \text{and} \quad i_{k+1} = i_k + 1 \quad \text{for} \quad p'_k = n$$

(A8.3.11)

and again there is just one significant multiplication. It follows from the principles of mathematical induction that in this case:

$$\mu_k = Q + 1 \qquad \text{for all } k \ge 1 \tag{A8.3.12}$$

The multiplication pattern for the general situation corresponding to equation (A8.3.3) can now be derived by induction on m. When m=0, the multiplication pattern is derived above as:

$$\left.\begin{aligned} \mu_k &= Q + 0 \quad \text{for} \quad in + \nu \le k \le (i+1)n \\ &= Q + 1 \quad \text{for} \quad (i+1)n + 1 \le k \le (i+1)\, n + \nu \end{aligned}\right\}$$

(A8.3.13)

where i is a positive integer. The multiplication pattern for $m > 0$ is therefore assumed to be:

$$\left.\begin{aligned} \mu_k &= Q + m \quad \text{for} \quad in + \nu \le k \le (i+1)n \\ &= Q + m + 1 \quad \text{for} \quad (i+1)n + 1 \le k \le (i+1)\, n + \nu \end{aligned}\right\}$$

(A8.3.14)

Because the convolution (C) for m+1 is related to that for m by:

$$C_{m+1} = C_m + \sum_{p=1}^{n} a_{mn+\nu+p}\, \underline{e}(k - mn - \nu - p) \tag{A8.3.15}$$

and because the number of significant multiplications in the last term is specified by equation (A8.3.12) as 1, then for m+1:

$$\left.\begin{aligned} \mu_k &= Q + (m+1) \quad \text{for} \quad in + \nu \le k \le (i+1)n \\ &= Q + (m+1) + 1 \quad \text{for} \quad (i+1)n + 1 \le k \le (i+1)\, n + \nu \end{aligned}\right\}$$

(A8.3.16)

By the principle of mathematical induction, equation (A8.3.14) applies for all integer values m, and this result appears as equation (8.3.6) in the main text. Attention is drawn to the obviously periodic nature of the multiplication pattern.

References

1. LUCKY, R.W; SALZ, J. and WELDON, E.J. 'Principles of Data Communication', McGraw Hill Publishers (1968)
2. ALTABER, J. et al. 'Truly Distributed Control, using one microprocessor per real time task', Paper 1 IEE Conference on Trends in on-line computer control systems' (1982)
3. SMITH, I.C. and WALL, D.N. 'Programmable Electronic Systems for Reactor Safety', Atom 395, 10-13, September (1989)
4. BHASKAR, K. 'Computer Security: Threats and Countermeasures', NCC/Blackwell Publishers (1993)
5. BODE, H.W. 'Network Analysis and Feedback Amplifier Design', D Van Nostrand Publishers (1956)
6. MACIEJOWSKI, J.M. 'Multivariable Feedback Design', Addison-Wesley Publishers (1989)
7. BELL, D.J. and MUNRO, N. 'Design of Modern Control Systems' IEE Series on Control Engineering 18 (1982)
8. AIZERMAN, M.A. 'On a Problem Concerning the In-the-Large Stability of Dynamic Systems', Uspekhi Mat. Nauk, 4, 187-188, (1949)
9. ZEMANSKY, M.W. and DITTMAN, R.H. 'Heat and Thermodynamics', McGraw Hill Publishers (1981)
10. DIEUDONNE, J. 'Treatise on Analysis', Academic Press (1993)
11. WIENER, N. 'The Fourier Integral and Certain of its Applications', Cambridge University Press (1933)
12. TITCHMARSH, E.C. 'The Theory of Functions', Oxford University Press (1939)

13. CARSLAW, H.S. 'Introduction to the theory of Fourier's Series and Integrals', 3rd Edition, Cambridge University Press (1930)
14. BROWN, W.C. 'Vector Spaces and Matrices', M. Dekker Publishers (1991)
15. JONES, F. 'Lebesgue Integration on Euclidean Space', Jones and Bartlett Publishers (1993)
16. LIGHTHILL, M.J. 'Introduction to Fourier Analysis and Generalised Functions', Cambridge Monographs on Mechanics and Applied Mathematics, Cambridge University Press (1959)
17. GOLD, B and RADER, C.M. 'Digital Processing of Signals', McGraw Hill Publishers (1969)
18. 'Selected Papers in Digital Signal Processing II', edited by the Digital Signal Processing Committee of the IEEE Acoustics, Speech and Signal Processing Society, IEEE Press (1976)
19. 'Papers on Digital Signal Processing', edited by A.V. Oppenheim, MIT Press (1969)
20. 'Analog Digital Conversion Handbook', Digital Equipment Corporation (1964)
21. 'The Analogic Data Conversion Systems Digest', Analogic Corporation (1978)
22. GORDON, B.M. 'Linear Electronic Analog/Digital Conversion Architectures, Their Origins, Parameters, Limitations and Applications', IEEE Trans. Circuits and Systems 25, No 7, 6-33 (1978)
23. BARKER, R.H. 'A Transducer for Digital Data Transmission Systems', Proc. IEE 103, Part B, No 7, 42-51 (1956)
24. MURRAY, W. 'Computer and Digital System Architecture', Prentice Hall Publishers (1990)
25. KERSHAW, J.D. 'Digital Electronics', 4th Edition, Chapman and Hall Publishers (1992)
26. 'Integrated Circuits Data Book', Burr-Brown Corporation 33 (1989), 33b (1990)
27. EPSTEIN, H.C. et al. 'Economical, High-Performance Optical Encoders', Hewlett-Packard Journal, 99-106, October (1988)
28. RAGAZZINI, J.R. and FRANKLIN, G.F. 'Sampled Data Control Systems', McGraw Hill Publishers (1958)

29. BIRKHOFF, G. and Mac LANE, S. 'A Survey of Modern Algebra', Macmillan Publishers (1953)
30. JACKSON, L et al. 'An Approach to the Implementation of Digital Filters', IEEE Trans. Audio and Electroacoustics 17, No 2, 104-108 (1969)
31. SCALF, H.L. and PETERSON, S.C. 'The ASIC Guidebook', Chapman and Hall (1992)
32. HWANG, K 'Computer Arithmetic - principles, architecture and design', J.Wiley Publishers (1979)
33. KULISCH, U 'Mathematical Foundations of Computer Arithmetic', IEEE Trans. Computers 26, 610-620 (1977)
34. 'Digital Signal Processing Handbook', Advanced Micro Devices Inc (1976)
35. EDWARDS, R. et al. 'Comparison of Noise Performances of Programming Methods in the Realisation of Digital Filters', Proc. of the Symposium on Computer Processing in Communications, Microwave Research Institute Symposia Series 14, 295-311 (1969), Brooklyn Polytechnic Press
36. JACKSON, L.B. 'Roundoff-Noise Analysis for Fixed-Point Digital filters Realized in Cascade or Parallel Form', IEEE Trans. Audio and Electroacoustics 18, No 2, 107-122 (1970)
37. JURY, E.I. 'Sampled-Data Control Systems', J. Wiley Publishers (1958)
38. KNOPP, K. 'Theory and Application of Infinite Series', Dover Publishers (1990)
39. RABINER, L.R. et al. 'Some Comparisons between FIR and IIR Digital Filters', Bell System Technical Journal 53, No 2, 305-331 (1973)
40. DOP, L.M.G. and EDWARDS, A.J. 'An Experimental Study of Molten Fuel-Sodium Interactions', BNES International Conference on Fast Reactor Core and Fuel Structure Behaviour, Inverness (1990)
41. KNOWLES, J.B. and EDWARDS, R. 'Effect of a Finite Word Length Computer in a Sampled-data Feedback System', Proc. IEE 112, No 6, 1197-1207 (1965)
42. GOLD, B and RADER, C.M. 'Effects of Quantization Noise in Digital Filters', Spring Joint Computer Conference, Proc. AFIPS 28, 213-219 (1966)

43. KNOWLES, J.B. and OLCAYTO, E.M. 'Coefficient Accuracy and Digital Filter Response', IEEE Trans. Circuit Theory 15, 31-41 (1968)

44. AVENHAUS, E. 'On the Design of Digital Filters with Coefficients of Limited Word Length', IEEE Trans. Audio and Electroacoustics 20, 206-212 (1972)

45. GOLDEN, R.M. and KAISER, J.F. 'Design of Wideband Sampled-data Filters', Bell System Technical Journal 43, 1533-1546 (1964)

46. HARDY, G.H. 'A Course of Pure Mathematics', 10th Edition, Cambridge University Press (1958)

47. CLARSEN, T.A.C.M. et al. 'Some Remarks on the Classification of Limit Cycles in Digital Filters', Philips Research Reports, 28, 297-305 (1973)

48. CLARSEN, T.A.C.M. et al. 'Second-order Digital Filter with only One Magnitude Truncation Quantiser and having practically No Limit Cycles', IEE Electronic Letters 9, 531-532 (1973)

49. CLARSEN, T.A.C.M. et al. 'Frequency Domain Criteria for the Absence of Zero-Input Limit Cycles in Non-linear Discrete-Time Systems, with applications to Digital Filters', IEEE Trans. Circuit Theory 22, 232-239 (1975)

50. JURY, E.I. 'Analysis and Synthesis of Sampled-Data Control Systems', Trans AIEE 73, Part 1, 332-346 (1954)

51. KNOWLES, J.B. and EDWARDS, R. 'Ripple Performance and Choice of Sampling Frequency for a Direct Digital Control System', Proc. IEE 133, 1885-1892 (1966)

52. BARKER, R.H. 'The Theory of Pulse Monitored Servomechanisms and their use for Prediction', Signals Research and Development Establishment, Christchurch, England, Report No 1046 (1950)

53. BARKER, R.H. 'The Pulse Transfer function and its Application to Sampling Servo-Systems', Proc. IEE 99, Part IV, 302-317 (1952)

54. MASSERA, J.L. 'Contributions to Stability Theory', Annals of Mathematics 64, 182-205 (1956)

55. CESARI, L. 'Asymptotic Behaviour and Stability Problems in Ordinary Differential Equations', 2nd Edition, Springer Verlag (1963)
56. 'Contributions to the Theory of Non-Linear Oscillations', Volumes I to V edited by S Lefschetz, Princeton Press (1950 - 1960)
57. HAGEDORN, P. 'Non-Linear Oscillations', Oxford University Press (1988)
58. BAKER, G.L. and GOLLUP, J.P. 'Chaotic Dynamics: An Introduction', Cambridge University Press (1990)
59. KNOWLES, J.B. 'Simulation and Control of Electrical Power Stations' Research Studies Press (1990)
60. EVANS, W.R. 'Graphical Analysis of Control Systems', Trans. AIEE 67, 547-551 (1948)
61. TRUXAL, J.G. 'Control System Synthesis', McGraw Hill Publishers (1955)
62. NYQUIST, H. 'Regeneration Theory', Bell System Technical Journal 11, 126-147 (1932)
63. CUMMINS, J.D. 'A Note on the Errors and Signal to Noise Ratio of Binary Cross-Correlation Measurements of System Impulse Response', AEEW-R329 (1964)
64. CUMMINS, J.D. 'The Simultaneous Use of Several Pseudo Random Binary Sequences in the Identification of Linear Multivariable Dynamic Systems', AEEW-R507 (1965)
65. GUILLEMIN, E.A. 'Synthesis of Passive Networks', J. Wiley Publishers (1957)
66. WEINBERG, L. 'Network Analysis and Synthesis', McGraw Hill Publishers (1962)
67. KAISER, J.F. 'Digital Filters', Chapter 7 of 'System Analysis by Digital Computer', edited by JF Kaiser and FF Kuo, J.Wiley Publishers (1966)
68. JOHNSON, G.W., LINDORF, D.P. and NORDLING, C.G.A 'Extension of Continuous Data System Design Techniques to Sampled-Data Control Systems', Trans AIEE 74, Part II, 252-259 (1955)
69. SALZER, J.M. 'Frequency Analysis of Digital Computers operating in Real Time', Proc. IRE 42, 457-466 (1954)

70. SCHMIDT, S.F. 'Application of Continuous System Design Concepts to the design of Sampled-Data Systems', Trans AIEE 78, Part II, 74-79 (1959)

71. NYQUIST, H. 'Certain Topics in Telegraph Transmission', Journal AIEE 47, 214-216, (1928)

72. TAYLOR, P.L. 'Servomechanisms', Longmans Publishers (1960)

73. MARCY, H.T. 'Parallel Circuits in Servomechanisms', Trans AIEE 65, 521-529 (1946)

74. JAMES, H.M., NICHOLS, N.B and PHILLIPS, R.S. 'Theory of Servomechanisms', McGraw Hill Publishers (1947)

75. BELSEY, F.H. Metropolitan Vickers Research Report C565 (1945)

76. LIVERSAGE, J.H. 'Backlash and Resilience within Closed Loop of Automatic Control', Automatic and Manual Control, Edited by Tustin A 343-372, Pitman Press (1952)

77. SHINSKEY, F.G. 'Feedforward Control Applied', Journal Instrument Society of America, 61-65, November (1963)

78. WOOLVERTON, P.F. and MURRILL, P.W. 'An Evaluation of Four Ideas in Feedforward (Distillation) Control', Instrument Technology 14 No 1, 35-40 (1967)

79. BUTTERFIELD, M.H. and GALL, C.J. 'Dynamic Model Verification by Comparison with Plant Tests on the Winfrith SGHWR', Proc. BNES Conference on Boiler Dynamics and Control in Nuclear Power Stations, 3.1-3.7 (1973)

80. HALL, A.C. 'Application of Circuit Theory to the Design of Servomechanisms', J. Franklin Institute 242, 279-307 (1946)

81. DZUNG, L.S. 'The Stability Criterion', Automatic and Manual Control, Edited by Tustin A, 3-23, Pitman Press (1952)

82. ROSENBROCK, H.H. 'Design of Multivariable Control systems using the Inverse Nyquist Array', Control Systems Centre Report No 48, UMIST, Manchester, England (1969)

83. MUNRO, N. and BOWLAND, B.J. 'The UMIST Computer-Aided Control system Design Suite', User's Guide, Control Systems Centre, UMIST, Manchester England (1980)

84. OWENS, D.H. 'Feedback and Multivariable Systems', 7 IEE Control Engineering Series, Peter Peregrinus Publishers (1978)

85. MUNRO, N. 'Computer aided design I: The inverse Nyquist array design method' edited by O'Reilly, J., Multivariable control for industrial applications, Peter Peregrinus Publishers (1987)
86. WIENER, N. 'Extrapolation, Interpolation and Smoothing of Stationary Time Series', J.Wiley Publishers (1949)
87. CRAMER, H. 'Mathematical Methods of Statistics', Princeton Press (1946)
88. DAVENPORT, W.B. and ROOT, W.L. 'Random Signals and Noise', McGraw Hill Publishers (1958)
89. DOOB, J.L. 'Stochastic Processes', J.Wiley Publishers (1953)
90. Chambers English Dictionary, published by W and R Chambers with Cambridge University Press (1988)
91. RACK, A.J. 'Effect of Space Charge and Transit Time on the Shot Noise in Diodes', Bell System Technical Journal 17, 592-619 (1938)
92. SIEGERT, A.J.F. 'Passage of Stationary Processes through Linear and Non-Linear Devices', IRE Trans. PGIT 36, 4-25 (1954)
93. BENDAT, J.S. 'Principles and Applications of Random Noise Theory', J.Wiley Publishers (1958)
94. BARTLETT, M.S. 'The Critical Community Size for Measels in the United States', Journal of the Royal Statistical Society 123, Part 1, 37-44 (1960)
95. League of Red Cross and Red Crescent Societies. 'Report on Assessment Mission to the Areas Affected by the Chernobyl Disaster' (1990)
96. JORDAN, J. et al. 'Correlation-based Measurement Systems', Harwood Publishers (1989)
97. COOLEY, J.W. et al. 'Applications of the Fast Fourier Transform to Computation of Fourier Integrals, Fourier Series and Convolution Integrals', IEEE Trans. Audio and Electroacoustics 15, 79-84 (1967)
98. TAYLOR, J.R. and ZAFIRATOS, C.D. 'Modern Physics for Scientists and Engineers' Prentice Hall Publishers (1991)

99. NIEDERREITER, H. 'Random Number Generation and Quasi-Monte Carlo Methods', US Society for Industrial and Applied Mathematics (1992)

100. PERRY, J.L. et al. 'A Digital Hardware Realization of a Random Number Generator', IEEE Trans. Audio and Electroacoustics 20, No 4, 236-240 (1972)

101. FARKAS, H. and KRA, I. 'Riemann Surfaces', Springer Publishers (1992)

102. GUILLEMIN, E.A. 'The Mathematics of Circuit Analysis', J.Wiley Publishers (1951)

103. KALMAN, R.E. 'A New Approach to Linear Filtering and Prediction Problems', Trans. ASME, J. of Basic Engineering 82D, 35-45 (1960)

104. KALMAN, R.E. and BUCY, R.S. 'New Results in Linear Filtering and Prediction Theory', Trans. ASME, J. of Basic Engineering 83D, 95-108 (1961)

105. MORRIS, A.S. and STIRLING, M.J.H. 'Model Tuning using the Extended Kalman Filter', Electronics Letters' 15, 201-202 (1979)

106. WOJCIK, P.J. 'Online Estimation of Signal and Noise Parameters with Application to Adaptive Kalman Filtering', Circuits Systems and Signal Processing, 10, 137-152 (1991)

107. ANNIBAL, P.S. 'Statistical Analysis and Kalman Filtering applied to Nuclear Materials Accountancy', AEA Technology Winfrith SRDP R-174 (1990)

108. SPIVAK, M. 'Calculus on Manifolds', Benjamin Publishers (1965)

109. COURANT, R. 'Differential and Integral Calculus', Volume 2, Blackie and Sons Publishers (1962)

110. THOMAS, L.C. 'The Biquad: Part I Some Practical Design Considerations :Part II A Multipurpose Active Filtering System', Proc. IEEE Trans. on Circuit Theory 18, 350-361 (1971)

111. BODE, H.W. and SHANNON, C.E. 'A Simplified Derivation of Linear Least Squares Smoothing and Prediction Theory', Proc. IRE 38, 417-425 (1950)

112. GOLDMAN, S. 'Frequency Analysis, Modulation and Noise', McGraw Hill Publishers (1948)

113. DAVIS, W.D.T. 'Generation and properties of maximum length sequences', Control 302-304 June (1966)

114. KNOWLES, J.B. 'A Contribution to Computer Control', Ph.D thesis UMIST (1962)

115. KNOWLES, J.B. and TSUI, H.T. 'Correlating Devices and Their Estimation Errors', Journal of Applied Physics 38, No 2, 607-612 (1967)

116. SKLANSKY, J and RAGAZZINI, J.R. 'Analysis of errors in sampled data feedback systems', Trans. AIEE 74, Part II, 65-71 (1955)

117. STEWART, R.M. 'Statistical Design and Evaluation of Filters for the Restoration of Sampled Data', Proc. IRE 44, 253-257 (1956)

118. PATEL, S.M. 'Ripple Power in Sampled Data Control Systems', M.Sc Dissertation, UMIST (1966)

119. GERSHO, A. 'Principles of Quantization', IEEE Trans. Circuits and Systems 25, 427-436 (1978)

120. KENDALL, M.G. and STUART, A. 'Advanced Theory of Statistics', Volumes I and II Griffin Publishers (1958)

121. WIDROW, B. 'A Study of Rough Amplitude Quantization by Means of Nyquist Sampling Theory', Trans. IRE on Circuit Theory CT3, No 4, 266-276 (1956)

122. KOSYAKIN, A.A. 'The Statistical Theory of Amplitude Quantization of a Stationary Signal', Automatic and Remote Control 22, No 6, 624-630 (1961)

123. WATTS, D.G. 'A General Theory of Amplitude Quantization with Applications to Correlation Determination', IEE Monograph No 481M (1961)

124. KNOWLES, J.B. 'Word-Length Limitations on the performance of Linear Digital Filters', AEEW-M1119 (1973)

125. ERDELYI, E.T. 'Tables of Integral Transforms', Volumes 1 and 2, McGraw Hill Publishers (1954)

126. GRIFFITHS, J.B. 'Theory of Classical Dynamics', Cambridge University Press (1985)

127. BENNETT, W.R. 'Spectra of Quantized Signals', Bell System Technical Journal 27, 446-472, (1948)

128. KNOWLES, J.B. and EDWARDS, R. 'Autocorrelation function of the Quantisation-Error Process for a Sinusoid', IEE Electronics Letters 4, No 9, 180-182 (1968)

129. SORNMOONPIN, O. 'Investigation of Quantisation Errors', M.Sc Dissertation, UMIST (1966)

130. KNOWLES, J.B. and EDWARDS, R. 'A Simplified Analysis of Computational Errors in a Feedback System incorporating a Digital Computer', Symposium on Direct Digital Control, The Society of Instrument Technology (1965)

131. KNOWLES, J.B. and EDWARDS, R.'Computational Error Effects in a Direct Digital Control System', Automatica 4, 7-29 (1966)

132. EDWARDS, R. 'Degenerative Effects in Direct Digital Control Systems', Ph.D Thesis, UMIST (1966)

133. SANDBERG, I.W. and KAISER, J.F. 'A Bound on Limit Cycles in Fixed Point Implementation of Digital Filters', IEEE Trans. Audio and Electroacoustics 20, 110-112 (1972)

134. WEST, J.C. 'A System Ulitizing Coarse and Fine Position Measuring Elements Simultaneously in Remote-Position-Control Servo Mechanisms', Proc. IEE, 99, Part 2, 135-143 (1952)

135. JACKSON, L.B. 'An Analysis of Roundoff Noise in Digital Filters', Sc D dissertation, Stevens Institute of Technology, Castle Point, Holboken, NJ, USA (1969)

136. AVENHAUS, E. 'On the design of Digital Filters with Coefficients of Limited Wordlength', IEEE Trans. Audio and Electroacoustics 20, 206-212 (1972)

137. CHAN, D.S.K. and RABINER, L.R. 'Analysis of Quantisation Errors in the Direct Form for Finite Impulse Response Digital Filters', IEEE Trans. Audio and Electroacoustics 21, 354-366 (1973)

138. 'Operational Amplifiers', edited by Graeme J.G. et al., McGraw Hill Publishers (1989)

139. GEIGER, R.L. and BUDAK, A. 'Active Filters with Zero Amplifier Sensitivity', IEEE Trans. Circuits and Systems 26, 277-288 (1979)

140. KEATS, A.B. and KNOWLES, J.B. 'Data Transmission up to 4800 bits/sec over Local Post Office Lines without using a Modem', AEEW-R1069 (1976)

141. BOOTON, R.C. et al. 'Non-linear Servomechanisms with Random Inputs', MIT Dynamic Analysis and Control Laboratory Report No 70 (1953)

142. KNOWLES, J.B. 'A Comprehensive, yet computationally simple, Direct Digital Control System design technique', Proc. IEE 125, 1383-1395 (1978) also AEEW-R1172 (1977)

143. TOU, J.T. 'Digital and Sampled Data Control Systems', McGraw Hill Publishers (1959)

144. CHESTNUT, H. and MEYER, R.W. 'Servomechanisms and Regulating System Design', Volume II, J.Wiley Publishers (1955)

145. SMITH, C.H. et al. 'Characteristics of Sampling Servo Systems', Automatic and Manual Control - Proc. of the Cranfield Conference, Butterworth Scientific Publishers (1951)

146. BERGEN, A.R. and RAGAZZINI, J.R. 'Sampled Data Processing Techniques for Feedback Control Systems', Trans. AIEE 73, Part 2, 236-247 (1954)

147. SEMELKA, F.W. 'Time Domain Synthesis of Sampled Data Control Systems', Ph. D Thesis, University of California Berkeley (1960)

148. LINVILL, W.K. 'Sampled Data Control Systems studied through a comparison with Amplitude Modulation', Trans. AIEE 70, Part II, 1779-1788 (1951)

149. LINVILL, W.K. and SALZER, J.M. 'Analysis of Control Systems involving Digital Computers', Proc. IRE 41, 901-908 (1953)

150. CRUSCA, F. and ALDEEN, M. 'Multivariable Frequency-Domain Techniques for the Systematic Design of Stabilizers for Large-Scale Power-Systems', IEEE Trans. on Power-Systems 6, 1133-1139 (1991)

151. MUNRO, N. and ENGELL, S. 'Regulator Design for the F100 Gas Turbofan Engine', Proc. IEE International Conference on Control and its Applications, Warwick University, 380-387 (1981)

152. KIDD, P.T. et al. 'Multivariable Control of a Ship Propulsion System', Proc. of Sixth Ship Control Systems Symposium, Ottowa Canada (1981)

153. KNOWLES, J.B. Internal Document at AEE Winfrith (1983) also as Chapter 2 of 'Real Time Computer Control', edited by Bennett S and Linkens DA, Peter Peregrinus Publishers Ltd.

154. KRANC, G.M. 'Compensation of an error sampled system by a multirate controller', Trans. AIEE 76, Part II, 149-159 (1957)

155. KNOWLES, J.B. and EDWARDS, R. 'Finite Wordlength Effects in Multirate Direct Digital Control Systems', Proc. IEE 112, 2376-2384 (1965)

156. KNOWLES, J.B. and EDWARDS, R. 'Aspects of Subrate Digital Control Systems', Proc. IEE 113, 1893-1901 (1966)

157. KNOWLES, J.B. and EDWARDS, R. 'Critical Comparison of Multirate and Single rate Direct Digital Control System Performance', Computational Methods Session II of the Joint Automatic Control Conference, Colorado, USA (1969)

158. LOEVE, M. 'Probability Theory' 2nd Edition, Van Nostrand Publishers (1960)

Index

Abel's lemma	6,44,360
A-D converter systems	39-42
Aizerman conjecture	4
all pass controller	176
amplitude quantisation frequency	211,219-220
analytic region	142,357,359-361
antialiasing filter	26,117,264-266
antiresonance filter	see notch network
autocorrelation function or sequence	123,135,152,156,160-164,171
backlash	105,276
band limited and non-band limited	24-26,98-99
bandwidth -	
of characteristic function	217-219
of closed loop system	98-104,189,268,271,304
of notch network (filter)	277,282
ratio (sampling frequency: input bandwidth)	98,176,197,223
baseband	24,72,117,175,190,264,372
bilinear transformation	117-120
binary point	27

Birkoff ergodic hypothesis	124
bit -	
least significant	27,136,182
most significant	29
Bode diagram	116,283
Borel - Lebesgue theorem	352
box car circuit	see zero order hold
branch cut	67,148
break frequency	118
Bromwich contour	16
canonical programming	59-61
cascade programming	58,250,260,296,322,336
Cauchy's theorem	358,367
central limit theorem	124,247
characteristic equation	84
characteristic frequency	88
characteristic function -	
bandwidth of	217-219,221-230,369-370
definition and existence of	209
idealised for rounding errors	211-215,364-368
chatter	see control error component due to ripple
Chebyshev polynomials	229
Chernobyl	129
clamp circuit	see zero order hold
class of a control system	270
coarse-fine control modes	292
coefficient of excess, skew	247
coefficient quantisation -	
in an arbitrary transfer function	252-260
in a notch network	279-283
commercial programming convention	27
complex poles, advantages in compensation	272-273
continuous data	18
control error	173-175,231,233

control error component due to a finite wordlength -
advantages of complex poles 236-238
derivative action 239,249
distribution of scalar gain 252,292
Normality 247-248
overflow 61,252
programming technique 239,249-251
control error component due to ripple 72-74,174-201,268-269, 287
convolution integral or summation 14,63-64,150,154,159, 247
correlation coefficient -
definition 129
Markov process 220
quantisation errors 225-230
sinusoid 224
critical point 108,111
cross correlation function or sequence 126,154,158,160, 171
cyclic permuting code 40

D-A converters -
electronic circuits 35-37
phase lag produced 37
phase retard compensation 283
damping factor 88
destabilising effect of sampling data 91-92
deterministic embedding 298
deterministic function or sequence 43,122,161,169
δ-function 17
difference equation see recurrence equation
direct programming technique 55-56,234,242,249, 253, 295, 318,334

discrete Fourier or Laplace transformation	see pulse Fourier or Laplace transformation
divided reset	105,261
dominant poles	88,92,103
drift	246,323-324,339-340
δ-unit	37,63,178,182
dynamic stability	see stability
eigen value	11
encirclement theorem	106,302
energy spectrum of continuous data	13
energy spectrum for a sequence	97,357-358
ensemble -	
definition	121
average	see mean
equivalent open loop controller	177-178,184,311-312,336
ergodic	124-125,168,321
expectation (statistical)	see mean
extension of inequalities	349,356
fail safe	113
feedforward	106,261,276
fictitious sampling	63,301
filters -	see antialising and anti-resonance filter
continuous data	19,264
digital filters form the bilinear transformation	117-119
digital filters recursive and non-recursive	50
high order digital filters	56,252,258-259
idealised low pass	21,98
Kalman	151,264
	see shaping filter

Wiener	151,198,264
finite settling time system	288
final value theorem	48,109
fixed point arithmetic	27
folding	see antialiasing filter
Fourier integral (transformation)	12-14
Fourier series	7-12,23,180
Fubini's theorem	152,257
gain margin	114,116,272,274,275, 285,288
Gaussian	see Normal
gear box	105,276
Gibbs phenomenon	10
glitches	35
Gray code	40
Green's function	see weighting function
hidden oscillations	76
hysteresis	see backlash
idealised low pass filter	21,98
impulse response	see weighting function
initial value theorem	44,109
irreversible process	4
instability	see stability
intermodulation	26,165
Inverse Nyquist -	
array	116
diagram	115-117,267,292
inversion integral of the z-Transformation -	
one-sided	45-47,299,304,382
two-sided	136-140, 244, 249,256

Kolmogorov's theorem 321

Laplace -
transfer functions 15-17
transformation integrals 14-15
least frequency envelope -
existence for single rate systems 95
non-existence for multirate systems 305
left-shifts or shifts-left -
description 56
comparison with multiple additions 294
Leibnitz's theorem 209
limit cycle noise (oscillations) 50,61,247
linearly independent (uncorrelated) 128
line power spectrum 12
linear system -
definition 10-11,80-81
important practical non-linearities 3-4,105,261,292-296,347
importance to control engineering 3,79-80
Normal stochastic inputs 124
pulse real frequency response function 93-97
pulse transfer function 50-51,61-65
stability of 80-84,299
stochastic frequency response measurement 163-166
stochastic input-output relations 151-166
transfer function 15-17

Markovian process 134,220,224
M-circles 114,116
mean(statistical) 123,131

meromorphic function 44,106,211,351,353,357, 361
minor feedback loop 84,105,261
Morera's theorem 359
multiplication errors -
  source of 33-34
  statistics 206-231
multiplication pattern 319-322
multirate system, definition 297

Newton-Raphson algorithm 329
norm, definition of 4-5
non-recursive filter or difference equation 49-50
Normality of finite wordlength control errors 247
Normal process 124
normally summable, definition 5-6
notch network(filter) 252,276-283
Nyquist Diagram theory for -
  single rate systems 105-115
  multirate systems 302-305
  subrate systems 328-332
Nyquist's (Shannon's) sampling theorem 98

order of a filter or controller 49
orthogonal functions 7-8
oscillatory structural mode 276,279
overflow 28,56,211,252,292-296

P+I and P+D controllers 271
parallel programming 56-58,250,259,296,322, 333,336
Parceval's theorem 8,13,169,217

peak magnification factor 101-105,114,275
periodogram 131,165
phase advance compensation 270-272
phase lag in D-A conversion 37
phase margin 114,116,284-285
phase retard compensation 283-286
Plancheral's theorem 13
Planck-Boltzmann 133
pointwise(simply) convergent series 9,44
power spectrum of a stochastic process 130-134
prefilter see antialiasing filter
prewarped 118
principal value 146
probability density function -
    definition 121-122
    quantisation(rounding) errors 216
    relationship to Characteristic functions 209
prototypical pulse transfer function 287
pulse amplitude modulation 22-24,179-183,261
pulse Fourier Transformation 70
pulse Laplace Transformation 70
pulse power spectrum of a stochastic process 134-140
pulsed RC networks 201-204

quantisation error statistics -
    effect of additive noise 220-230
    for a few probable states 208-209
    idealised 210-216,364-368
quantisation frequency see amplitude quantisation frequency
quantiser 207-208
quietness 50

randomly phased embedding 307-310
random switch-on instant 179-180,183,308, 371
rank see transfer function
real frequency response function -
continuous data system 19
effect of coefficient quantisation 252-260
multirate sampling system 304-305
single rate sampling system 93-95
subrate sampling system 328-332
realisation of a stochastic process 121
real time 20
recurrence(difference or recursion) equation 32, 49-50, 55, 58-59, 82,234,242,332-333
recursive filter or difference equation 49-50
regression, linear 127-128
representative member of an ensemble 125
residues, method of 15,46,66,69,93,145-147, 299,353,359
resilient drive shaft 105,195,206,276
resonant frequency 101-105,114,276-277
response time 88,90,100-105
Riemann surface 67,148
ripple -
comparison of single and multirate systems 324-325
effect of input spectrum bandwidth 195-198
effect of input spectrum's cut-off rate 195-198
higher order data holds 193
multirate system analysis 311-315
single rate system analysis 72-74,189-193
subrate system analysis 336-337
Wiener filter approach 198-199

robustness 292
Root Locus method 84-87
rounding(finite wordlength) control error component -
multirate system analysis 318-322
single rate system analysis 233-236,242-244
subrate system analysis 334-336

sampling frequency -
choice of 26,72-74,98,193-199,268-269
interaction with compensation 193-194,267-268
interaction with stability 91-92,101-102,103-104
interaction with wordlength requirement 273
Wiener filter considerations 198-199
sample and hold amplifier 39
sampling as pulse amplitude modulation see pulse amplitude modulation
sampling theorem 98
scalar gain 85,87,106,108,252,296
scatter diagram 127
scientific programming convention 27
shaft encoder 40-42
shaping filter 136,167,170,244-245,250,322,336
sidebands -
multirate system 297,311,316
single rate system 24,72,117,175,190,193,264,372
subrate system 337
significant multiplication 319
sign - magnitude convention 27
single pulse sequence 51,82,136,170,245,258
sinusoidal sequence 43,94-95
Smelka's synthesis 291-292

spectrum -
- energy, for continuous deterministic data 13
- energy, for deterministic sequences 97
- line power, for periodic continuous data 12
- power, for continuous stochastic data 130-133
- pulse power, for stochastic sequences 134-135
- quantisation errors 220-225

stability -
- adequate 84,189
- conditional 113
- effect of overflow or saturation 292
- effect of time delays 92
- global 78,80
- linear system 80-84,299
- margins 114,116,207,252,285, 328

stage, of a digital or continuous data compensator 271
staleness factor 286
stationary or steady state solution 77,93,149-151,284-285, 304-305
state vector 59,81
staticisor register 37,71,182,284
statistical expectation see mean
statistically independent 128
stereographic projection 138,147
stiffness of a compensator(controller) 48,198,272
stochastic process 122
strict sense stationary 122-123
structural resonance see oscillatory structural mode
subrate system, definition 327

total steady state control error 174
transfer function -
 combined in cascade 53
 combined in parallel 53
 Laplace, development 15-16
 loop 54-55
 overall 54-55
 prototypical 287
 pulse, development 50-51,63-65
 pulse Fourier, definition 70-71
 pulse Laplace, definition 69-70
 rank 86-87,192-193,371
 stray 255
 unity feedback, definition 54
transient overshoot 88,90,101-105
triangular inequality, definition 231
two's complement arithmetic -
 addition 29
 additions versus shift-lefts 294
 advantages of 28
 definition 28
 fixed point 27
 immunity to intermediate overflows 31-32
 least significant bit 27
 most significant bit 29
 multiplication 33-34
 overflow, definition 28
 scaling to avoid overflow 293-295
 subtraction 29
 width of quantisation, definition 27
 wordlength, definition 27

uncorrelated see linearly independent
unit circle 48,83,95,96,138,169, 249, 279, 357

unit step function, definition 14
unit step sequence, definition 43

weighting function or sequence 20,63,150,159,170
white noise 133,136,142,167-168, 220, 241-242

wide sense stationary, definition 123
width of quantisation, definition 27
window function 117
w-plane 119

zeros, different values aid compensation 274
zero order hold -
    comparison with higher order holds 193
    transfer function 38
    real frequency response 38
z-Transformation, definition -
    one-sided 43-44
    two-sided 135